VOLUME ONE HUNDRED AND FIFTY

# CURRENT TOPICS IN DEVELOPMENTAL BIOLOGY

## Cell-Cell Signaling in Development

# CURRENT TOPICS IN DEVELOPMENTAL BIOLOGY

*"A meeting-ground for critical review and discussion of developmental processes"*
**A.A. Moscona and Alberto Monroy (Volume 1, 1966)**

VOLUME ONE HUNDRED AND FIFTY

# Current Topics in DEVELOPMENTAL BIOLOGY

## Cell-Cell Signaling in Development

Edited by

**THOMAS KORNBERG**
*Cardiovascular Research Institute,*
*University of California,*
*San Francisco, CA, United States*

ELSEVIER

ACADEMIC PRESS
An imprint of Elsevier

Academic Press is an imprint of Elsevier
50 Hampshire Street, 5th Floor, Cambridge, MA 02139, United States
525 B Street, Suite 1650, San Diego, CA 92101, United States
The Boulevard, Langford Lane, Kidlington, Oxford OX5 1GB, United Kingdom
125 London Wall, London, EC2Y 5AS, United Kingdom

First edition 2022



**Notices**
Knowledge and best practice in this field are constantly changing. As new research and experience broaden our understanding, changes in research methods, professional practices, or medical treatment may become necessary.

Practitioners and researchers must always rely on their own experience and knowledge in evaluating and using any information, methods, compounds, or experiments described herein. In using such information or methods they should be mindful of their own safety and the safety of others, including parties for whom they have a professional responsibility.

To the fullest extent of the law, neither the Publisher nor the authors, contributors, or editors, assume any liability for any injury and/or damage to persons or property as a matter of products liability, negligence or otherwise, or from any use or operation of any methods, products, instructions, or ideas contained in the material herein.

ISBN: 978-0-12-820155-8
ISSN: 0070-2153

For information on all Academic Press publications
visit our website at https://www.elsevier.com/books-and-journals

*Publisher:* Zoe Kruze
*Developmental Editor:* Naiza Mendoza
*Production Project Manager:* Vijayaraj Purushothaman
*Cover Designer:* Miles Hitchen

Typeset by STRAIVE, India

# Contents

# Contributors

**J. Fernando Bazan**
*ħ* bioconsulting llc, Stillwater, MN, United States

**Giovanna M. Collu**
Department of Cell, Developmental, & Regenerative Biology, Graduate School of Biomedical Sciences, Icahn School of Medicine at Mount Sinai, New York, NY, United States

**Ana-Citlali Gradilla**
Tissue and Organ Homeostasis, Centro de Biología Molecular "Severo Ochoa" (CSIC-UAM), Nicolás Cabrera 1, Universidad Autónoma de Madrid, Cantoblanco, E-28049, Madrid, Spain

**Isabel Guerrero**
Tissue and Organ Homeostasis, Centro de Biología Molecular "Severo Ochoa" (CSIC-UAM), Nicolás Cabrera 1, Universidad Autónoma de Madrid, Cantoblanco, E-28049, Madrid, Spain

**Zena Hadjivasiliou**
Department of Physics and Astronomy; Institute for the Physics of Living Systems, University College; Mathematical and Physical Biology Laboratory, The Francis Crick Institute, London, United Kingdom

**Ginger Hunter**
Department of Biology, Clarkson University, Potsdam, NY, United States

**William D. Jones**
Department of Cell and Developmental Biology, University of Pennsylvania Perelman School of Medicine, Philadelphia, PA, United States

**Yildiz Koca**
Department of Cell, Developmental, & Regenerative Biology, Graduate School of Biomedical Sciences, Icahn School of Medicine at Mount Sinai, New York, NY, United States

**Andres M. Lebensohn**
Laboratory of Cellular and Molecular Biology, Center for Cancer Research, National Cancer Institute, National Institutes of Health, Bethesda, MD, United States

**Blaise Louis**
Centre for Stem Cells and Regenerative Medicine, King's College London, London, United Kingdom

**Lorenz Mittermeier**
Program in Cancer and Stem Cell Biology, Duke-NUS Medical School, Singapore, Singapore

**Marek Mlodzik**
Department of Cell, Developmental, & Regenerative Biology, Graduate School of Biomedical Sciences, Icahn School of Medicine at Mount Sinai, New York, NY, United States

**Mary C. Mullins**
Department of Cell and Developmental Biology, University of Pennsylvania Perelman School of Medicine, Philadelphia, PA, United States

**Victor A. Negri**
Centre for Stem Cells and Regenerative Medicine, King's College London, London, United Kingdom

**Rajat Rohatgi**
Department of Biochemistry; Department of Medicine, Stanford University School of Medicine, Stanford, CA, United States

**Vasiliki Salameti**
Centre for Stem Cells and Regenerative Medicine, King's College London, London, United Kingdom

**Mihaela Serpe**
Neurosciences and Cellular and Structural Biology Division, Eunice Kennedy Shiver National Institute of Child Health and Human Development, National Institutes of Health, Bethesda, MD, United States

**Rosario Vicidomini**
Neurosciences and Cellular and Structural Biology Division, Eunice Kennedy Shiver National Institute of Child Health and Human Development, National Institutes of Health, Bethesda, MD, United States

**David M. Virshup**
Program in Cancer and Stem Cell Biology, Duke-NUS Medical School, Singapore, Singapore

**Fiona M. Watt**
Centre for Stem Cells and Regenerative Medicine, King's College London, London, United Kingdom; Directors' Research Unit, European Molecular Biology Laboratory, Heidelberg, Germany

**Sebastiaan Zijl**
Centre for Stem Cells and Regenerative Medicine, King's College London, London, United Kingdom

# Preface

There is a hierarchy among cells such that some cells direct their neighbors to adopt particular fates. As we now understand that the genes that direct developmental pathways are expressed only in the cells they control, and as we also understand the basic mechanics of cell-specific gene expression, a major question that remains in developmental biology is this: How cells communicate with each other to adopt their particular fates? Progress in the field over the past several decades has identified the important signaling proteins—bone morphogenic protein, Wnt, Hedgehog, fibroblast growth factor, and Notch—that cells use to communicate developmental information. These signaling proteins are used universally in tissues and across evolution. Recent findings have now identified the mechanisms by which the signaling proteins disperse to client cells after leaving the cells that make them and provide a basis to understand how cells collate and interpret the information the signaling proteins convey. These exciting topics are addressed in this volume.

THOMAS KORNBERG
Cardiovascular Research Institute
University of California
San Francisco, CA, United States

CHAPTER ONE

# Hedgehog on track: Long-distant signal transport and transfer through direct cell-to-cell contact

**Ana-Citlali Gradilla* and Isabel Guerrero***
Tissue and Organ Homeostasis, Centro de Biología Molecular "Severo Ochoa" (CSIC-UAM), Nicolás Cabrera 1, Universidad Autónoma de Madrid, Cantoblanco, E-28049, Madrid, Spain
*Corresponding authors: e-mail address: acgradilla@cbm.csic.es; iguerrero@cbm.csic.es

## Contents



## Abstract

The function of Hedgehog (Hh) as a morphogen results from its long-distance distribution from producing to neighboring receiving cells within the developing tissue. This signal distribution enables, for example, the formation of a concentration gradient eliciting distinct cellular responses that will give rise to spatial patterning. Hh is a lipid modified protein and its dispersion is better guaranteed through cytonemes, cell protrusions that allow direct cell membrane contact and signal transfer at a distance. Hh and its receptor Patched (Ptc) meet at cytoneme contacts in a way that reminds synapses. Both Hh and Ptc require a recycling process prior to presentation in cytonemes. Increasing research on the role of cytonemes in Hh signaling is revealing cellular mechanisms that link signal transport through dynamic cytonemes with concurrent regulation of cell adhesion. The equilibrium between these two processes is being unveiled as crucial to both patterned morphogen distribution and signal transfer. In addition, these discoveries are pushing forward our understanding of the role of extracellular elements involved in the Hh pathway, such as the Hh coreceptors Ihog and Boi and the glypicans Dally and Dally-like protein (Dlp).

*Current Topics in Developmental Biology*, Volume 150
ISSN 0070-2153
https://doi.org/10.1016/bs.ctdb.2022.03.002

## Abbreviations

| | |
|---|---|
| **AMPs** | adult muscle precursors |
| **Aop** | Anterior open |
| **ASP** | air sac primordium |
| **Bnl** | Branchless |
| **Boi** | Brother of Ihog |
| **Btl** | Breathless |
| **CAM** | cell adhesion molecule |
| **Cut** | cut |
| **Disp** | Disptached |
| **Dlp** | Dally like protein |
| **En** | Engrailed |
| **FGF** | fibroblast growth factor |
| **Fn** | Fibronectine |
| **Fur1** | Furin1 |
| **GPI** | glycosylphosphatidylinositol |
| **GRASP** | GFP reconstitution across synaptic partners |
| **Hh** | Hedgehog |
| **Ihog** | Interference hedgehog |
| **MVBs** | multi vesicular bodies |
| **Myo10** | Myosin 10 |
| **Ptc** | Patched |
| **Ser** | Serrate |
| **Shf** | Shifted |
| **Syb** | Synaptobrevin |
| **Syt** | Synaptotagmin |

## 1. Introduction

Spatiotemporal cell determination and differentiation within morphogenetic fields, essential for tissue development, are largely established by morphogens that disperse with defined profiles. As a morphogen, Hedgehog (Hh) has a different distribution depending on the cellular context, varying from the formation of continuous concentration gradients to the signaling of distant target cells after bypassing intermediate non-target tissue. Therefore, understanding how this spatial arrangement is achieved is vital to comprehend signaling mechanisms. The discovery of specialized cellular structures, such as the signaling membrane protrusions called cytonemes, has provided a plausible though still controversial mechanism to achieve distribution of the membrane anchored Hh through direct membrane-to-membrane contacts.

Lipid modifications in Hh hamper its free dispersion through the aqueous extracellular space and are required for its full long-distance signaling function. Signal transfer through direct cell-to-cell membrane contact benefit from this membrane anchoring, as it could grant signal-controlled delivery (Bischoff et al., 2013; Chen, Huang, Hatori, & Kornberg, 2017; Gradilla et al., 2014) (Fig. 1). Thus, dynamics of cytoneme extension and retraction, as well as that of signal transfer, should be crucial aspects to ensure correct Hh signaling. In this line of thought, recent models predicting and testing gradient formation have found that cytoneme dynamics and their contacts can robustly achieve graded signaling (Aguirre-Tamaral & Guerrero, 2021). Furthermore, it is through cytoneme-mediated distribution that gradient formation is sustained notwithstanding topography changes in different tissue contexts, and maintains gradient shape independently of morphogen quantity (Hatori, Wood, Barbosa, & Kornberg, 2021).

Despite cytonemes playing a crucial role in several signaling pathways (Ali-Murthy & Kornberg, 2017; González-Méndez, Gradilla, & Guerrero, 2019; Zhang & Scholpp, 2019) little is known about their formation and signaling specificity. Here, we review the latest research regarding relevant cytoneme regulation mechanisms. As other actin-based protrusive structures, cytonemes are influenced by both intracellular and extracellular cues, as well as by biochemical and biophysical features. Cell adhesion might play a key role in Hh cytoneme extension and dynamics. Furthermore, known fundamental Hh pathway components such as the Hh coreceptors Interference hedgehog

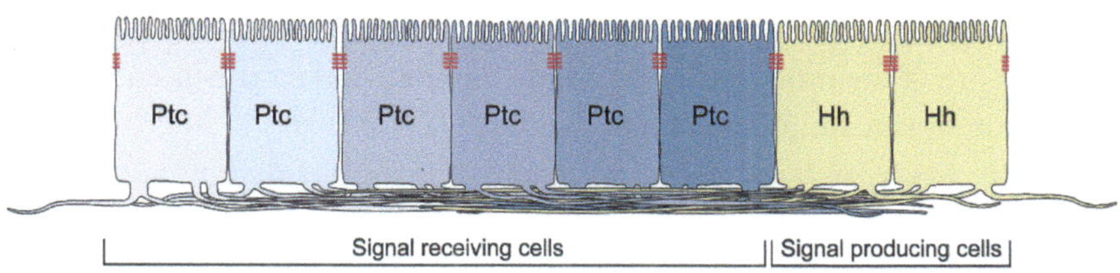


**Fig. 1** Cytonemes coordinate cell signaling to establish morphogenetic gradients. (A) Hh sending (pale green) and receiving (blue) cells communicate with each other, extending cytonemes that emerge from the basal surface of a developing epithelium The distant receiving cells are highlighted with paler colors. Cytonemes from both signal source and receptor cells contact with each other for signal transfer. Graded distribution of the signal is given by the proportional number of signal-transferring contacts according to the distance between cells. This mechanism of cellular communication enables Hh gradient formation and patterning of tissues. (For interpretation of the references to colour in this figure legend, the reader is referred to the web version of this article.)

(Ihog) and Brother of ihog (Boi) (Yao, Lum, & Beachy, 2006), and the GPI-membrane-anchored glypicans Dally and Dally-like protein (Dlp) (Desbordes & Sanson, 2003; Han, Belenkaya, Wang, & Lin, 2004; Lum et al., 2003; Williams et al., 2010) are unveiling their potential as cytoneme regulators (Bilioni et al., 2013; González-Méndez, Seijo-Barandiarán, & Guerrero, 2017; Simon et al., 2021; Yang et al., 2021), providing further comprehension of their mediation on Hh signaling.

Another aspect is the intracellular distribution of Hh and other Hh pathway components during cytoneme-mediated signaling. Polarized distribution of Hh has been proposed to occur through Hh packaging *via* Multi Vesicular Bodies (MVBs) to form extracellular vesicles (Gradilla et al., 2014; Matusek et al., 2014; Hurbain et al., 2022). This polarization has been reported to be directed either to basal membranes headed for cytoneme-dependent distribution (Callejo et al., 2011; Chen et al., 2017; Gradilla et al., 2014) or to apical membranes for cytoneme-independent dispersal (D'Angelo, Matusek, Pizette, & Thérond, 2015; Matusek et al., 2014; Hurbain et al., 2022). In addition, polarization of the main Hh receptor Patched (Ptc) (Chen et al., 2017; González-Méndez et al., 2017) and other Hh pathway components has also been described (Bilioni et al., 2013; Callejo et al., 2011; Simon et al., 2021; Stewart et al., 2018). Nevertheless, the detailed mechanisms for signal transport along cytonemes, transfer and reception at contact sites are still far from being solved.

This review is an effort to put together the novelties concerning cytoneme as regulators of signal presentation and reception in the hope of understanding the signaling mechanisms both during development and tissue homeostasis. Better knowledge in this field might result in molecular tools to potentially modify aberrant signaling for targeted treatment of cancer and other signaling dependent diseases.

## 2. Dynamic cytonemes and gradient formation models

Cytonemes have been described as long and thin actin-based cellular protrusions that transfer morphogen signals from cell to cell; by electron microscopy, they measure from 7 to 300 μm in length and between 20 and 200 nm in diameter (Sanders, Llagostera, & Barna, 2013; Wood et al., 2021). *In vivo* imaging of cytonemes has allowed visualization of their dynamics and shown a strong correlation between cytoneme extension and Hh gradient, both in space and time. Accordingly, experimental blocking of cytoneme extension tightly reflects on Hh gradient reduction (Bischoff et al., 2013). Thus, experimental measurements of cytoneme features and dynamics should be relevant to comprehend Hh gradient formation (Fig. 1). In this line of thought, recent mathematical modeling based on computational analysis of gradient establishment has been able to identify key aspects of cytoneme-mediated distribution and signaling (Aguirre-Tamaral & Guerrero, 2021). Cytoneme features and dynamics, including elongation and retraction timing as well as contact probability, have been integrated into a software tool (Cytomorph) that can reproduce the distribution of morphogens. This tool has been able to identify cytoneme length and cell size as crucial traits for gradient formation, as well as corroborate cytoneme-cytoneme contact as a preferential way to achieve the Hh gradient in *Drosophila* (Aguirre-Tamaral & Guerrero, 2021) (Fig. 2).

Cytomorph has also been used to compare the model based on cytoneme-mediated distribution with that based on free diffusion. Interestingly, both models can reproduce the Hh gradient in the wing disc (Fig. 2C); however, when tested in a different tissue, only the cytoneme model is able to reproduce the Hh graded distribution (Fig. 2D). The diffusion model, on the other hand, requires readjusting the diffusion constant depending on the tissue, while in the cytoneme model the molecular properties of the morphogen in different tissue contexts remain unaltered (Aguirre-Tamaral & Guerrero, 2021).

Another attempt to compare both models, distribution through simple diffusion or through cytonemes, has been developed by modeling signaling in a simple monolayer of cells. Fancher and Mugler (2020) have found that modeling graded distribution through cytonemes fits better for short-range gradients, significantly diminishing noise, while large-range gradient shapes are better attained by simple diffusion. Thus, the authors propose that some signaling contexts might utilize cytonemes while others could be based on free diffusion, or a combination of both (Fancher & Mugler, 2020). However, these models are not taking into account potential tissue context-dependent obstacles either for diffusion or for cytonemes. Introducing cytoneme dynamics into gradient modeling has been essential to corroborate earlier experimental observations and to show that the combination of different cytoneme behaviors is what attains gradient shape (Aguirre-Tamaral & Guerrero, 2021) (Fig. 2). This allows the coexistence of short- and long-range gradient shapes and its adjustment to different tissues. This *in silico* model has also revealed its potential to provide robustness to

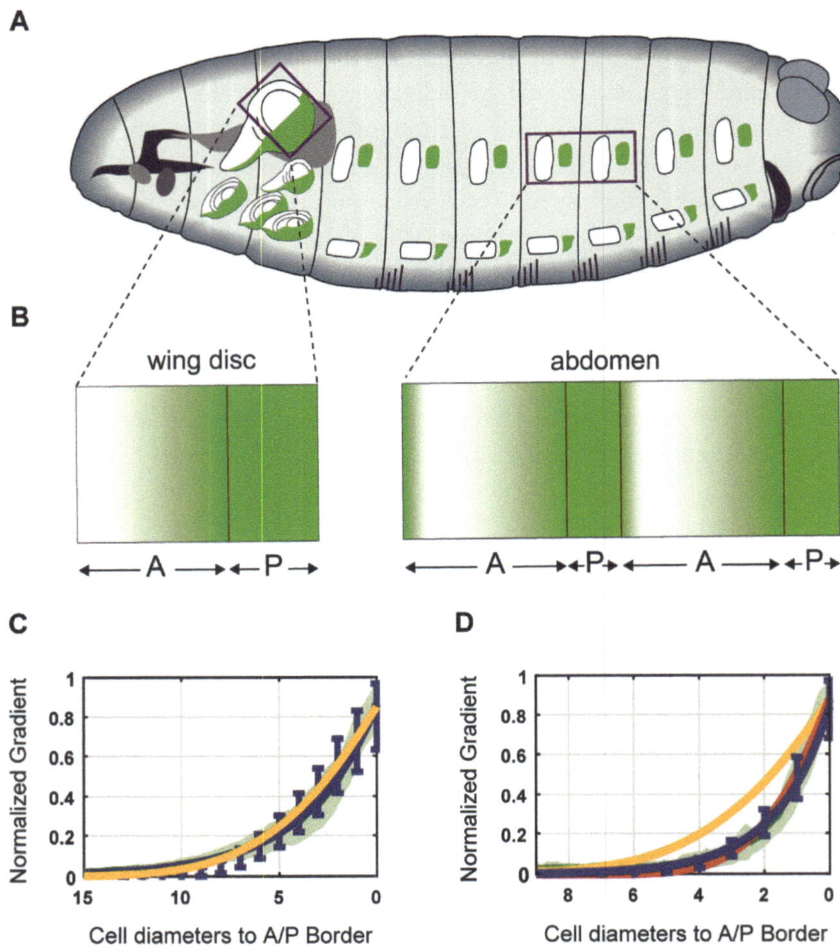


**Fig. 2** Experimental and simulated Hh gradients in *Drosophila* wing disc and abdominal epithelia. (A) Drawing of a *Drosophila* larva depicting thoracic imaginal discs and abdominal histoblast nests, with their A and P compartments marked in white and green, respectively. Purple boxes outline the wing pouch of the wing and the dorsal histoblast nests of two segments. (B) The rectangles depict Hh-expressing (green) in the P compartments and the signaling gradient (pale green) in the A compartments of the wing disc epithelium (left) and in two contiguous abdominal segments. (C) Plot representing a comparison between the quantified experimental gradient in the wing disc (green) and the gradients predicted by cytoneme model (blue) and diffusion-degradation model (yellow). Note that both graphs are identical. (D) Plot representing a comparison between the quantified experimental gradient of abdominal histoblast (green) and the gradients predicted by cytoneme model (blue) and diffusion-degradation model (yellow); note that the diffusion-degradation model when considering a diffusion coefficient 3 times smaller (red) is similar to cytoneme model (For interpretation of the references to colour in this figure legend, the reader is referred to the web version of this article) (Aguirre-Tamaral & Guerrero, 2021)

gradient formation as, for instance, a theoretical reduction in the number of signal producing cells could be compensated by an increase in the number of cytonemes per cell.

In some tissues, signaling can also be produced throughout cell migration. Modeling of this sort of scenario has also found cytoneme-mediated distribution as more advantageous, acquiring steeper gradients in shorter times, in the case of Wnt signaling during zebrafish neural plate development (Rosenbauer et al., 2020). These results based on a search and capture model, support previous model simulations of Wnt signaling gradient formation through dynamic cytonemes (Bressloff & Kim, 2019). The consideration of contact-mediated cell communication between juxtaposed cells has been shown to provide robustness and refinement within an *in silico* morphogenetic prediction, where free dispersion of the morphogen would act first giving a broader initial cue for positional information (Kuyyamudi, Menon, & Sinha, 2021).

Experimentally, the Hh gradient has been shown to remain unchanged after increasing morphogen production, providing further evidence for the tuning potential of cytoneme-mediated morphogen distribution (Hatori et al., 2021). In contrast, a model considering a free diffusion of an augmented amount of morphogen should have resulted in either a wider gradient distribution range or required additional regulatory mechanisms to remain unchanged. Therefore, both experimental and *in silico* results continue advocating for a cytoneme mediated mechanism of gradient formation, specially highlighting the importance of cytoneme dynamics in signaling regulation.

## 3. Cytoneme-mediated Hh signaling between different cell types

Signaling between different tissues and cell types is determinant to generate appropriate cell numbers and identities (Ribatti & Santoiemma, 2014). Actually, in the developing *Drosophila* imaginal wing disc, the notum primordium regulates specific responses in three close target tissues: the epithelial wing disc cells, the tracheal air sac primordium (ASP) and the myoblasts, which are the adult muscle precursors (AMPs) that generate flight muscles (Bate & Arias, 1991). To obtain a regulated response in each tissue, signaling from a common pool of a group of ligands would be unfeasible; cytoneme-mediated cell signaling is a more consistent mechanism for an optimal transport and reception. In fact, signaling seems to be cytoneme-dependent in all three tissues (Huang & Kornberg, 2015) and responsible for the uptake, among other signals, of Hh and FGF (Branchless (Bnl) in Drosophila) working cooperatively (Hatori & Kornberg, 2020).

Hh produced by the epithelial wing disc is distributed between ASP and AMPs, activating target genes that are common as well as tissue specific ones. Hh from the disc regulates the expression of En, Ser, Cut and Aop in the ASP and, as expected, cytonemes from the ASP tip contain motile punctae that include Hh and its receptor Ptc. Some targets, such as Cut and Aop, regulated by Hh signaling in ASP, also depend on Bnl signal transduction (Hatori et al., 2021). Moreover, using single cell RNA sequencing data (scRNAseq), Hh can activate the specific targets *neurotactin* and *midline* in the AMPs, to specify a subset of myoblasts. Ptc expression was observed primarily in a group of precursors of direct adult muscles, but also in two more dorsal smaller groups of cells, precursors of indirect muscles (Everetts, Worley, Yasutomi, Yosef, & Hariharan, 2021). The proximity of these Ptc-expressing mesodermal cells to Hh-secreting epithelial cells suggests that they are responding *via* cytonemes to Hh secreted by the disc cells rather than to a circulating Hh pool available to the whole AMP.

Controlling the coordinated cytoneme growth between the ASP and the wing disc, a positive feedback mechanism has been described for Hh signaling activation from the disc toward the ASP, since Hh reception can increase the length of the ASP tip cytonemes (Hatori et al., 2021). Similarly, another positive feedback mechanism has been reported for Bnl/FGF signaling is cytoneme contacts. These contacts determine graded signaling and the number and length of cytonemes through a regulatory feedback loop not only at receiving cells (Du, Sohr, Yan, & Roy, 2018), but also at ligand producing cells (Du, Sohr, & Roy, 2021).

In contrast, in the Drosophila ovary, where cytonemes are also involved in the communication between different cell types, after experimentally blocking Hh reception in a group of cells, the cytonemes keep growing until they touch cells able to receive Hh (Rojas-Rios, Guerrero, & Gonzalez-Reyes, 2012). These results are in agreement with others in which blocking the presentation of the receptor Ptc at plasma membrane, provokes an extension of the Hh gradient in the wing disc (González-Méndez et al., 2020, 2017; Torroja, Gorfinkiel, & Guerrero, 2004). Therefore, further investigation is needed to clarify the feed-back auto regulatory loop of signal reception and cytoneme promotion.

Finally, in accordance with the many signals being concurrently driven and coordinated by cytonemes during morphogenesis, imaging by electron microscopy of the wing disc has discovered a larger number of cytoneme structures than had previously been described through confocal microscopy. These cytonemes all share a distinctive shape, characterized by alternating thin and wide sections; they contain ribosomes in some wide sections, and they present membranous compartments of different sizes and shapes along their lengths (Wood et al., 2021). Thus, synchronization of several pathway-specific elements with the intracellular actin machinery and extracellular clues might well be pivotal to the formation of specialized signaling filopodia.

## 4. Cytoneme establishment and regulation: The weight of cell adhesion in Hh signaling

The knowledge of the regulatory mechanisms for cytoneme establishment is the basis to understand morphogen signaling. As cellular membrane projections, cytonemes would require intracellular mechanisms for their nucleation at the plasma membrane, as well as continuous coordination between intra and extracellular constituents to allow elongation and retraction. In what follows, we will describe several cases of known molecular components of the Hh pathway that also have an extracellular role, mediating cell adhesion, signal release and reception (reviewed in Beachy, Hymowitz, Lazarus, Leahy, & Siebold, 2010; González-Méndez et al., 2019).

Besides their function in Hh delivery and reception, the Hh coreceptors Ihog and Boi have a role as cell adhesion molecules (CAM) (Hsia et al., 2017; Simon et al., 2021; Williams et al., 2010; Yang et al., 2021; Yao, Lum, et al., 2006; Yao, Munson, Webb, & Lis, 2006; Zheng et al., 2010). Similarly, glypicans in the extracellular matrix (ECM) have several functions: they interact with the Hh reception complex during the signaling process (Desbordes & Sanson, 2003; Han et al., 2004; Lum et al., 2003; Williams et al., 2010) as well as during Hh delivery and transport (Bilioni et al., 2013; Callejo et al., 2011; Han et al., 2004). In addition, Ihog has a role in cytoneme stabilization that allows their visualization after tissue fixation (Bischoff et al., 2013; González-Méndez et al., 2017). Recent research regarding Ihog functional domains and their involvement in cytoneme stabilization has found that, contrary to expected, it is the extracellular fragment that is essential for the stabilization function (Simon et al., 2021; Yang et al., 2021). The transmembrane protein Ihog presents extracellularly two fibronectin type III (Fn) domains and four Inmunoglobulin (Ig) domains (Yao, Lum, et al., 2006; Yao, Munson, et al., 2006) (Fig. 3).

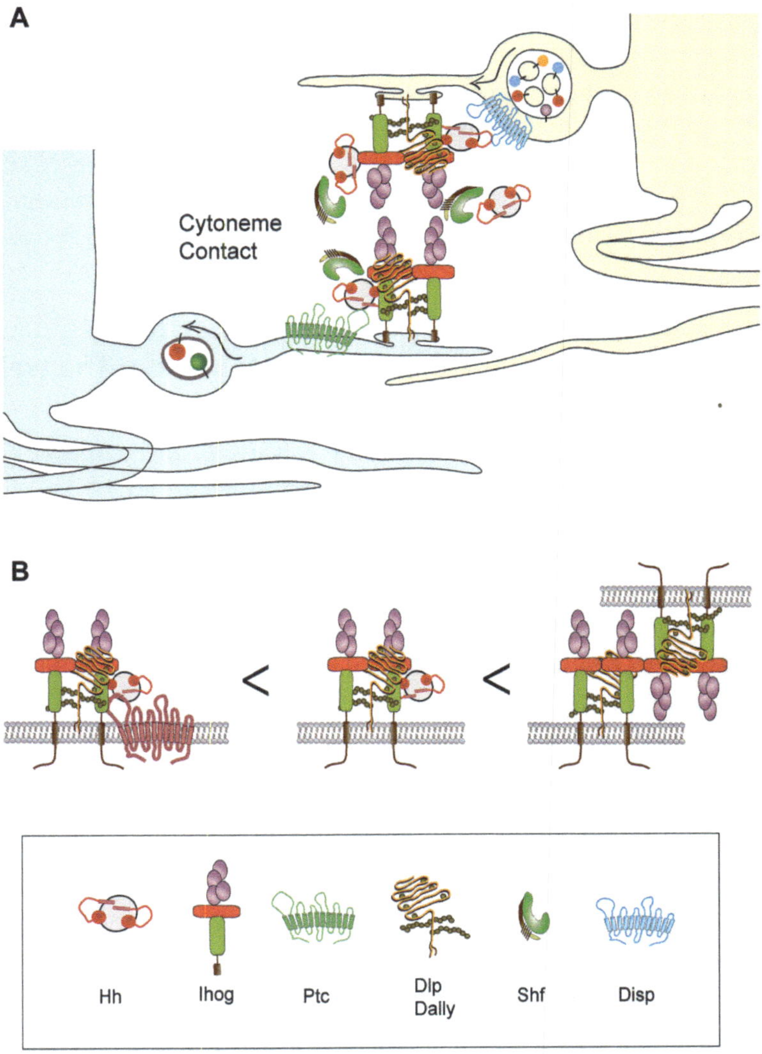


**Fig. 3** Morphogenetic synaptic model. (A) Hh signaling conceived as a synapse-like process between Hh producing and receiving cells. Adhesion proteins and glypicans from both type of cytonemes mediate this close cell-to-cell contact and facilitate the binding of Hh to its receptor complex, either from cytoneme membranes or from exovesicles released at the site. Hh travels in multivesicular bodies (MVBs) along cytonemes and is released in exosomes to interact at the contact sites. The Hh coreceptors Ihog (shown as adhesion protein) and the glypican Dlp and Dally are recruited for ligand/receptor interaction. The soluble factor Shifted (Shf) helps the Hh release from the producing cell and/or Hh presentation to the receiving cell cytonemes. The transmembrane proteins Dispatched (Disp), required for Hh release from the producing cells, and the Hh receptor Patched (Ptc) in the receiving cytonemes also interact with some of the other proteins during the release and reception

Experimental deletion of the fibronectin domains has revealed that their presence is necessary for cytoneme stabilization as well as for signaling, and binding Hh, Ptc and the glypicans Dally and Dlp (Simon et al., 2021; Williams et al., 2010; Yang et al., 2021; Yao, Lum, et al., 2006; Yao, Munson, et al., 2006; Zheng et al., 2010).

Interestingly, further analysis of the Ihog domains involved in cytoneme stabilization has shown that although the two FNIII domains participate in interactions with glypicans, ligand and receptor, they do not do it through the same interacting subdomains (McLellan et al., 2008; Simon et al., 2021; Yang et al., 2021; Zheng et al., 2010). The Fn2 subdomain interacts with Ptc, while the Fn1 binds both Hh and glypicans, but through different aminoacids (Fig. 3). In addition, the Ihog/glypican activity in cytoneme stabilization had been previously described as an interaction *in trans* between contacting cytonemes (González-Méndez et al., 2017). In summary, the research shows that the Hh coreceptor Ihog functions as a CAM (Hsia et al., 2017), inducing cytoneme stabilization by interacting with glypicans through its extracellular domain (González-Méndez et al., 2017; Yang et al., 2021) (Fig. 3A).

Furthermore, an *in trans* homophilic Ihog–Ihog interaction has also been described as key for cytoneme stabilization (Yang et al., 2021) (Fig. 3B). It has been proposed that Ihog is able to bind to itself in pairs, being capable of acting as a cross-linker-like molecule inducing cytoneme stabilization. It was also suggested that the Ihog–Ihog homophilic interaction between producing and receiving cells in the connection of cytonemes *in trans* can be switched into a heterophilic Ihog–Hh interaction, which dominates and can even displace the *in trans* Ihog-Ihog binding (Fig. 3B). Thus, it has been speculated that at cytoneme contacts, the competition for Ihog by Hh and Ptc, also known to interact with the Ihog Fn2 domain (Zheng et al., 2010), allows signal reception and cytoneme retraction in a

processes. (B) Model of the reception process regulated by the adhesion properties of the coreceptor Ihog at the contact site between cytonemes. Ihog and the glypicans Dlp and Dally are also recruited at cytonemes for ligand/receptor interaction. Right panel: Ihog–Ihog interaction induces high membrane adhesion facilitating the contact between presenting and receiving cytonemes. Medium panel: In the presence of Hh, the cell adhesion induced by Ihog-Ihog homodimers in the P compartment diminishes. Left panel: At the contact, the presence of Ptc, Hh and other proteins (not shown) diminishes the density of the Ihog-Ihog homodimers, decreasing the membrane adhesion and facilitating the releases of Hh for reception. The symbol $<$ represents lower to higher membrane affinities.

dynamic manner (Yang et al., 2021) (Fig. 3A.) Equilibrium between these two types of interactions, as well as the influence of extracellular molecular cues such as glypicans, might direct Hh cytoneme dynamics and directionality (Fig. 3B).

Similarly, in the Bnl/FGF signaling between the disc and the ASP, it was recently discovered that Bnl is anchored to the source cell surface by a glycosylphosphatidylinositol (GPI) moiety (Du et al., 2021). This lipid modification ensures Bnl attachment to basal cytoneme membranes, facilitating direct contact with cytonemes protruding from the receiving ASP cells and bearing the receptor Breathless (Btl/FGFR) (Du et al., 2018). In this way, Bnl also acts as a CAM protein self-promoting cytoneme projection in the ASP and in the disc toward each other to form signaling contacts (Du et al., 2021) as it has been proposed for Ihog in Hh signaling.

On the other hand, despite Ihog and Boi having been long considered as functionally redundant (Camp, Currie, Labbé, van Meyel, & Charron, 2010; Yao, Lum, et al., 2006; Yao, Munson, et al., 2006; Zheng et al., 2010), ectopic Ihog has a role in cytoneme stabilization that Boi cannot replace (Simon et al., 2021). The lack of function of either Ihog or Boi affects Hh gradient response differently, Ihog the one having a pre-eminent role in controlling the Hh gradient. Furthermore, Ihog, but to a lesser extent Boi, binds glypicans at the basal side of the epithelium, where cytonemes are located, and it has been proposed that Ihog specifically affects Hh cytoneme-mediated gradient formation through its interaction with glypicans (Simon et al., 2021). In contrast, the vertebrate orthologs of Ihog and Boi, CDON and BOC, both appear to have a relevant role on cytoneme stabilization, BOC possibly being more prevalent at cytonemes (Hall et al., 2021; Sanders et al., 2013). Altogether, the role of the adhesion protein Ihog in cytoneme-mediated Hh signaling, both in producing and receiving cells, is revealing an orchestrated mechanism between intracellular, inter-cellular and extracellular protein interactions.

## 5. Transport through the extending protrusion

In agreement with signal transfer through direct membrane-membrane contact, visualization of signaling cytonemes has revealed the presence of several pathway-specific elements at cytoneme tips and at cytoneme contacts. However, the mechanisms for the localization or transport of these molecules along extending cytonemes are still mostly unexplored. Recent research in mammalian cells suggested that active transport of Sonic-Hh (SHh) molecules along cytonemes might be driven by the actin motor Myosin 10 (Myo10) (Hall et al., 2021). SHh and Myo10 travel at similar velocities, and this transport appears to be mediated by the Myo10 cargo binding domain, since cells expressing a mutant form for this domain fail to enrich cytonemes with SHh and cannot induce a robust signal response in mammalian tissue culture cell. Apico/basal polarized mouse embryonic fibroblasts lacking Myo10 also present a reduced number of basal cytonemes (Hall et al., 2021), in agreement with a reported role for Myo10 in filopodia outgrowth (Bohil, Robertson, & Cheney, 2006). We wonder if myosins like Myo10 are responsible in *Drosophila* for the transport of Hh pathway elements along cytonemes, contributing to their allocation.

An elegant method to visualize transport within long cellular extensions using engineered multiheaded myosin motors was developed by Zhang et al. (2021); these motors move bidirectionally and faster than naturally occurring dimeric myosin motors. Engineered optogenetic motors move backwards and forwards under the control of blue light pulses. These tools have been proven as capable of transporting actin-regulators that can modify cytoneme occurrence *in vivo* by increasing or decreasing their number (Zhang et al., 2021). The tools were actually used to manipulate filopodia formation within a regenerating limb in the axolotl, demonstrating a requirement for SHh signaling in *in vivo* regeneration together with the appearance of newly generated cytoneme networks. These motors were able to transport GFP-tagged proteins allowing the visualization of their transport *in vivo,* even for the transmembrane proteins Patched1 (Zhang et al., 2021) and Dispatched1 (Disp1) (Hall et al., 2021), required for the release of SHh from the producing cells (reviewed in Hall, Cleverdon, & Ogden, 2019). Furthermore, enhancement of cytoneme occurrence has also been found after Disp ectopic expression (Hall et al., 2019; Hatori et al., 2021). Additional research is needed to understand the mechanisms for transport and allocation of signaling-pathway proteins along cytonemes, but Myosins are likely to have a pivotal role regulating cytoneme establishment and cargo transport.

In addition, intracellular transport has been proposed to reach the point of Hh cargo at the base of cytonemes (Callejo et al., 2011). There is strong evidence toward a prevalence of basal cytoneme occurrence, thus intracellular shipping mechanisms of signaling pathway elements to basal membranes have to be relevant (reviewed in Gradilla, Sanchez-Hernandez, Brunt, & Scholpp, 2018). In the *Drosophila* developing wing disc, both

Hh and Ptc have been found to recycle from the apical to the basal membrane through a MVB mediated mechanism (González-Méndez et al., 2020; Gradilla et al., 2014) (Fig. 4). This vesicle recycling process of Hh and Ptc is dependent on the ESCRT complex needed for the extracellular vesicle biogenesis, while their final presentation at basal membranes relies on membrane-fusion Snare proteins, such as Synaptobrevin (Syb) (Chen et al., 2017; González-Méndez et al., 2020; Gradilla et al., 2014; Huang, Liu, & Kornberg, 2019). Similar apical-to-basal recycling mechanisms have been described for other signaling proteins such as Wingless (Wg), which requires the ubiquitin ligase Godzilla, known to interact with the Snare protein Syb (Yamazaki et al., 2016).

Interestingly, signal exocytosis following an endocytosis within the same cell seems to initiate these intracellular vesicle-sorting processes toward a defined cell membrane destination (reviewed in Gradilla et al., 2018). In agreement, a novel endosome-mediated pathway for Hh recycling in the *Drosophila* wing disc has recently been described. The recycling vesicles, named Hherisomes, have been defined as tubular endosomes and their generation is enhanced by the expression of the GTPase Rab11 (Pizette, Matusek, Herpers, Thérond, & Rabouille, 2021). Hherisomes have been speculated not to be a step toward degradation but to serve as Hh recycling vehicles toward cytoneme-mediated Hh signaling, since they seem to promote high levels of Hh responses. In addition, Hherisomes have been defined as vesicles independent from Disp function and clearly different from MVB. Therefore, additional research is needed to investigate whether this vesicle recycling mechanism co-exists with previously proposed mechanisms, if they could be related or if they rather have completely separate roles in Hh recycling.

MVB-dependent recycling to basal membranes has also been suggested for other pathway elements (González-Méndez et al., 2020; Gradilla et al., 2014; Hall et al., 2019). Basal polarization of components such as Ihog or Disp agrees with their proposed role in both Hh secretion and cytoneme dynamics regulation (Bischoff et al., 2013; Callejo et al., 2011; Gradilla et al., 2014; Hall et al., 2019). In contrast, an apical pool of secreted Hh has been suggested to contribute to long-range gradient formation in the *Drosophila* wing disc (D'Angelo et al., 2015; Gore et al., 2021; Matusek et al., 2014; Hurbain et al., 2022). A recycling mechanism of Hh and its coreceptor Ihog for this apical signaling pool has as well been proposed (Matusek et al., 2014). These authors observed an MVB mediated biogenesis of extracellular vesicles during Hh apical secretion, similar to the one

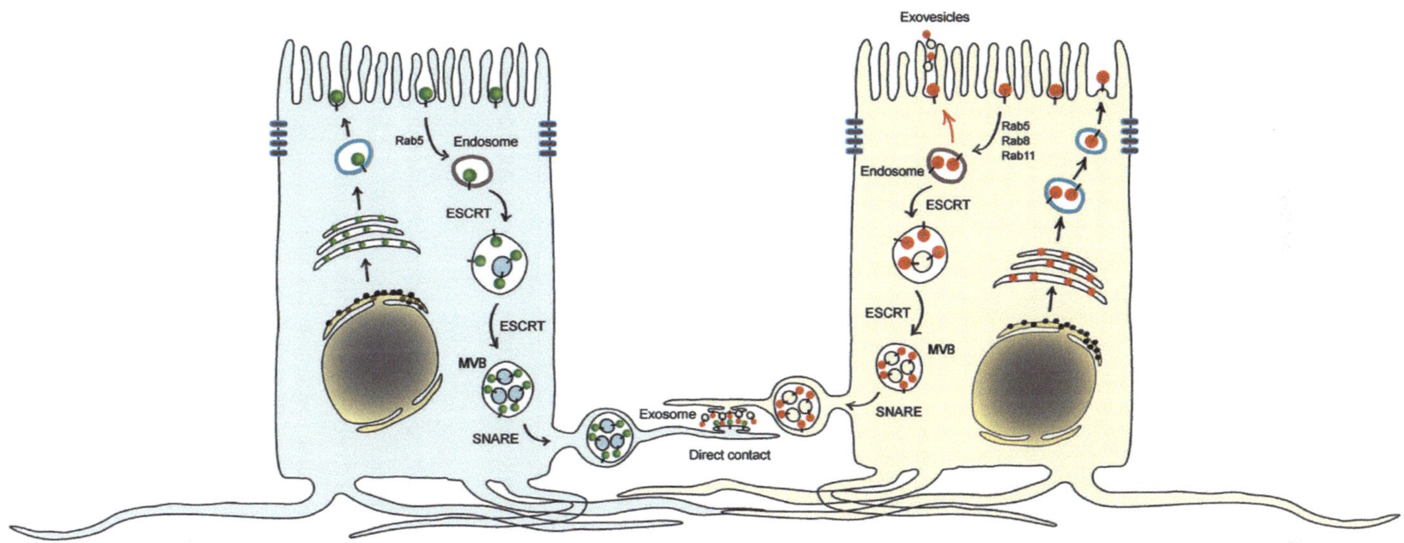


**Fig. 4** Basolateral polarization of Hh and Ptc for cytoneme-mediated signaling. Mechanisms for either formation or loading of basal polarized cytonemes in epithelial cells are still not well characterized. The basolateral secretion of Hedgehog (Hh) and Ptc has been defined and it is achieved *via* an MVB-mediated recycling mechanism from the apical to the basolateral side of the wing disc epithelium. Hh and Ptc recycles after an endocytic process using the ESCRT complex to form multivesicular bodies (MVB). Finally, Hh and Ptc are presented in cytonemes as exovesicles (EVs) for Hh and at the plasma membrane for Ptc using the SNARE complex.

described for Hh basal trafficking. Additionally to the proposed role of Rab8 dependent endocytosis and recycling function for Hh basolateral secretion (Callejo et al., 2011), Rab 8 was also suggested to participate in apical signaling (Gore et al., 2021) (Fig. 4).

The cues that might determine the mechanisms for intracellular trafficking and final destination of these proteins remain hidden. An example of a molecular cue to define intracellular recycling is the enzymatic cleavage needed for Disp to be functional. Disp is cleaved by Furin1 (Fur1) before recycling to reach its final destination at the basolateral plasma membranes (Stewart et al., 2018). Bnl/FGF has also been shown to go through a previous endoproteolytic cleavage by Fur1 at the Golgi network, leaving a functional truncated C-terminal Bnl fragment that localizes to basal membranes (Du et al., 2018). This posttranslational modification is essential for proper directed signaling during the ASP development, and only the cleaved and GPI modified Bnl localizes to delivering cytonemes.

An active Hh also requires cleavage and lipid modifications (N-palmitoylation and cholesterol addition at the C-terminal end) (Bumcrot, Takada, & McMahon, 1995; Lee, von Kessler, Parks, & Beachy, 1992; Pepinsky et al., 1998; Porter et al., 1995; Porter, Young, & Beachy, 1996). However, to date the potential role of these lipid modifications in intracellular trafficking is unclear. Nevertheless, Hh lipid modifications guarantee the anchoring to membranes and it is expected to also warrant Hh vesicle-mediated trafficking to cytonemes (Callejo et al., 2011).

## 6. Signal transfer, reception and retraction

Hh distribution and gradient formation through cytonemes is also highly regulated at signal reception (González-Méndez et al., 2020, 2017). In *Drosophila* epithelia (wing disc and abdominal histoblasts) reception occurs at contacts between cytonemes that carry Hh-loaded exosovesicles, and cytonemes that bear the receptor Ptc (González-Méndez et al., 2017; Gradilla et al., 2014) (Figs. 3 and 4). Hh cytoneme-cytoneme contacts are reminiscent of a synapsis connection, due to the presence of classic synaptic molecules such as Syb and Synaptotagmin (Syt) (Chen et al., 2017; González-Méndez et al., 2020; Huang et al., 2019). Using the GRASP technique (Feinberg et al., 2008), in *Drosophila* the morphogenetic contacts, also known as morphogenetic synapses, maintain a distance between signal-presenting and receiving membranes similar to that described for neuronal synapsis (Chen et al., 2017; González-Méndez et al., 2017; Huang et al., 2019; Roy, Huang, Liu, & Kornberg, 2014).

Moreover, we recently found that among other SNARE proteins, Syb is required for the correct placement of the receptor Ptc at the receiving cytoneme membrane, and therefore for correct Hh signaling (González-Méndez et al., 2020) (Figs. 3A and 4). In addition, mebrane potential differences have been found at Hh receiving cells upon signal activation, which promotes depolarization and enhances further signaling (Emmons-Bell & Hariharan, 2021; Spannl et al., 2020). Moreover, the structure of both transmembrane proteins Ptc and Disp have been described as RND transporters-like, able to transport cholesterol depending on Na+ gradient and controlling its availability at membranes which modulates Hh pathway activation (Qi, Di Minin, Vercellino, Wutz, & Korkhov, 2019; Zhang et al., 2018; Cannac et al., 2020). However, signal transferring of the lipidated and strongly membrane-attached Hh, either from vesicles or cytoneme membranes towards receiving cytonemes is still an unanswered question.

It is then possible that a mechanism is required for lipid-modified Hh to be released from the presenting membranes for its entrance into the receiving cells. Recently, a mechanism for transferring SHh from producing to receptor mammalian cells has been proposed, and implies successive SHh lipid-dependent steps (Wierbowski et al., 2020). This mechanism is in accordance with the proposed stepwise interaction of the Hh binding partners previous to transfer (Bilioni et al., 2013; Gradilla & Guerrero, 2013). Like SHh, the *Drosophila* Hh requires Disp1 for release, and Ihog/Boi (CDON/BOC) for reception. Although the ortholog of the secreted protein SCUBE2 is absent, the *Drosophila* Shifted (Shf) is also secreted and necessary for Hh release and long-distance signaling (Glise et al., 2005; Gorfinkiel, Sierra, Callejo, Ibañez, & Guerrero, 2005). In addition, Shf, like SCUBE2, is predicted by structural analysis to bind Hh through its lipid modifications (Liepinsh, Bányai, Patthy, & Otting, 2006.; Malinauskas, Aricescu, Lu, Siebold, & Jones, 2011). SCUBE2 interacts with SHh and its coreceptors BOC and CDON in mammalian cells (Wierbowski et al., 2020) and Shf in *Drosophila* with Hh, Ihog, Boi and glypicans (Gorfinkiel, Sierra, Callejo, Ibañez and Guerrero, 2005; Avanesov & Blair, 2013; Bilioni et al., 2013; Sánchez-Hernández, Sierra, Ortigão-Farias, & Guerrero, 2012).

Furthermore, it has been suggested that the Drosophila GPI anchored Dlp might perform for Hh reception a function similar to that of the vertebrate coreceptor GAS1 (Kim, Saunders, Hamaoka, Beachy, & Leahy, 2011; Wierbowski et al., 2020; Williams et al., 2010). Interestingly, Dlp has been recently shown to bind the Wg lipid moiety and proposed to be a potential Wg-lipid shield, handing off Wg to its receptor (McGough et al., 2020). Thus, Dlp could also act to tunnel lipid modified Hh to be presented to its receptor. In accordance, we have demonstrated that all these proteins, Hh, Disp, Dlp, Ihog and Shf, allocate within basal cytonemes for activation of Hh signaling (Bilioni et al., 2013; Callejo et al., 2011; González-Méndez et al., 2017). On the other hand, a shedding of lipids before reception has also been reported during Hh release, since the unprocessed palmitoylated peptides sterically inhibit Shh binding sites responsible for its interaction with Ptc (Schürmann et al., 2018). Nevertheless, both mechanisms for Hh release are compatible with the hypothesis of Hh signal transfer at cytoneme contact sites at a morphogenetic synapse (Bilioni et al., 2013; Callejo et al., 2011; González-Méndez et al., 2020, 2017) (Figs. 3A and 4).

## 7. Concluding remarks

Hh signaling is fundamental for development and adult tissue maintenance, and aberrant Hh signaling has been also linked to tumor progression. Lipid modified-Hh transfer through direct cell-to-cell membrane contact benefits from its anchoring to membranes, as it could grant signal-controlled delivery. *In silico* modeling of cytoneme-mediated signaling shows that dynamics of cytoneme extension and retraction, as well as that of signal transfer, are crucial aspects to ensure for correct Hh graded signaling.

On the other hand, recent research on the basis of cytoneme regulation points to the importance of pathway-specific key elements. These elements can coordinate the intracellular mechanisms for protrusion formation with extracellular guidance cues and cell adhesion components. In this line, the function of Ihog as both Hh co-receptor and cell adhesion protein is relevant to cytoneme behavior and cytoneme-mediated signal transfer. Equilibrium between Ihog interactions, including the Hh reception complex and glypicans of the ECM, might be directing Hh cytoneme dynamics and orientation. Furthermore, there is increasing evidence for intracellular trafficking to allocate Hh pathway elements at and along cytonemes, a requirement also shared with synaptic processes. However, more research is needed to clarify how the final signal transfer is achieved.

Therefore, cytoneme regulation is crucial to sustain spatial regulation of signaling, contributing to a robust mechanism of graded distribution and facilitating communication between different cell types during organogenesis. In addition, it has been shown in the *Drosophila* wing epithelium that genetic removal of cytonemes is sufficient to prevent tumor progression, improve adult survival and restore apico-basal cell polarity (Fereres, Hatori, Hatori, & Kornberg, 2019). Research directed to further elucidate cytoneme-mediated Hh signaling would then be crucial not only to understand developmental mechanisms, but also to improve our knowledge of cancer progression, including potential interactions between tumor cells and the surrounding normal tissues.

## Acknowledgments

We are grateful to Pedro Ripoll for his comments on the manuscript and to members of Guerrero's lab for their insights and discussions. This work was supported by grants BFU2017-83789-P, PID2020-114533GB-C21 and TENTACLES consortium RED2018-102411-T to I.G. from the Spanish Ministry of Science, Innovation and Universities and by institutional grants from the Fundación Areces and Banco de Santander to the CBMSO.

## References

Aguirre-Tamaral, A., & Guerrero, I. (2021). Improving the understanding of cytoneme-mediated morphogen gradients by in silico modeling. *PLoS Computational Biology*, *17*(8), e1009245. https://doi.org/10.1371/JOURNAL.PCBI.1009245.

Ali-Murthy, Z., & Kornberg, T. B. (2017). Erratum: Correction: Bicoid gradient formation and function in the Drosophila pre-syncytial blastoderm. *eLife*, *6*, e26811. https://doi.org/10.7554/eLife.26811.

Avanesov, A., & Blair, S. S. (2013). The drosophila WIF1 homolog shifted maintains glypican-independent hedgehog signaling and interacts with the hedgehog co-receptors Ihog and Boi. *Development*, *140*(1), 107–116. https://doi.org/10.1242/DEV.078444.

Bate, M., & Arias, A. M. (1991). The embryonic origin of imaginal discs in drosophila. *Development*, *112*(3), 755–761. https://doi.org/10.1242/DEV.112.3.755.

Beachy, P. A., Hymowitz, S. G., Lazarus, R. A., Leahy, D. J., & Siebold, C. (2010). Interactions between hedgehog proteins and their binding partners come into view. *Genes & Development*, *24*(18), 2001–2012. https://doi.org/10.1101/GAD.1951710.

Bilioni, A., Sánchez-Hernández, D., Callejo, A., Gradilla, A. C., Ibáñez, C., Mollica, E., et al. (2013). Balancing hedgehog, a retention and release equilibrium given by dally, Ihog, Boi and shifted/DmWif. *Developmental Biology*, *376*(2), 198–212. https://doi.org/10.1016/j.ydbio.2012.12.013.

Bischoff, M., Gradilla, A., Seijo, I., Andrés, G., Rodríguez-navas, C., González-méndez, L., et al. (2013). Cytonemes are required for the establishment of a normal hedgehog morphogen gradient in drosophila epithelia. *Nature Cell Biology*, *15*(11), 1269–1281. https://doi.org/10.1038/ncb2856.

Bohil, A. B., Robertson, B. W., & Cheney, R. E. (2006). Myosin-X is a molecular motor that functions in filopodia formation. *Proceedings of the National Academy of Sciences of the United States of America*, *103*(33), 12411–12416. https://doi.org/10.1073/PNAS.0602443103.

Bressloff, P. C., & Kim, H. (2019). Search-and-capture model of cytoneme-mediated morphogen gradient formation. *Physical Review E*, *99*(5), 052401. https://doi.org/10.1103/PhysRevE.99.052401.

Bumcrot, D. A., Takada, R., & McMahon, A. P. (1995). Proteolytic processing yields two secreted forms of sonic hedgehog. *Molecular and Cellular Biology*, *15*(4), 2294–2303. https://doi.org/10.1128/MCB.15.4.2294.

Callejo, A., Bilioni, A., Mollica, E., Gorfinkiel, N., Andrés, G., Ibáñez, C., et al. (2011). Dispatched mediates hedgehog basolateral release to form the long-range morphogenetic gradient in the drosophila wing disk epithelium. *Proceedings of the National Academy of Sciences of the United States of America, 108*(31), 12591–12598. https://doi.org/10.1073/pnas.1106881108.

Camp, D., Currie, K., Labbé, A., van Meyel, D. J., & Charron, F. (2010). Ihog and Boi are essential for hedgehog signaling in drosophila. *Neural Development, 5*(1). https://doi.org/10.1186/1749-8104-5-28.

Cannac, F., Qi, C., Falschlunger, J., Hausmann, G., Basler, K., & Korkhov, V. M. (2020). Cryo-EM strcture of the Hedgehog release protein Dispatched. *Science Advances, 6*(16), eaay7928. https://doi.org/10.1126/sciadv.aay7928.

Chen, W., Huang, H., Hatori, R., & Kornberg, T. B. (2017). Essential basal cytonemes take up hedgehog in the drosophila wing imaginal disc. *Development (Cambridge), 144*(17), 3134–3144. https://doi.org/10.1242/dev.149856.

D'Angelo, G., Matusek, T., Pizette, S., & Thérond, P. P. (2015). Endocytosis of hedgehog through dispatched regulates long-range signaling. *Developmental Cell, 32*(3), 290–303. https://doi.org/10.1016/J.DEVCEL.2014.12.004.

Desbordes, S. C., & Sanson, B. (2003). The glypican Dally-like is required for hedgehog signalling in the embryonic epidermis of drosophila. *Development, 130*(25), 6245–6255. https://doi.org/10.1242/dev.00874dev.00874 [pii].

Du, L., Sohr, A., & Roy, S. (2021). Glypiation enables FGF to self-regulate its tissue-specific dispersion and interpretation through cytonemes. *BioRxiv*. https://doi.org/10.1101/2021.02.23.432493.

Du, L., Sohr, A., Yan, G., & Roy, S. (2018). Feedback regulation of cytoneme-mediated transport shapes a tissue-specific FGF morphogen gradient. *eLife*, 7. https://doi.org/10.7554/ELIFE.38137.

Emmons-Bell, M., & Hariharan, I. (2021). Membrane potential regulates Hedgehog signalling in the Drosophila wing imaginal disc. *EMBO Reports, 22*(4), e51861. https://doi.org/10.15252/embr.202051861.

Everetts, N. J., Worley, M. I., Yasutomi, R., Yosef, N., & Hariharan, I. K. (2021). Single-cell transcriptomics of the drosophila wing disc reveals instructive epithelium-to-myoblast interactions. *eLife, 10*. https://doi.org/10.7554/ELIFE.61276.

Fancher, S., & Mugler, A. (2020). Diffusion vs. direct transpor in the precision of morphogen readout. *eLife, 9*(e58981), 1–27, e58981. https://doi.org/10.7554/eLife.58981.

Feinberg, E. H., VanHoven, M. K., Bendesky, A., Wang, G., Fetter, R. D., Shen, K., et al. (2008). GFP reconstitution across synaptic partners (GRASP) defines cell contacts and synapses in living nervous systems. *Neuron, 57*(3), 353–363. https://doi.org/10.1016/j.neuron.2007.11.030.

Fereres, S., Hatori, R., Hatori, M., & Kornberg, TB. (2019). Cytoneme-mediated signaling essential for tumorigenesis. *PLOS Genetics, 15*(9), e1008415. https://doi.org/10.1371/journal.pgen.1008415.

Glise, B., Miller, C. A., Crozatier, M., Halbisen, M. A., Wise, S., Olson, D. J., et al. (2005). Shifted, the drosophila ortholog of Wnt inhibitory factor-1, controls the distribution and movement of hedgehog. *Developmental Cell, 8*(2), 255–266. https://doi.org/S1534-5807(05)00005-5 [pii]10.1016/j.devcel.2005.01.003.

González-Méndez, L., Gradilla, A. C., & Guerrero, I. (2019). The cytoneme connection: Direct long-distance signal transfer during development. *Development (Cambridge), 146*(9). https://doi.org/10.1242/dev.174607.

González-Méndez, L., Gradilla, A.-C., Sánchez-Hernández, D., González, E., Aguirre-Tamaral, A., Jiménez-Jiménez, C., et al. (2020). Polarized sorting of patched enables cytoneme-mediated hedgehog reception in the Drosophila wing disc. *EMBO Journal, 39*(11). https://doi.org/10.15252/embj.2019103629.

González-Méndez, L., Seijo-Barandiarán, I., & Guerrero, I. (2017). Cytoneme-mediated cell-cell contacts for hedgehog reception. *eLife*, *6*. https://doi.org/10.7554/eLife.24045.

Gore, T., Matusek, T., D'Angelo, G., Giordano, C., Tognacci, T., Lavenant-Staccini, L., et al. (2021). The GTPase Rab8 differentially controls the long- and short-range activity of the hedgehog morphogen gradient by regulating hedgehog apico-basal distribution. *Development*, *148*(5). https://doi.org/10.1242/DEV.191791.

Gorfinkiel, N., Sierra, J., Callejo, A., Ibañez, C., & Guerrero, I. (2005). The drosophila ortholog of the human Wnt inhibitor factor shifted controls the diffusion of lipid-modified hedgehog. *Developmental Cell*, *8*(2), 241–253. https://doi.org/10.1016/j.devcel.2004.12.018.

Gradilla, A.-C., González, E., Seijo, I., Andrés, G., Bischoff, M., González-Mendez, L., et al. (2014). Exosomes as hedgehog carriers in cytoneme-mediated transport and secretion. *Nature Communications*, *5*, 5649. https://doi.org/10.1038/ncomms6649.

Gradilla, A. C., & Guerrero, I. (2013). Hedgehog on the move: A precise spatial control of hedgehog dispersion shapes the gradient. *Current Opinion in Genetics & Development*. https://doi.org/S0959-437X(13)00065-8 [pii]10.1016/j.gde.2013.04.011.

Gradilla, A.-C., Sanchez-Hernandez, D., Brunt, L., & Scholpp, S. (2018). From top to bottom: Cell polarity in hedgehog and Wnt trafficking. *BMC Biology*, *16*(1), 1–11. https://doi.org/10.1186/S12915-018-0511-X.

Hall, E. T., Cleverdon, E. R., & Ogden, S. K. (2019). Dispatching sonic hedgehog: Molecular mechanisms controlling deployment. *Trends in Cell Biology*, *29*(5), 385–395. https://doi.org/10.1016/J.TCB.2019.02.005.

Hall, E. T., Dillard, M. E., Stewart, D. P., Zhang, Y., Wagner, B., Levine, R. M., et al. (2021). Cytoneme delivery of sonic hedgehog from ligand-producing cells requires myosin 10 and a dispatched-boc/cdon co-receptor complex. *eLife*, *10*, 1–68. https://doi.org/10.7554/eLife.61432.

Han, C. H., Belenkaya, T. Y., Wang, B., & Lin, X. (2004). Drosophila glypicans control the cell-to-cell movement of hedgehog by a dynamin-independent process. *Development*, *131*(3), 601–611. https://doi.org/10.1242/dev.00958.

Hatori, R., & Kornberg, T. B. (2020). Hedgehog produced by the *Drosophila* wing imaginal disc induces distinct responses in three target tissues. *Development*, *147*(22). https://doi.org/10.1242/dev.195974.

Hatori, R., Wood, B. M., Barbosa, G., & Kornberg, T. B. (2021). Regulated delivery controls drosophila hedgehog, wingless and decapentaplegic signaling. *eLife*, *10*. https://doi.org/10.7554/ELIFE.71744.

Hsia, E. Y. C., Zhang, Y., Tran, H. S., Lim, A., Chou, Y. H., Lan, G., et al. (2017). Hedgehog mediated degradation of Ihog adhesion proteins modulates cell segregation in drosophila wing imaginal discs. *Nature. Communications*, *8*(1). https://doi.org/10.1038/s41467-017-01364-z.

Huang, H., & Kornberg, T. B. (2015). Myoblast cytonemes mediate Wg signaling from the wing imaginal disc and Delta-notch signaling to the air sac primordium. *eLife*, *4*(MAY). https://doi.org/10.7554/ELIFE.06114.

Huang, H., Liu, S., & Kornberg, T. B. (2019). Glutamate signaling at cytoneme synapses. *Science*, *363*(6430), 948–955. https://doi.org/10.1126/science.aat5053.

Hurbain, I., Macé, A.-S., Romao, M., Prince, E., Sengmanigov, L., Ruel, L., & D'Angelo, G. (2022). Microvilli-derived extracellular vesicles carry Hedgehog morphogenic signals for Drosophila wing imaginal disc development. *Current Biology*, *32*(2), 361–373. https://doi.org/10.1016/j.cub.2021.11.023.

Kim, M. S., Saunders, A. M., Hamaoka, B. Y., Beachy, P. A., & Leahy, D. J. (2011). Structure of the protein core of the glypican dally-like and localization of a region important for hedgehog signaling. *Proceedings of the National Academy of*

*Sciences of the United States of America*, *108*(32), 13112–13117. https://doi.org/1109877108 [pii]10.1073/pnas.1109877108.

Kuyyamudi, C., Menon, S. N., & Sinha, S. (2021). Contact-mediated cellular communication supplements positional information to regulate spatial patterning during development. *Physical Review E*, *103*(6), 062409. https://doi.org/10.1103/PhysRevE.103.062409.

Lee, J. J., von Kessler, D. P., Parks, S., & Beachy, P. A. (1992). Secretion and localized transcription suggest a role in positional signaling for products of the segmentation gene hedgehog. *Cell*, *71*(1), 33–50. https://doi.org/10.1016/0092-8674(92)90264-D.

Liepinsh, E., Bányai, L., Patthy, L., & Otting, G. (2006). NMR structure of the WIF domain of the human Wnt-inhibitory Factor-1. *Journal of Molecular Biology*, *357*(3), 942–950. https://doi.org/10.1016/j.jmb.2006.01.047.

Lum, L., Yao, S., Mozer, B., Rovescalli, A., Von Kessler, D., Nirenberg, M., et al. (2003). Identification of hedgehog pathway components by RNAi in drosophila cultured cells. *Science*, *299*(5615), 2039–2045. https://doi.org/10.1126/science.1081403.

Malinauskas, T., Aricescu, R., Lu, W., Siebold, C., & Jones, Y. (2011). 8 8 6 VOLUME 18 NUMBER 8 AUGUST 2011 nature structural & molecular biology. Nature Publishing Group. https://doi.org/10.1038/nsmb.2081.

Matusek, T., Wendler, F., Polès, S., Pizette, S., D'Angelo, G., Fürthauer, M., et al. (2014). The ESCRT machinery regulates the secretion and long-range activity of hedgehog. *Nature*, *516*(7529), 99–103. https://doi.org/10.1038/nature13847.

McGough, I. J., Vecchia, L., Bishop, B., Malinauskas, T., Beckett, K., Joshi, D., et al. (2020). Glypicans shield the Wnt lipid moiety to enable signalling at a distance. *Nature*, *585*(7823), 85–90. https://doi.org/10.1038/s41586-020-2498-z.

McLellan, J. S., Zheng, X., Hauk, G., Ghirlando, R., Beachy, P. A., & Leahy, D. J. (2008). The mode of hedgehog binding to Ihog homologues is not conserved across different phyla. *Nature*, *455*(7215), 979–983. https://doi.org/10.1038/nature07358.

Pepinsky, R. B., Zeng, C., Wen, D., Rayhorn, P., Baker, D. P., Williams, K. P., et al. (1998). Identification of a palmitic acid-modified form of human sonic hedgehog. *The Journal of Biological Chemistry*, *273*(22), 14037–14045. http://www.ncbi.nlm.nih.gov/pubmed/9593755.

Pizette, S., Matusek, T., Herpers, B., Thérond, P. P., & Rabouille, C. (2021). Hherisomes, Hedgehog specialized recycling endosomes, are required for high level Hedgehog signaling and tissue growth. *Journal of Cell Science*, *134*(10). https://doi.org/10.1242/JCS.258603.

Porter, J. A., von Kessler, D. P., Ekker, S. C., Young, K. E., Lee, J. J., Moses, K., et al. (1995). The product of hedgehog autoproteolytic cleavage active in local and long-range signalling. *Nature*, *374*(6520), 363–366. https://doi.org/10.1038/374363a0.

Porter, J. A., Young, K. E., & Beachy, P. A. (1996). Cholesterol modification of hedgehog signaling proteins in animal development. *Science*, *274*(5285), 255–259. https://doi.org/10.1126/science.274.5285.255.

Qi, C., Di Minin, G., Vercellino, I., Wutz, A., & Korkhov, V. M. (2019). Structural basis of sterol recognition by human hedgehog receptor PTCH1. *Science Advances*, *5*(9), eaaw6490. https://doi.org/10.1126/sciadv.aaw6490.

Ribatti, D., & Santoiemma, M. (2014). Epithelial-mesenchymal interactions: A fundamental developmental biology mechanism. *International Journal of Developmental Biology*, *58*(5), 303–306. https://doi.org/10.1387/IJDB.140143DR.

Rojas-Rios, P., Guerrero, I., & Gonzalez-Reyes, A. (2012). Cytoneme-mediated delivery of hedgehog regulates the expression of bone morphogenetic proteins to maintain germline stem cells in drosophila. *PLoS Biology*, *10*(4), e1001298. https://doi.org/10.1371/journal.pbio.1001298PBIOLOGY-D-11-02864 [pii].

Rosenbauer, J., Zhang, C., Mattes, B., Reinartz, I., Wedgwood, K., Schindler, S., et al. (2020). Modeling of Wnt-mediated tissue patterning in vertebrate embryogenesis. *PLoS Computational Biology*, *16*(6), e1007417. https://doi.org/10.1371/JOURNAL.PCBI.1007417.

Roy, S., Huang, H., Liu, S., & Kornberg, T. B. (2014). Cytoneme-mediated contact-dependent transport of the drosophila decapentaplegic signaling protein. *Science*, *343*(6173). https://doi.org/10.1126/SCIENCE.1244624.

Sánchez-Hernández, D., Sierra, J., Ortigão-Farias, J. R., & Guerrero, I. (2012). The WIF domain of the human and drosophila Wif-1 secreted factors confers specificity for Wnt or hedgehog. *Development*, *139*(20), 3849–3858. https://doi.org/10.1242/DEV.080028.

Sanders, T. A., Llagostera, E., & Barna, M. (2013). Specialized filopodia direct long-range transport of SHH during vertebrate tissue patterning. *Nature*, *497*(7451), 628–632. https://doi.org/10.1038/nature12157.

Schürmann, S., Steffes, G., Manikowski, D., Kastl, P., Malkus, U., Bandari, S., et al. (2018). Proteolytic processing of palmitoylated hedgehog peptides specifies the 3-4 intervein region of the drosophila wing. *eLife*, *7*. https://doi.org/10.7554/eLife.33033.

Simon, E., Jiménez-Jiménez, C., Seijo-Barandiarán, I., Aguilar, G., Sánchez-Hernández, D., Aguirre-Tamaral, A., et al. (2021). Glypicans define unique roles for the hedgehog co-receptors Boi and Ihog in cytoneme-mediated gradient formation. *eLife*, *10*. https://doi.org/10.7554/ELIFE.64581.

Spannl, S., Buhl, T., Nellas, I., Zeidan, S. A., Iyer, K. V., Khaliullina, H., & Eaton, S. (2020). Glycolysis regulates Hedgehog signalling via the plasma membrane potential. *EMBO Journal*, *39*(21), e101767. https://doi.org/10.15252/embj.2019101767.

Stewart, D. P., Marada, S., Bodeen, W. J., Truong, A., Sakurada, S. M., Pandit, T., et al. (2018). Cleavage activates dispatched for sonic hedgehog ligand release. *eLife*, *7*. https://doi.org/10.7554/ELIFE.31678.

Torroja, C., Gorfinkiel, N., & Guerrero, I. (2004). Patched controls the hedgehog gradient by endocytosis in a dynamin-dependent manner, but this internalization does not play a major role in signal transduction. *Development*, *131*(10), 2395–2408. http://www.ncbi.nlm.nih.gov/entrez/query.fcgi?cmd=Retrieve&db=PubMed&dopt=Citation&list_uids=15102702.

Wierbowski, B. M., Petrov, K., Aravena, L., Gu, G., Xu, Y., & Salic, A. (2020). Hedgehog pathway activation requires coreceptor-catalyzed, lipid-dependent relay of the sonic hedgehog ligand. *Developmental Cell*, *55*(4), 450–467. e8. https://doi.org/10.1016/j.devcel.2020.09.017.

Williams, E. H., Pappano, W. N., Saunders, A. M., Kim, M. S., Leahy, D. J., & Beachy, P. A. (2010). Dally-like core protein and its mammalian homologues mediate stimulatory and inhibitory effects on hedgehog signal response. *Proceedings of the National Academy of Sciences of the United States of America*, *107*(13), 5869–5874. https://doi.org/10.1073/pnas.1001777107.

Wood, B. M., Baena, V., Huang, H., Jorgens, D. M., Terasaki, M., & Kornberg, T. B. (2021). Cytonemes with complex geometries and composition extend into invaginations of target cells. *Journal of Cell Biology*, *220*(5). https://doi.org/10.1083/JCB.202101116.

Yamazaki, Y., Palmer, L., Alexandre, C., Kakugawa, S., Beckett, K., Gaugue, I., et al. (2016). Godzilla-dependent transcytosis promotes Wingless signalling in Drosophila wing imaginal discs. *Nature Cell Biology*, *18*(4), 451–457. https://doi.org/10.1038/ncb3325.

Yang, S., Zhang, Y., Yang, C., Wu, X., El Oud, S. M., Chen, R., et al. (2021). Competitive coordination of the dual roles of the hedgehog co-receptor in homophilic adhesion and signal reception. *eLife*, *10*. https://doi.org/10.7554/ELIFE.65770.

Yao, S., Lum, L., & Beachy, P. (2006). The ihog cell-surface proteins bind hedgehog and mediate pathway activation. *Cell, 125*(2), 343–357. http://www.ncbi.nlm.nih.gov/entrez/query.fcgi?cmd=Retrieve&db=PubMed&dopt=Citation&list_uids=16630821.

Yao, J., Munson, K. M., Webb, W. W., & Lis, J. T. (2006). Dynamics of heat shock factor association with native gene loci in living cells. *Nature*. https://doi.org/10.1038/nature05025.

Zhang, Y., Bulkley, D. P., Xin, Y., Roberts, K., Asarnow, D. E., Sharma, A., & Beachy, P. A. (2018). Structural basis for cholesterol transport-like activity of the Hedgehog receptor Patched. *Cell, 175*(5), e14. 1352–1364. https://doi.org/10.1016/j.cell.2018.10.026.

Zhang, Z., Denans, N., Liu, Y., Zhulyn, O., Rosenblatt, H. D., Wernig, M., et al. (2021). Optogenetic manipulation of cellular communication using engineered myosin motors. *Nature Cell Biology, 23*(2), 198–208. https://doi.org/10.1038/s41556-020-00625-2.

Zhang, C., & Scholpp, S. (2019). Cytonemes in development. In *Vol. 57. Current opinion in genetics and development* (pp. 25–30). Elsevier Ltd. https://doi.org/10.1016/j.gde.2019.06.005.

Zheng, X., Mann, R. K., Sever, N., Zheng, X., Mann, R. K., Sever, N., et al. (2010). Genetic and biochemical definition of the hedgehog receptor. *Genes and Development, 24*(1), 57–71. https://doi.org/10.1101/gad.1870310.

CHAPTER TWO

# Receptor control by membrane-tethered ubiquitin ligases in development and tissue homeostasis

**Andres M. Lebensohn[a,*], J. Fernando Bazan[b,*], and Rajat Rohatgi[c,d,*]**
[a]Laboratory of Cellular and Molecular Biology, Center for Cancer Research, National Cancer Institute, National Institutes of Health, Bethesda, MD, United States
[b]ℏ bioconsulting llc, Stillwater, MN, United States
[c]Department of Biochemistry, Stanford University School of Medicine, Stanford, CA, United States
[d]Department of Medicine, Stanford University School of Medicine, Stanford, CA, United States
*Corresponding authors: e-mail address: andres.lebensohn@nih.gov; jfbazan@gmail.com; rrohatgi@stanford.edu

## Contents



*Current Topics in Developmental Biology*, Volume 150
ISSN 0070-2153
https://doi.org/10.1016/bs.ctdb.2022.03.003

## Abstract

Paracrine cell-cell communication is central to all developmental processes, ranging from cell diversification to patterning and morphogenesis. Precise calibration of signaling strength is essential for the fidelity of tissue formation during embryogenesis and tissue maintenance in adults. Membrane-tethered ubiquitin ligases can control the sensitivity of target cells to secreted ligands by regulating the abundance of signaling receptors at the cell surface. We discuss two examples of this emerging concept in signaling: (1) the transmembrane ubiquitin ligases ZNRF3 and RNF43 that regulate WNT and bone morphogenetic protein receptor abundance in response to R-spondin ligands and (2) the membrane-recruited ubiquitin ligase MGRN1 that controls Hedgehog and melanocortin receptor abundance. We focus on the mechanistic logic of these systems, illustrated by structural and protein interaction models enabled by AlphaFold. We suggest that membrane-tethered ubiquitin ligases play a widespread role in remodeling the cell surface proteome to control responses to extracellular ligands in diverse biological processes.


## Abbreviations

| | |
|---|---|
| **AGRP** | agouti-related protein |
| **AML** | acute myeloid leukemia |
| **ASGR** | asialoglycoprotein receptor |
| **ASP** | agouti signaling protein |
| **ATRN** | attractin |
| **BMP** | bone morphogenetic protein |
| **BMPR1A** | bone morphogenetic protein receptor type-1A |
| **BR** | basic region |
| **CLD** | cyclophilin-like domain |
| **CRD** | Cys-rich domain |
| **CUB** | complement C1r/C1s, uegf, bmp1 |
| **DEP** | dishevelled, egl-10, pleckstrin |
| **DIR** | dishevelled-interaction region |
| **DIX** | dishevelled, axin |
| **DUB** | deubiquitylating enzyme |
| **DVL** | dishevelled |
| **E1** | ubiquitin activating enzyme |
| **E2** | ubiquitin conjugating enzyme |
| **E3** | ubiquitin ligase enzyme |
| **ECD** | extracellular domain |
| **EGFL** | epidermal growth factor-like |
| **ER** | endoplasmic reticulum |
| **ERAD** | ER-associated degradation |
| **ESCORT-I** | endosomal sorting complex required for transport-I |
| **FU** | furin-like repeat |
| **FZD** | frizzled |
| **GDU1** | glutamine dumper-1 |
| **GPC** | glypican |
| **GPCR** | G protein-coupled receptor |
| **GPI** | glycosylphosphatidylinositol |

| | |
|---|---|
| **GRAIL** | gene related to anergy in lymphocytes |
| **HECT** | homologous to the E6-AP carboxyl terminus |
| **Hh** | Hedgehog |
| **HRR** | His-rich region |
| **HS** | heparan sulfate |
| **HSPG** | heparan sulfate proteoglycan |
| **ICD** | intracellular domain |
| **KO** | knock-out |
| **KSHV** | Kaposi's sarcoma herpesvirus |
| **LDLR** | low-density lipoprotein receptor |
| **LGR** | Leu-rich repeat G protein-coupled receptor |
| **LRP** | low-density lipoprotein receptor-related protein |
| **LRR** | Leu-rich repeat |
| **MARCH** | membrane-associated RING-CH |
| **MEGF8** | multiple EGFL domains protein 8 |
| **MGRN1** | mahogunin RING finger 1 |
| **MHC-I** | class I major histocompatibility complex |
| **MIR2** | modulator of immune recognition 2 |
| **MMM** | MGRN1, MEGF8 and MOSMO |
| **PA** | protease associated |
| **PD-L1** | programmed death-ligand 1 |
| **PDZ** | postsynaptic density 95, discs large, zona occludens-1 |
| **PE** | β-propeller/epidermal growth factor-like |
| **PROTAC** | proteolysis-targeting chimera |
| **PRR** | Pro-rich region |
| **PTCH1** | patched 1 |
| **RBR** | RING-between-RING |
| **RING** | really interesting new gene |
| **RSPO** | R-spondin |
| ***S. pombe*** | *Saccharomyces pombe* |
| **scFv** | single-chain variable fragment |
| **SDC** | syndecan |
| **SHH** | Sonic Hedgehog |
| **SMO** | smoothened |
| **SNARE** | SNAP receptor |
| **SRR** | Ser-rich region |
| **SUFU** | suppressor of fused |
| **TETAMS** | tetra-amelia with lung hypo/aplasia syndrome |
| **TFR1** | transferrin receptor 1 |
| **TM** | transmembrane |
| **TMD** | TM domain |
| **TSG101** | tumor suppressor gene 101 |
| **TSP** | thrombospondin type 1 |
| **Ub** | ubiquitin |
| **WNT/PCP** | WNT/planar cell polarity |
| **WNT/STOP** | WNT-dependent stabilization of proteins |
| **WT** | wild type |

## 1. Introduction

The fates of signaling receptors and other membrane proteins are regulated by ubiquitylation during all stages of their life cycle: protein quality control in the endoplasmic reticulum (ER), sorting, trafficking and expulsion into exosomes, endocytic clearance, and lysosomal degradation (Foot, Henshall, & Kumar, 2017). The attachment of ubiquitin (Ub) and polyubiquitin chains to Lys residues on a target protein is carried out by the intricate interplay between three conserved families of enzymes, whose structures and molecular mechanisms have been revealed by various approaches (Cappadocia & Lima, 2018; Hershko, Ciechanover, & Varshavsky, 2000). First, Ub is activated by its attachment to a ubiquitin activating enzyme (E1) through a thioester linkage. Second, the activated Ub is transferred to a Cys on a ubiquitin conjugating enzyme (E2). Finally, ubiquitin ligase enzymes (E3s) facilitate the transfer of Ub from the E2 to specific Lys residues on a substrate protein or to a previously conjugated Ub in a growing polyubiquitin chain. Given the presence of seven Lys residues that decorate the Ub surface, linear or branched chains containing various combinations of Ub linkages can be attached to the substrate, and this topologically diverse "Ub code" can drive different outcomes (Komander & Rape, 2012; Kwon & Ciechanover, 2017). E3s provide the crucial substrate specificity to the ubiquitylation reaction, and sometimes this recognition event requires the assembly of large multiprotein complexes (Harper & Schulman, 2021; Morreale & Walden, 2016; Zheng & Shabek, 2017). The really interesting new gene (RING) E3s comprise the largest family (~600 members), characterized by the presence of a compact RING domain nucleated by two bound $Zn^{2+}$ ions (Deshaies & Joazeiro, 2009). RING domains recruit a Ub-charged E2 and position it optimally for transfer of Ub to a substrate that is captured by a separate recognition module (Metzger, Pruneda, Klevit, & Weissman, 2014). Homologous to the E6-AP carboxyl terminus (HECT) E3s (~29 members) and RING-between-RING (RBR) E3s (13 members) mediate Ub transfer through a two-step process involving a thioester intermediate between Ub and a catalytic Cys on the E3 itself prior to the transfer of Ub onto the substrate (Zheng & Shabek, 2017). An additional level of regulation is afforded by ~100 deubiquitylating enzymes (DUBs) that remove Ub from proteins (Clague, Urbé, & Komander, 2019).

In this chapter we focus on Ub modifications performed by RING E3s that have recognizable transmembrane (TM) helices. Approximately 50 of the ~600 annotated RING E3s fall into this class (Fenech et al., 2020; Li et al., 2008; Neutzner et al., 2011). However, the actual number of membrane-tethered E3s may be significantly larger, since cytoplasmic E3s can be recruited to the plasma membrane by stable association with a TM co-receptor, and such complexes are difficult to predict by sequence analysis alone. We use the term "membrane-tethered" to refer to both classes of E3s—those that are anchored to the membrane by an intrinsic TM domain and those that are recruited by non-covalent association with a TM protein. Much of the research in this area has been on TM E3s that function in the ER as part of the ER-associated degradation (ERAD) system and other protein quality control pathways (Fenech et al., 2020; Foot et al., 2017; Sardana & Emr, 2021). For example, TM E3s such as the prototype yeast protein Hrd1 ubiquitylate misfolded ER proteins that are retro-translocated through a pore-like assembly to the cytoplasm, tagging them for proteasomal degradation (Phillips & Miller, 2021). However, a growing number of structurally distinct membrane-tethered E3s have been shown to function outside of the ER to regulate the abundance of signaling receptors at the cell surface, and consequently the sensitivity of cells to signaling ligands. We will describe two such systems that function in developmental signaling pathways to control tissue patterning and morphogenesis, as well as in stem cell self-renewal, tissue homeostasis and regeneration. We anticipate that regulation of signaling strength in target cells—the cells exposed to signaling ligands—by membrane-tethered E3s will emerge as a general control mechanism in signaling pathways beyond those discussed in this chapter.

The recognition mechanisms that these membrane-tethered E3s employ to bind their targets and position their RING domains for effective Ub transfer to the cytoplasmic chains of substrate receptors remain largely unknown. We take advantage of the recent advances in the prediction of protein folds and protein-protein interactions by deep learning-based programs like AlphaFold and RoseTTaFold (Baek et al., 2021; Bryant, Pozzati, & Elofsson, 2021; Evans et al., 2021; Jumper et al., 2021; Tunyasuvunakool et al., 2021) to create models of multimodular TM E3 complexes. These methods are useful to generate hypotheses for how E3s recognize substrates through extracellular, TM and intracellular contacts, and how they may themselves be regulated by ligands. *We note that all the structures shown in the figures represent AlphaFold models unless indicated otherwise.*

# 2. Classification of membrane-tethered E3s

Excluding the E3s involved in protein quality control pathways in the ER, membrane-tethered E3s fall into three broad architectural classes (Fig. 1).

## 2.1 MARCH family TM E3s

Homologs of the membrane-associated RING-CH (MARCH) proteins were first identified as gene products that allow viruses to evade the host immune response by downregulating class I major histocompatibility complex (MHC-I) proteins (reviewed in Bauer, Bakke, & Preben Morth, 2017). MARCH proteins have been implicated in regulating the cell surface expression and trafficking of many single-pass TM proteins that play a role in T-cell activation: class I and II MHC proteins (antigen presentation), ICAM-1 (cell-cell adhesion), CD4 (T-cell co-receptor), CD86 (co-stimulatory signal), and cytokine receptors. Eleven MARCH family members have been recognized by the close similarity of their distinctive RING domains. Seven of these (MAR1–4, 8–9 and 11) contain a tight hairpin composed of two TM helices that follows an N-terminal RING module, two of them (MAR5–6) have more complex arrangements of multiple angled TM stretches, and two outliers (MAR7 and 10) have a single C-terminal TM helix (Figs. 1A and 2A). The compact RING-TM-TM portion of the major group of MARCH TM E3s is predicted to form their only structured part, although their cytoplasmic chains, composed largely of long disordered segments at both the N- and C-termini, likely carry cryptic modification sites and short interaction motifs. This conserved, ~160 residue-long RING-TM-TM module is capable of both recruiting a Ub-charged E2 via its juxtamembrane RING domain and recognizing the substrate to catalyze Ub transfer. Therefore, substrate recognition likely involves intra-membrane binding of one or multiple TM helices in the substrate to the MARCH TM hairpin motif.

Modeling of the shared RING-TM-TM module reveals that the MARCH RING domain is bipartite, built primarily by the canonical $Zn^{2+}$-binding motif located just before the TM hairpin, but completed by a conserved β-strand that immediately follows the second TM helix (Fig. 2A). As a result, the MARCH RING domain is closely juxtaposed to the hairpin TM structure at the level of the inner leaflet of the plasma membrane, and may be uniquely responsive to structural rearrangements within the TM hairpin motif upon substrate recognition inside the plasma membrane (Trenker et al., 2021). The more complex TM architectures of MAR5 and MAR6 still display this bipartite RING domain structure, but with some

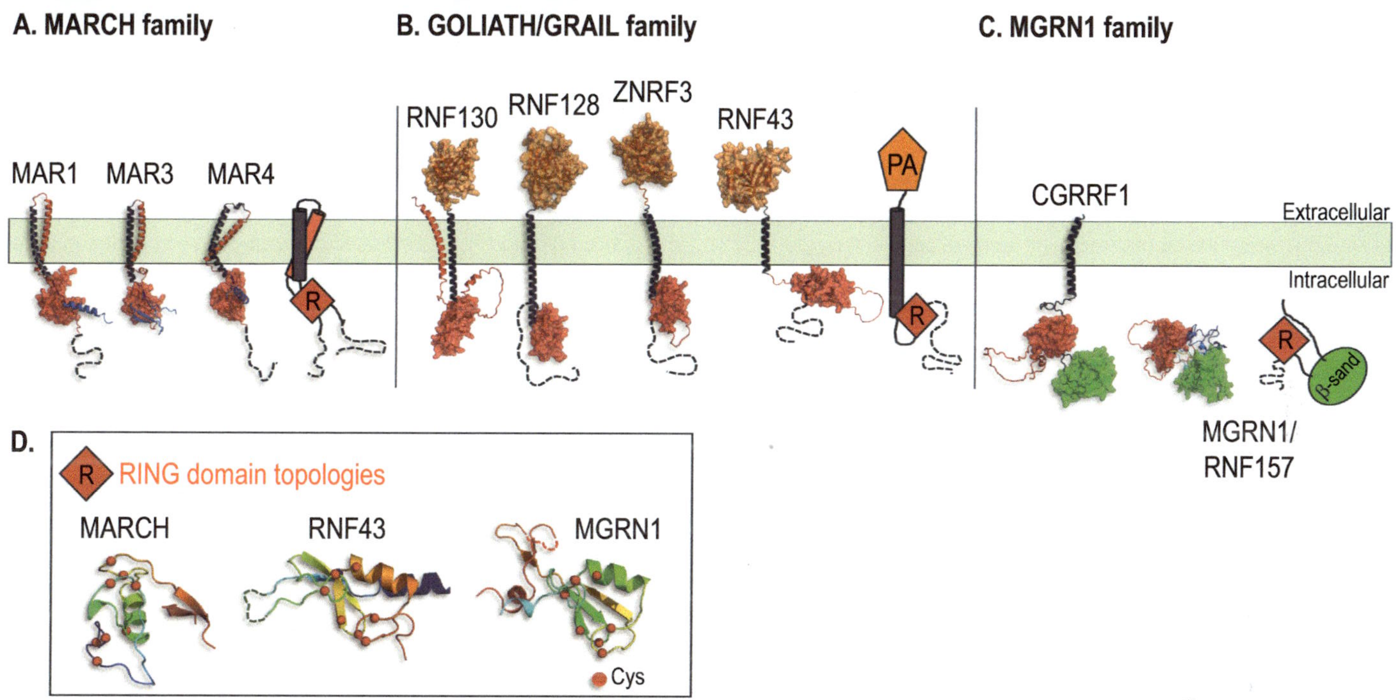


**Fig. 1** Structural models of the three main classes of membrane-tethered E3s. (A–C) AlphaFold models of representative members of the MARCH (A), GOLIATH/GRAIL (B) and MGRN1 (C) E3 families, with cartoons used throughout the figures to represent each family. (A) In the MARCH family, substrate recognition is accomplished by two closely linked TM helices (gray and red) folded as a hairpin, and Ub transfer is catalyzed by a tightly associated RING domain. (B) Members of the GOLIATH/GRAIL family contain an extracellular PA domain that can bind to ligands and serve in substrate recognition. (C) The MGRN1 family is characterized by a RING domain juxtaposed to a putative substrate-binding β-sandwich domain (β-sand, green). MGRN1 and RNF157 lack TM helices, but are recruited to the membrane by interactions with single pass TM proteins (see Figs. 6 and 7), while CGRRF1 is tethered to the membrane by a single TM helix. (D) Topologies of the RING domains in one representative member of each of the three E3 families shown. In this and all subsequent figures, the RING domain is shown as a red space-filling model in the structural representations and as a red diamond labeled "R" in the cartoons. All the structures shown in the figures represent AlphaFold models, unless indicated otherwise with a Protein Data Bank (PDB) ID shown in italics. All structures are drawn to the same scale within each figure, except for structures shown in boxes. Dotted lines denote unstructured segments of the proteins for which folds could not be predicted. Molecular graphics were generated with PyMOL (www.pymol.org).

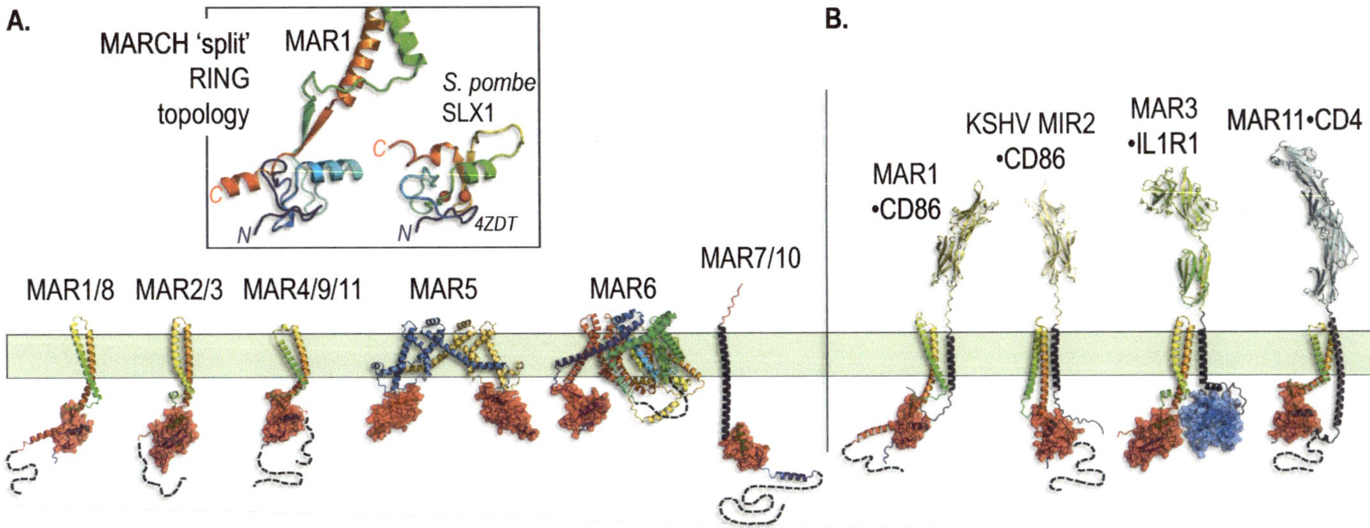


**Fig. 2** The MARCH family TM E3s and their substrate recognition mechanisms. (A) AlphaFold models of representatives of the 11 MARCH family members (MAR1/8, MAR2/3, MAR4/9/11 and MAR7/10 have similar structures, so only one of each group is shown in the figure). The unique "split" RING topology is highlighted in the box (see main text for description). For comparison, the bipartite RING domain of MAR1 is shown next to the RING domain of the *Saccharomyces pombe* (*S. pombe*) protein SLX1 (PDB ID 4ZDT) (Lian, Xie, & Qian, 2016). (B) Models of MARCH family members bound to their substrates highlight the importance of interactions between TM helices within the plane of the membrane. CD86 can be targeted by both MAR1 and the viral homolog MIR2 with slightly divergent folds and mechanisms.

variations. MAR5 adopts a predicted dimer fold that completes each of the two RING domains with a polypeptide chain from their respective partner subunits (Fig. 2A). MAR6 reunites the N-terminal portion of the RING domain with a β-strand that follows the C-terminal TM helix, with an intervening 610-residue sequence that crosses the plasma membrane multiple times (Fig. 2A).

The clearest indication that the distinctive MARCH hairpin TM structure is responsible for substrate recognition comes from a comparative study describing the engagement and Ub modification of CD86, a protein that provides co-stimulatory signals to T cells, by two distinct MARCH-class E3s: human MAR1 and modulator of immune recognition 2 (MIR2), a viral MARCH homolog from Kaposi's sarcoma herpesvirus (KSHV) (Fig. 2B) (Trenker et al., 2021).

## 2.2 GOLIATH/GRAIL family TM E3s

Members of the GOLIATH/GRAIL family of TM E3s have a common domain architecture: an N-terminal extracellular protease associated (PA) domain connected by a linker of varying length to a single TM helix, closely followed by a cytoplasmic RING domain (Figs. 1B and 3A). In the human proteome, we find 12 PA-TM-RING E3s (RNF13, 43, 128, 130, 133, 148–150, 167, 204, 215, and ZNRF3) and two outlier members that lack the PA domain (RNF24 and RNF122). In some of these E3s, the RING domain is predicted to pack against the last two turns of an extended TM helix, restricting their conformational flexibility (Fig. 3A). One point of variability between the members of this family revealed by AlphaFold modeling is the seamless extension of the TM helix into an amphipathic cytoplasmic helix, which forms a rigid scaffold that positions the RING domain at different distances from the plasma membrane. The distance ranges from practically no extension of the TM helix (as in the case of RNF43, in which the RING domain is connected through a linker to a short cytoplasmic extension of the helix) to 5 helical turns (for ZNRF3) or even 8 helical turns (for RNF130, also known as GOLIATH). RNF130 has a second, C-terminally distal TM helix that packs against the canonical TM helix (Fig. 3A) in a manner reminiscent of some MARCH family E3s (Fig. 2).

The best studied of the PA-TM-RING proteins is RNF128, also known as gene related to anergy in lymphocytes (GRAIL) (reviewed in Whiting, Su, Lin, & Garrison Fathman, 2011). RNF128 suppresses T-cell responsiveness and cytokine transcription by ubiquitylating and down-regulating multiple cell surface molecules involved in T-cell activation,

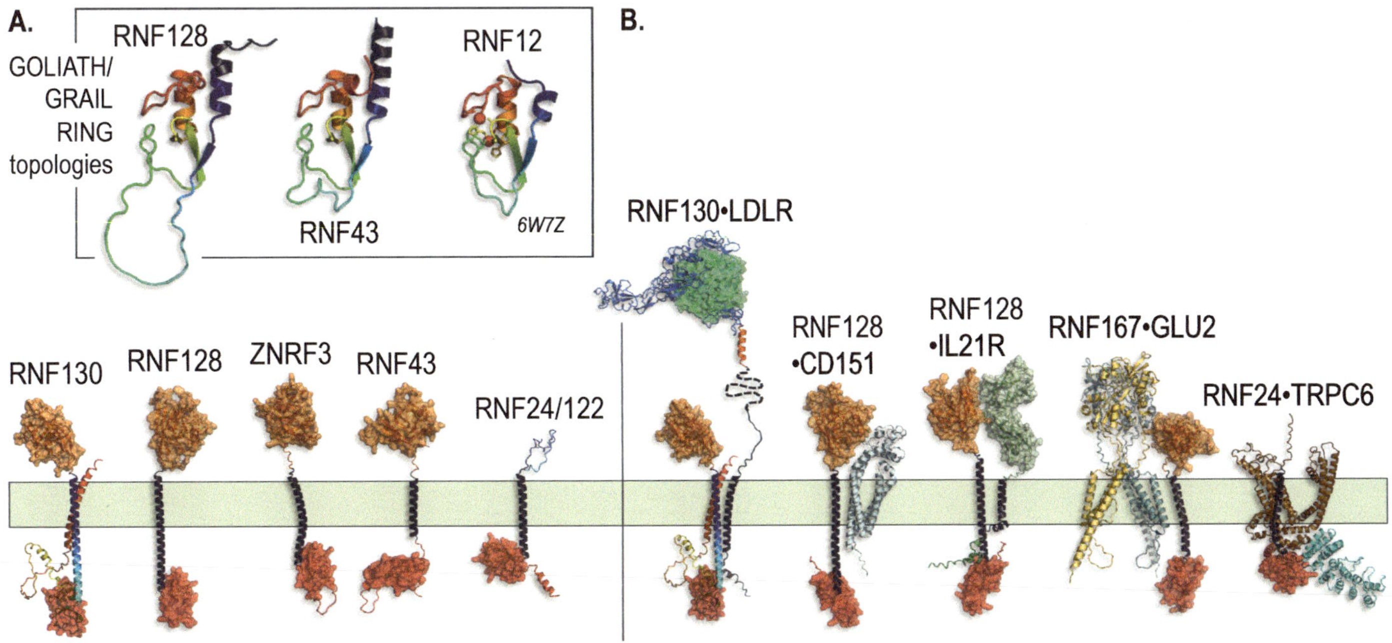


**Fig. 3** The GOLIATH/GRAIL family TM E3s and their substrate recognition mechanisms. (A) AlphaFold models of GOLIATH/GRAIL family members (RNF24/122 have similar structures, so only one of them is shown). The RING domain topologies for RNF128 and RNF43 are highlighted in the box. While no structures of the RING domain of GOLIATH/GRAIL family members have been solved, the RING domain most closely resembles that of the crystal structure of RNF12 (PDB ID 6W7Z) (Middleton, Zhu, & Day, 2020), shown for comparison. (B) AlphaFold models of GOLIATH/GRAIL family members interacting with their substrates suggest the importance of recognition events that span extracellular, TM and intracellular domains. The PA domain (orange) of RNF128 binds to the extracellular domains of substrates (Lineberry, Leon, Soares, & Garrison Fathman, 2008).

including CD83, CD81, CD151 and CD40L (Lineberry et al., 2008; Su, Iwai, Lin, & Garrison Fathman, 2009). The *Drosophila* GOLIATH family members have been shown to ubiquitylate the SNAP receptor (SNARE) protein VAMP3: loss-of-function mutations in GOLIATH or GODZILLA in flies result in the accumulation of membrane proteins in Rab5-positive giant endosomes due to defects in recycling endosome trafficking (Yamazaki et al., 2013). ZNRF3 and RNF43, which belong to a distinct branch of the GOLIATH/GRAIL family, regulate the cell surface abundance of WNT and bone morphogenetic protein (BMP) signaling receptors (Hao et al., 2012; Koo et al., 2012; Lee, Seidl, Sun, Glinka, & Niehrs, 2020). PLR-1, a relative of this family in *C. elegans*, also regulates the density of multiple receptors for WNT ligands on the plasma membrane (Moffat, Robinson, Bakoulis, & Clark, 2014). While the mechanism of substrate recognition by these PA-TM-RING proteins remains incompletely understood, studies of RNF128 suggest that the PA domain directly binds to the extracellular domains of substrate TM proteins, recruiting them for ubiquitylation by the cytoplasmic RING domain (Fig. 3B) (Lineberry et al., 2008). Thus, substrate recognition and ubiquitylation are segregated on opposite sides of the plasma membrane. However, ZNRF3 and RNF43 may require a cytoplasmic adaptor protein for substrate recognition, as discussed later (Jiang, Charlat, Zamponi, Yang, & Cong, 2015).

## 2.3 MGRN1 family membrane-recruited E3s

E3s lacking a TM helix can nevertheless be tightly tethered to the plasma membrane via direct, non-covalent association with an integral membrane protein. Mahogunin RING finger 1 (MGRN1, also known as RNF156) and its vertebrate-specific paralog RNF157 are the only examples of such E3s described to date (Fig. 1C). These E3s are associated with two single-pass TM proteins to regulate Hedgehog and melanocortin receptor signaling (He, Eldridge, Jackson, Gunn, & Barsh, 2003; Kong et al., 2020). Interestingly, MGRN1 and RNF157 are related to CGRRF1 (also known as RNF197), an E3 that is anchored to the membrane by a single N-terminal TM segment but lacks an extracellular domain, and has been implicated in ERAD (Fig. 1C) (Glaeser et al., 2018). It is likely that other cytoplasmic E3s also associate with TM partners to ubiquitylate membrane proteins, but the cytoplasmic sequence motifs or cryptic structural modules in the co-receptors that drive complex formation have not been cataloged.

In the following sections we elaborate in depth on how membrane-tethered E3s control signaling receptors on the cell surface. We focus on one example of TM E3s—ZNRF3 and RNF43—and one example of a membrane-recruited E3—MGRN1—within the context of the developmental and tissue homeostasis signaling systems in which they have been best characterized.

## 3. The R-spondin-ZNRF3/RNF43 signaling system tunes WNT and BMP receptor abundance

The ZNRF3 and RNF43 PA-TM-RING E3s have been most extensively studied in the context of the R-spondin (RSPO) system, a signaling module that tunes the abundance of cell surface receptors in the WNT (Hoppler & Moon, 2014) and BMP (Derynck & Miyazono, 2017) pathways by regulated ubiquitylation, endocytosis and lysosomal degradation (see review by Niehrs, 2012 for a timeline of the discovery and initial characterization of the RSPO system). Recent work has also uncovered ZNRF3/RNF43-independent roles for RSPOs as WNT pathway agonists (Carmon, Gong, Yi, Thomas, & Liu, 2014) and antagonists (Reis & Sokol, 2021), and in regulating other signaling pathways including TGFβ (Zhou et al., 2017), ERK/FGF (Reis & Sokol, 2020; Zhang et al., 2017), EGFR (Stevens & Williams, 2021; Yue et al., 2021), MAPK (Zheng et al., 2020) and estrogen receptor regulation via cAMP-PKA signaling (Geng et al., 2020). Since these systems do not use ZNRF3/RNF43, which is the focus of this chapter, we will not discuss them further.

The four members of the RSPO family of secreted glycoproteins (RSPO1–4) were discovered in close succession and immediately linked to activation of WNT/β-catenin signaling or stabilization of β-catenin (Chen et al., 2002; Kamata et al., 2004; Kazanskaya et al., 2004; Kim et al., 2005). The first report describing RSPO2 also suggested it may negatively regulate TGF-β signaling, but it was unclear if this was a secondary consequence of WNT signaling modulation or an independent effect (Kazanskaya et al., 2004). RSPOs were later also linked to regulation of β-catenin-independent WNT signaling, in particular the WNT/planar cell polarity (WNT/PCP) pathway (Ohkawara, Glinka, & Niehrs, 2011). However, the precise mechanism of WNT/β-catenin signaling regulation by RSPOs remained unclear, and their receptors unknown.

Leu-rich repeat G protein-coupled receptor 5 (LGR5) was discovered as a common WNT target gene in normal intestinal crypts and in colon cancer, and was later shown to be an exquisite marker of many types of WNT-driven adult stem cells (reviewed in Barker, Tan, & Clevers, 2013;

de Lau, Peng, Gros, & Clevers, 2014). LGR5 and its close paralogs LGR4 and LGR6 (throughout the chapter, we refer jointly to these three members of the LGR family as "LGRs") were independently identified as RSPO receptors by several groups (Carmon, Gong, Lin, Thomas, & Liu, 2011; de Lau et al., 2011; Glinka et al., 2011; Ruffner et al., 2012). Shortly thereafter, ZNRF3 and RNF43 were described as the effectors of RSPO signaling (Hao et al., 2012; Koo et al., 2012). ZNRF3 and RNF43 target WNT receptors for ubiquitylation and lysosomal degradation, and binding of RSPOs to both LGRs and ZNRF3/RNF43 prevents this process by promoting clearance of ZNRF3/RNF43 from the plasma membrane. Thus, the outcome of RSPO signaling through this mechanism is the accumulation of WNT receptors at the plasma membrane, which results in increased sensitivity of cells to WNT ligands. Additionally, the heparan sulfate proteoglycan (HSPG) syndecan 4 was also identified as an RSPO3 receptor involved in activation of WNT/PCP signaling (Ohkawara et al., 2011).

Experiments in cells and mice lacking LGR4/5/6 then led to the discovery that RSPO2 and RSPO3 can signal independently of LGRs (Lebensohn & Rohatgi, 2018; Szenker-Ravi et al., 2018). Similar findings were reported in 293T cells lacking LGR4 (Park et al., 2018). LGR-independent signaling was shown to be physiologically relevant, since mice lacking LGR4/5/6 did not exhibit many of the phenotypes observed in mice lacking RSPO2 or RSPO3, suggesting that RSPO2 and RSPO3 could still promote signaling in *Lgr4/5/6* triple knock-out (KO) mice (Szenker-Ravi et al., 2018). In the absence of LGRs, RSPOs were shown to use HSPGs such as glypicans (GPCs) and syndecans (SDCs) as alternative receptors to promote potentiation of WNT/β-catenin signaling through a mechanism that still required interactions between RSPOs and ZNRF3/RNF43, as well as internalization of RNF43 (Dubey et al., 2020; Lebensohn & Rohatgi, 2018).

More recently, RSPO2 and RSPO3 were shown to downregulate Type I BMP receptor levels through another LGR-independent mechanism (Lee et al., 2020). In this context, RSPO binding to ZNRF3/RNF43 and to the BMP receptor bone morphogenetic protein receptor type-1A (BMPR1A, also known as ALK3) promoted internalization and degradation of BMPR1A. This mechanism is very different from the way in which RSPOs regulate WNT receptor levels: binding of RSPOs to ZNRF3/RNF43 and BMPR1A directly downregulates BMPR1A levels, whereas binding of RSPOs to ZNRF3/RNF43, LGRs and/or HSPGs indirectly upregulates WNT receptors by preventing ZNRF3/RNF43 from inducing the ubiquitin-dependent internalization and lysosomal degradation of WNT receptors.

In the following sections, we first describe the system architecture of these different RSPO-ZNRF3/RNF43 signaling modalities, including the protein components, their relevant domains and interactions, and some of the post-translational regulation relevant to their signaling properties. We then discuss the mechanisms for each of the three signaling modalities, considering similarities and differences between them. Finally we discuss some physiological and pathological contexts in which these divergent signaling modalities operate, and consider the prospect of leveraging the modular nature of the RSPO-ZNRF3/RNF43 signaling system for therapeutic applications.

## 3.1 System architecture—Components, domains and interactions

The RSPO-ZNRF3/RNF43 signaling system includes five main interacting components: ligands, engagement receptors, effector receptors, target receptors and adaptors. While some of these components have been previously referred to using these terms (i.e., LGR4/5/6 have been called "engagement receptors" and ZNRF3/RNF43 "effector receptors" for RSPOs (Chen, Chen, Lin, Fang, & He, 2013; Xie et al., 2013)), here we define them as follows. Ligands comprise the four members of the RSPO family that initiate the signaling cascade. Engagement receptors are TM or membrane-tethered cell surface proteins that engage RSPO ligands. They include LGR4/5/6, HSPGs such as GPCs and SDCs, and the type I BMP receptor BMPR1A. Effector receptors are the TM E3s ZNRF3 and RNF43, which also engage RSPO ligands and transduce the signal by directly or indirectly modulating the abundance of cell surface receptors. Target receptors are the WNT receptors frizzled (FZD) and low-density lipoprotein receptor-related protein 6 (LRP6), and the type I BMP receptor BMPR1A. The final outcome of RSPO signaling is to effect changes in the cell surface abundance of target receptors, and in so doing, tune the sensitivity of cells to WNT and BMP ligands. BMPR1A is unique in that it is both an engagement receptor and a target receptor, since it binds RSPOs directly and its abundance on the cell surface is regulated by ZNRF3/RNF43. Finally, adaptors are proteins that mediate the specificity of ZNRF3/RNF43 towards their target receptors. Dishevelled (DVL) is the only such adaptor described so far. In the following sections we describe the domain structure of these components and the interactions relevant to RSPO-ZNRF3/RNF43 signaling (Figs. 4 and 5). We focus on the mammalian proteins, but descriptions of these components in other species can be found in the various reviews cited.

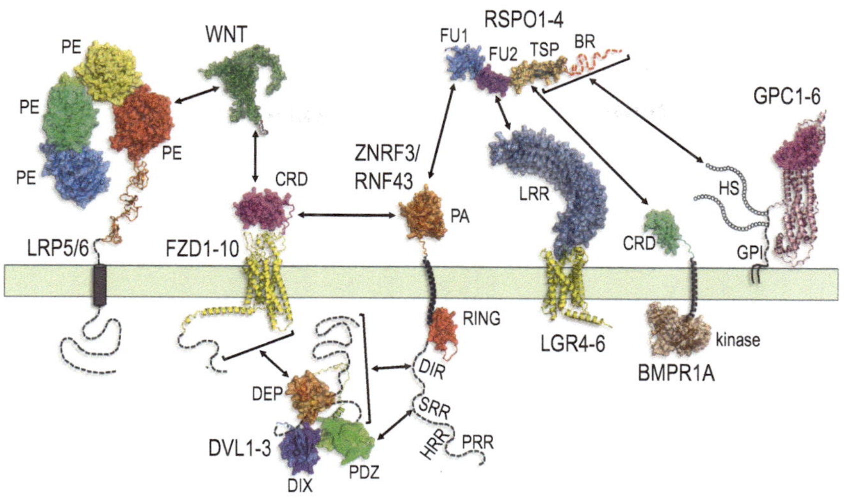


**Fig. 4** The RSPO-ZNRF3/RNF43 signaling system: components, domains and interactions. AlphaFold models of the major components of the RSPO-ZNRF3/RNF43 signaling system, indicating the domains and protein-protein interactions (double arrows) relevant for signal transduction. See main text for description. Dotted lines represent parts of the polypeptide chains for which the structure could not be predicted by AlphaFold. The HS chains and GPI anchor of GPC1–6 were drawn to represent their approximate sites of attachment to the polypeptide chain, but are not intended to depict their actual structures or dimensions.

### *3.1.1 Ligands: RSPOs*

RSPO1–4 are the four members of the RSPO subfamily of thrombospondin type 1 (TSP) repeat-containing proteins. All RSPOs contain two N-terminal tandem Cys-rich furin-like repeats connected by a flexible hinge, referred to as furin-like repeat 1 (FU1) and furin-like repeat 2 (FU2), followed by the TSP domain and a C-terminal region rich in basic amino acids (Lys and Arg), referred to as the basic region (BR) (Figs. 4 and 5A). This domain architecture is highly conserved among the four RSPOs (Kim et al., 2008; reviewed in de Lau, Snel, & Clevers, 2012), suggesting common functions. However, the length of the BR varies significantly between family members.

The FU1 domain of RSPOs interacts with the extracellular PA domain of ZNRF3 and RNF43 (Figs. 4 and 5B). Conserved residues in the RSPO FU1 domain and the ZNRF3 or RNF43 extracellular PA domain form an extensive interface comprising a mixture of hydrophobic and complementary charged interactions, as shown by a series of X-ray crystallographic structures (Chen et al., 2013; Peng et al., 2013; Zebisch et al., 2013;

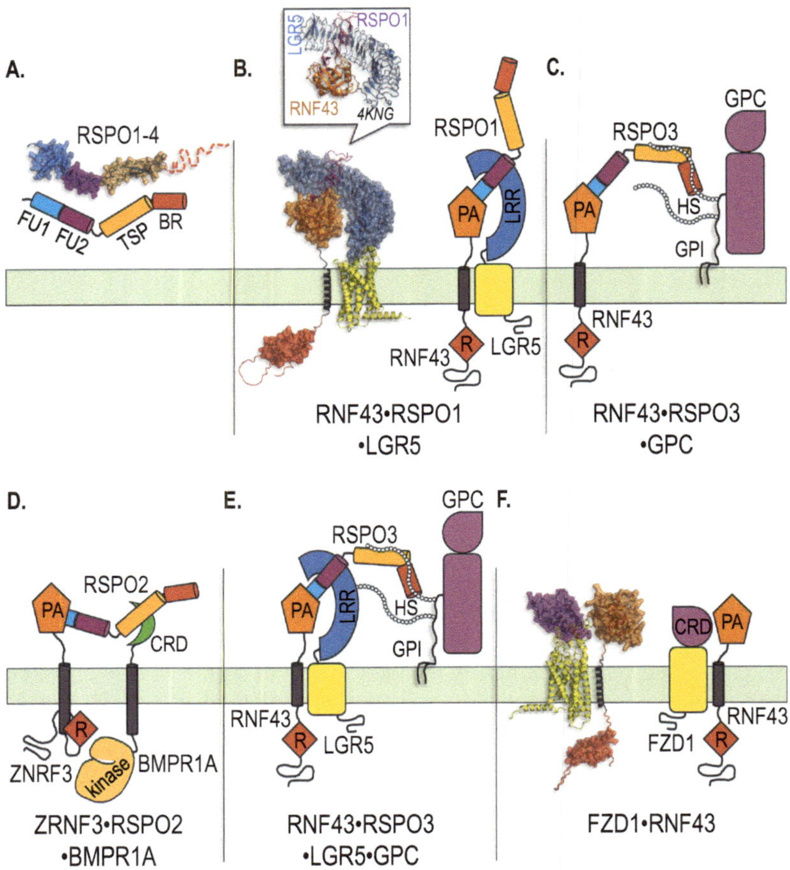


**Fig. 5** Protein complexes involved in ZNRF3/RNF43-mediated, LGR- and HSPG-dependent potentiation of WNT signaling, and BMPR1A-dependent inhibition of BMP signaling by RSPOs. (A) AlphaFold model and cartoon representation of RSPO1–4, showing the predicted modular architecture of the FU1, FU2, TSP and BR domains. (B) AlphaFold model and cartoon representation of the RNF43-RSPO1-LGR5 ternary complex that regulates WNT signaling by driving ZNRF3/RNF43 internalization and lysosomal degradation. In the model, a fragment of RSPO1 composed only of the FU1 and FU2 domains is shown, while in the cartoon representation full-length RSPO1 is shown to illustrate that the TSP/BR domains would extend into an open space not occupied by other polypeptides. The box shows the structure, solved by X-ray crystallography (PDB ID 4KNG), of the extracellular LRR domain of LGR5 and the PA domain of RNF43 bound to the RSPO1 FU1-FU2 fragment (Chen et al., 2013). Note that the crystal structure is nearly superimposable with the AlphaFold model. (C) and (D) Cartoon representations of the ternary complexes that mediate HSPG-dependent potentiation of WNT signaling (C) and BMPR1A-dependent inhibition of BMP signaling (D) by RSPOs. (E) Cartoon representation of a hypothetical quaternary complex that could promote simultaneous LGR- and HSPG-dependent potentiation of WNT signaling by RSPOs. While the existence

*(Continued)*

Zebisch & Yvonne Jones, 2015a). One distinctive feature of the FU1 domain, termed the "Met-finger" because it contains a Met residue at the tip, inserts into a hydrophobic pocket in the ZNRF3/RNF43 PA domain as a key determinant of the RSPO-ZNRF3/RNF43 interaction, and may account for the difference in the binding affinities between the four RSPO family members and ZNRF3/RNF43 (Zebisch et al., 2013). Point mutations in residues R66 and Q71 within the FU1 domain of RSPO1 (and corresponding residues in other RSPOs) abolish the interaction between RSPOs and ZNRF3/RNF43 (Xie et al., 2013; Zebisch et al., 2013), although there is some discrepancy between experiments about the extent to which these mutations impair potentiation of WNT signaling (Lebensohn & Rohatgi, 2018; Xie et al., 2013).

The FU2 domain of RSPOs interacts with the large Leu-rich repeat (LRR) array in the extracellular domain (ECD) of LGRs primarily through hydrophobic interactions, although charged interactions between residues in the FU1 domain and LGRs have also been described (Figs. 4 and 5B) (Chen et al., 2013; Peng et al., 2013; Wang et al., 2013; Xu, Xu, Rajashankar, Robev, & Nikolov, 2013; Zebisch & Yvonne Jones, 2015a). Point mutations in residues F106 and/or F110 within the FU2 domain of RSPO1 (and corresponding residues in other RSPOs) abrogate binding of RSPOs to LGRs and eliminate potentiation of WNT signaling (Peng, de Lau, Forneris, et al., 2013; Xie et al., 2013). RSPO proteins containing point mutations in these FU2 domain residues are therefore useful reagents to study LGR-independent modes of RSPO signaling (Dubey et al., 2020; Lebensohn & Rohatgi, 2018; Park et al., 2018; Szenker-Ravi et al., 2018).

The FU1 and FU2 domains used to be considered the "business end" of the mature RSPO proteins (de Lau et al., 2014), since a fragment comprising

**Fig. 5—Cont'd** of such a complex has not been confirmed experimentally, it is compatible with the spatial arrangement of the relevant domains in RSPO based on solved crystal structures (B), and is consistent with the ability of the TSP/BR domains of RSPO3, as well as HSPGs, to potentiate WNT/β-catenin signaling beyond the levels promoted by the FU1-FU2 fragment and LGRs alone (Dubey et al., 2020; Lebensohn & Rohatgi, 2018). (F) AlphaFold model and cartoon representation of a FZD1-RNF43 complex. The model suggests that the FZD1 CRD would interact with the PA domain of RNF43 and drive contacts between the TM helix of RNF43 and the 7TM of FZD1, potentially orienting the RING domain for ubiquitin transfer. In (B–F), ZNRF3 or RNF43 are arbitrarily shown for illustrative purposes, but both E3s are thought to mediate all of these signaling modalities. In (B–E) RSPO1, RSPO2 or RSPO3 are arbitrarily shown for illustrative purposes, but other RSPO ligands capable of mediating each of these signaling modalities are indicated in Table 1.

these two domains is necessary and sufficient to potentiate WNT signaling (Kazanskaya et al., 2004; Kim et al., 2008; Li et al., 2009). For this reason, all of the structural studies described were done with this minimal fragment. However, we now know that these two domains make only a partial contribution to the full repertoire of RSPO functions, since they are insufficient to signal through LGR-independent mechanisms (Dubey et al., 2020; Lebensohn & Rohatgi, 2018; Lee et al., 2020) and are significantly less potent than the full-length proteins when signaling through LGRs (Dubey et al., 2020; Lebensohn & Rohatgi, 2018).

The TSP and BR domains interact with heparin—a glycosaminoglycan polymer that mimics heparan sulfate (HS)—and with the HS chains of HSPGs such as GPCs and SDCs (Figs. 4 and 5C) (Bell et al., 2008; Chang et al., 2016; Glinka et al., 2011; Nam, Turcotte, Smith, Choi, & Yoon, 2006; Ohkawara et al., 2011; Ren et al., 2018). While the TSP and BR domains are often described separately, molecular modeling predicts that the positively charged surface of the TSP and BR domains forms a continuous binding interface for heparin (Ayadi, 2008; Dubey et al., 2020). Indeed, the TSP and BR domains can individually mediate binding of RSPOs to heparin (Nam et al., 2006) and RSPO constructs containing either the TSP or BR domain can induce HSPG-dependent potentiation of WNT signaling (Lebensohn & Rohatgi, 2018). The TSP domain is also required for binding of RSPO2 and RSPO3 to BMPR1A during downregulation of BMP signaling (Figs. 4 and 5D) (Lee et al., 2020).

### 3.1.2 Engagement receptors: LGRs, HSPGs, BMPR1A

LGR4/5/6 were identified and validated as RSPO engagement receptors through various independent approaches (Carmon et al., 2011; de Lau et al., 2011; Glinka et al., 2011; Ruffner et al., 2012). They are classified as the three Class B members of the LGR subgroup of the rhodopsin family of G protein-coupled receptors (GPCRs) (reviewed in Barker et al., 2013; de Lau et al., 2014) and mark stem cells in many embryonic and adult tissues (Kinzel et al., 2014; reviewed in Koo & Clevers, 2014; Leung, Tan, & Barker, 2018). They contain a large ECD consisting of 16–17 LRRs followed by a hinge region and the distinctive 7TM domain of rhodopsin-like GPCRs (Fig. 4). The concave face of the curved structure formed by the LRR array interacts with the FU2 domain of RSPOs (Figs. 4 and 5B) (Chen et al., 2013; Peng, de Lau, Forneris, et al., 2013; Wang et al., 2013; Xu et al., 2013; Zebisch & Yvonne Jones, 2015a). No G protein-coupled signaling activity triggered by binding to RSPO ligands has been reported (Carmon et al., 2011; de Lau et al., 2011), supporting the notion that LGRs transduce RSPO signals through other mechanisms discussed later.

HSPGs have also been implicated as engagement receptors for RSPOs (Dubey et al., 2020; Lebensohn et al., 2016; Lebensohn & Rohatgi, 2018; Ohkawara et al., 2011; Ren et al., 2018). HSPGs are a diverse class of cell surface and extracellular matrix glycoproteins decorated with HS glycosaminoglycan polysaccharide chains (reviewed in Christianson & Belting, 2014; Sarrazin, Lamanna, & Esko, 2011). Abundant carboxyl and sulfate groups on the HS chains make them polyanionic, promoting interactions with polybasic domains on proteins. While HSPGs broadly include the cell surface-associated GPCs and SDCs, the secreted extracellular matrix HSPGs (agrin, perlecan and type XVIII collagen) and the secretory vesicle proteoglycan serglycin, only GPCs and SDCs have been implicated as RSPO receptors. In mammals, there are six GPCs (GPCs1–6) and four SDCs (SDC1–4). GPCs are tethered to the plasma membrane through a glycosylphosphatidylinositol (GPI) anchor (Fig. 4), while SDCs are single-pass TM proteins. HS chains are attached to the protein core of GPCs close to the plasma membrane, and to the protein core of SDCs at more peripheral sites. Ligands can bind either to the protein core or to the HS chains of HSPGs. RSPOs interact with the HS chains of HSPGs through the TSP and BR domains (Figs. 4 and 5C) (Bell et al., 2008; Chang et al., 2016; Nam et al., 2006) and this interaction is required for RSPOs to potentiate WNT/β-catenin signaling (Dubey et al., 2020; Lebensohn & Rohatgi, 2018; Ren et al., 2018) as well as WNT/PCP signaling (Ohkawara et al., 2011).

The cell surface abundance of the type I BMP receptor BMPR1A can also be regulated by RSPOs (Lee et al., 2020). In this case, BMPR1A appears to be both the engagement and target receptor, since its own internalization is triggered when RSPO2 or RSPO3 bind to it and cross-link it with the effector receptors ZNRF3/RNF43. BMPR1A is one of seven type I TGFβ receptors in humans. It contains a small extracellular Cys-rich domain (CRD), a TM domain, and an intracellular juxtamembrane domain rich in Gly and Ser residues followed by a Ser kinase domain (Fig. 4) (reviewed in Heldin & Moustakas, 2016). The BMPR1A CRD binds with high affinity to the TSP domains of RSPO2 and RSPO3 (Figs. 4 and 5D) (Lee et al., 2020). The mechanism driving BMPR1A internalization following engagement of RSPOs is unknown (see Section 3.2.6).

### 3.1.3 Effector receptors: ZNRF3, RNF43

The closely related ZNRF3 and RNF43 proteins are members of the GOLIATH/GRAIL family of PA-TM-RING E3s (Fig. 3) (reviewed in de Lau et al., 2014; Hao, Jiang, & Cong, 2016; Zebisch & Yvonne Jones, 2015b). They were discovered as WNT/β-catenin target genes whose expression was correlated with that of *AXIN2* mRNA in primary tissue microarray data (Hao et al., 2012), or with LGR5-GFP abundance in LGR5+ small intestinal crypt stem cells (Koo et al., 2012). Both ZNRF3 and RNF43 contribute to a negative feedback mechanism that down-regulates WNT receptor levels (Hao et al., 2012; Koo et al., 2012). A genome-wide, forward genetic screen in haploid human cells designed to find attenuators of WNT/β-catenin signaling—genes that when deleted enhanced signaling in the presence of a low dose WNT ligand—later uncovered *ZNRF3* as the top hit (Lebensohn et al., 2016). Since the HAP1 cells in which this screen was conducted do not express *RNF43* mRNA (Lebensohn et al., 2016), these unbiased screen results suggest that *ZNRF3* is the most potent attenuator of WNT signaling in the genome, at least in haploid human cells.

The extracellular PA domain of ZNRF3 and RNF43 interacts with the FU1 domain of RSPOs (Figs. 4 and 5B) as discussed earlier. Comparison of ZNRF3 ECD structures in isolation and in complex with RSPO ligands did not reveal major conformational differences (Zebisch et al., 2013), suggesting that signal transduction upon binding of RSPOs is unlikely to be an autonomous property of the ZNRF3/RNF43 proteins, instead requiring other components of the system. The PA domain of the GOLIATH/GRAIL family member RNF128 (Fig. 3) interacts with transmembrane receptors such as CD40L and CD83 and targets them for ubiquitylation (Lineberry et al., 2008), suggesting that the PA domain of ZNRF3/RNF43 may do the same for the FZD family of WNT receptors. However, data regarding an interaction between the PA domain of ZNRF3/RNF43 and the ECD of FZD is conflicting, as we discuss later. The catalytic RING domain (Fig. 4) is required for ubiquitylation of the WNT receptor FZD, which leads to its internalization and lysosomal degradation, resulting in decreased sensitivity to WNT ligands (Hao et al., 2012; Koo et al., 2012). The RING domain also appears to be required for membrane clearance of BMPR1A (Lee et al., 2020). In addition to their defining PA, TM and RING domains, ZNRF3 and RNF43 have disordered cytoplasmic extensions containing a dishevelled-interaction region (DIR) followed by Ser-, His-, and Pro-rich regions (SRR, HRR and PRR, respectively) (Fig. 4). The DIRs of ZNRF3 and RNF43 interact with the C-terminal two thirds of DVL (Fig. 4) (Jiang et al., 2015). Another region of the RNF43 intracellular domain (ICD) located C-terminal to the DIR has also been postulated to interact with DVL2 (Fig. 4) (Tsukiyama et al., 2015). Phosphorylation/dephosphorylation of the ICD also regulates RNF43-mediated FZD ubiquitylation (Tsukiyama et al., 2020), as well as ZNRF3/RNF43 internalization (Chang, Kim, Glinka, Reinhard, & Niehrs, 2020; Kim, Reinhard, & Niehrs, 2021).

### 3.1.4 Target receptors: FZDs, LRP6, BMPR1A

FZD proteins (reviewed in Huang & Klein, 2004; MacDonald & He, 2012; Niehrs, 2012; Wang, Chang, Rattner, & Nathans, 2016) were the first WNT receptors to be identified (Bhanot et al., 1996). The 10 FZDs in humans (FZD1–10) are Class F members of the GPCR superfamily (reviewed in Malbon, 2004; Schulte & Bryja, 2007; Schulte & Wright, 2018). FZDs transduce both β-catenin-dependent and β-catenin-independent WNT signals, including those in the WNT/PCP, WNT/calcium (reviewed in Niehrs, 2012) and WNT-dependent stabilization of proteins (WNT/STOP) pathways (Acebron, Karaulanov, Berger, Huang, & Niehrs, 2014). FZDs contain an extracellular CRD followed by a linker region, a 7TM domain, and an ICD of variable length (Fig. 4). The FZD CRD interacts directly with WNT ligands (Fig. 4) via contacts at two opposing faces of the globular CRD, with the principal interaction involving the palmitate group of WNT docking into a hydrophobic groove in the CRD (Janda, Waghray, Levin, Thomas, & Christopher Garcia, 2012). Replacement of several conserved Lys residues throughout the intracellular loops of the FZD 7TM and the ICD with Arg residues abrogated changes in FZD levels in response to ZNRF3/RNF43 over-expression or depletion (Hao et al., 2012; Koo et al., 2012), suggesting that these Lys residues may be ubiquitylated by ZNRF3/RNF43. However, the relative contributions of these potential ubiquitylation sites to regulation of FZD levels by ZNRF3/RNF43 have not been determined.

LRP5 and LRP6, and the Drosophila ortholog Arrow, are WNT co-receptors required for WNT/β-catenin signaling but not for β-catenin-independent WNT/PCP signaling (reviewed in He, Semenov, Tamai, & Zeng, 2004; MacDonald & He, 2012). LRP5/6 are large (>1600 amino acids) single-pass TM proteins with an ECD formed by a closely packed set of four tandem β-propeller/epidermal growth factor-like (PE) repeats, followed by three low-density lipoprotein receptor (LDLR) type A repeats (Fig. 4). The LRP6 ECD interacts with WNT ligands in a manner that allows WNTs to simultaneously bind the FZD CRD (Fig. 4), bridging them into a ternary receptor complex that triggers cytoplasmic WNT signaling (Bourhis, Tam, Franke, Bazan, & Ernst, 2010; Chu et al., 2013; Hirai, Matoba, Mihara, Arimori, & Takagi, 2019; Tamai et al., 2000). While LRP6 internalization and degradation is regulated by ZNRF3/RNF43 (Chang et al., 2020; Giebel et al., 2021; Hao et al., 2012; Kim et al., 2021), the elements in LRP6 required for this regulation remain unknown.

The type I BMP receptor BMPR1A, discussed earlier, is unique among the targets of RSPO-ZNRF3/RNF43-dependent regulation in that it also engages RSPOs directly (Figs. 4 and 5D) (Lee et al., 2020), and can therefore be considered both a target and an engagement receptor. However, unlike in the case of WNT receptors, it is unclear whether regulation of cell surface BMPR1A abundance by RSPO-ZNRF3/RNF43-mediated endocytosis and lysosomal degradation involves BMPR1A ubiquitylation.

### 3.1.5 Adaptors: DVL

The three DVL proteins (DVL1–3 in humans) are crucial intracellular components of both β-catenin-dependent and β-catenin-independent WNT signaling pathways (reviewed in Gao & Chen, 2010; MacDonald & He, 2012; Sharma, Castro-Piedras, Simmons Jr, & Pruitt, 2018). They bind the cytoplasmic segments of FZD receptors and route WNT signals to the WNT/β-catenin or WNT/PCP pathways by forming distinct signaling complexes (reviewed in Gammons & Bienz, 2018; Mlodzik, 2016). DVLs interact with a diverse array of proteins through three highly conserved modules connected by flexible linkers that mediate their molecular functions: an N-terminal dishevelled, axin (DIX) domain, a central postsynaptic density 95, discs large, zona occludens-1 (PDZ) domain, and a C-terminal dishevelled, egl-10, pleckstrin (DEP) domain (Fig. 4). The DIX domain undergoes dynamic head-to-tail homo-polymerization (Kishida et al., 1999; Schwarz-Romond et al., 2007), which leads to formation of DVL assemblies (Schwarz-Romond, Metcalfe, & Bienz, 2007), and can also undergo hetero-polymerization with the related DAX domain of AXIN (Fiedler, Mendoza-Topaz, Rutherford, Mieszczanek, & Bienz, 2011; Kishida et al., 1999). The PDZ domain interacts with many proteins that mediate both WNT/β-catenin and WNT/PCP signaling, and may be involved in distinguishing between these two pathways (reviewed in Sharma et al., 2018). The interaction between the PDZ domain and a KTXXXW motif in the intracellular C-terminal tail of FZD recruits DVL to the WNT receptor complex and is crucial for transduction of WNT signals (Umbhauer et al., 2000; Wong et al., 2003). The DEP domain also targets DVL to the plasma membrane (reviewed in Consonni, Maurice, & Bos, 2014). It has a positively charged surface that likely interacts with phospholipids (Simons et al., 2009; Wong et al., 2000), and the DEP domain together with the C-terminal region of DVL interacts with a discontinuous motif in the FZD ICD (Tauriello et al., 2012).

DVL has been postulated as an adaptor required for recognition of FZD by ZNRF3/RNF43, a prerequisite step in promoting FZD degradation (Jiang et al., 2015). The three-way physical interaction between ZNRF3/RNF43, DVL and FZD is essential for the WNT/β-catenin inhibitory activity of ZNRF3/RNF43. This interaction is mediated by binding of the DVL DEP domain to FZD (Fig. 4), and by contacts between segments in the C-terminal two thirds of DVL (notably excluding the DIX, PDZ, and DEP domains) and the DIR of ZNRF3/RNF43 (Fig. 4). Accordingly, the DEP domain, but not the DIX or PDZ domains, are required for ZNRF3/RNF43-dependent FZD downregulation, and fusion of the DEP domain to ZNRF3/RNF43 eliminates the requirement of DVL to downregulate FZD levels (Jiang et al., 2015). An interaction between the PDZ domain of DVL and a region of the RNF43 ICD located C-terminal to the DIR is essential for inhibition of β-catenin-independent signaling through an undefined mechanism (Tsukiyama et al., 2015). This inhibition does not require the ubiquitin ligase activity of RNF43, or interactions between RNF43 and FZD, and does not result in downregulation of cell surface FZD.

## 3.2 RSPO-ZNRF3/RNF43 signaling mechanisms

In this section we describe three modalities of RSPO signaling mediated by ZNRF3/RNF43 that regulate the abundance of cell surface receptors, we contrast their salient features, and we discuss their plausible underlying molecular mechanisms. Other mechanisms through which ZNRF3/RNF43 control WNT/β-catenin signaling that do not impinge on the regulation of cell surface receptor levels will not be addressed here, but we refer the reader to the primary literature (Loregger et al., 2015; Spit et al., 2020).

### 3.2.1 LGR-dependent, ZNRF3/RNF43-mediated potentiation of WNT/β-catenin signaling by RSPO1-4

The first full picture of a mechanism driving potentiation of WNT/β-catenin signaling by RSPOs emerged with the discovery that ZNRF3 and RNF43 promote ubiquitylation-dependent internalization and lysosomal degradation of the WNT receptors FZD and LRP6 (Hao et al., 2012; Koo et al., 2012). Following internalization, RNF43 and FZD co-localize in RAB5+ early endosomes, and the final fate of FZD is lysosomal rather than proteasomal degradation, as surmised from the fact that the process can be inhibited by the lysosomal V-ATPase inhibitor bafilomycin A but not the proteasome inhibitor MG132 (Koo et al., 2012). Subsequent studies showed that at least three conditions contribute to ZNRF3/RNF43-mediated internalization and degradation of WNT receptors: (1) interaction of the adaptor protein DVL with both FZD and ZNRF3/RNF43 (Jiang et al., 2015), (2) phosphorylation of Ser residues in the SRR of ZNRF3/RNF43 (Tsukiyama et al., 2020) and (3) dephosphorylation of a 4Tyr motif in the DIR of ZNRF3 (Chang et al., 2020; Kim et al., 2021). Therefore, in the absence of RSPOs, clearance of FZD and LRP6 from the plasma membrane results in decreased sensitivity of cells to WNT ligands.

In the presence of RSPO ligands, binding of the FU1 domain of RSPO to the PA domain of ZNRF3/RNF43 and of the FU2 domain of RSPO to the LRRs of LGR4/5/6 results in formation of a ternary complex (Figs. 4 and 5B) (Chen et al., 2013; Moad & Pioszak, 2013; Xie et al., 2013; Zebisch & Yvonne Jones, 2015a). This molecular assembly triggers internalization of ZNRF3/RNF43, followed by lysosomal degradation, through a poorly understood process that requires the catalytic RING domain of ZNRF3/RNF43 (Hao et al., 2012) and can be counteracted through deubiquitylation of ZNRF3/RNF43 by the DUB USP42 (Giebel et al., 2021). As a consequence of ZNRF3/RNF43 clearance from the plasma membrane, ubiquitylation-dependent internalization and lysosomal degradation of FZD and LRP6 is diminished, leading to the accumulation of these WNT co-receptors on the cell surface (Hao et al., 2012). Therefore, the outcome of this RSPO signaling modality is to increase the sensitivity of cells to WNT ligands.

Because ZNRF3/RNF43 and LGRs do not interact directly with each other, the secreted RSPOs must engage both of them simultaneously through the adjacent FU1 and FU2 domains, respectively, acting as molecular cross-linkers (Fig. 5B) (Zebisch & Yvonne Jones, 2015b). The TSP and BR domains of RSPOs would appear to be dispensable for this mode of signaling, since they escape contact with either ZNRF3/RNF43 or LGRs. This is partially borne out by the fact that a fragment comprising only the FU1 and FU2 domains of RSPOs is sufficient to promote WNT/β-catenin signaling in cells and support the growth of small intestinal organoids (Kazanskaya et al., 2004; Kim et al., 2008; Li et al., 2009; Peng, de Lau, Forneris, et al., 2013). However, while this FU1-FU2 construct displays full signaling efficacy at sufficiently high concentrations, it is much less potent than the full-length protein containing the TSP/BR domains both in cells and in small intestinal organoids (Dubey et al., 2020; Kim et al., 2008; Lebensohn & Rohatgi, 2018), demonstrating that the TSP/BR domains contribute to signaling even in the presence of LGRs (Fig. 5E).

### *3.2.2 HSPG-dependent, ZNRF3/RNF43-mediated potentiation of WNT/β-catenin signaling by RSPO2/3*

Unexpectedly, RSPO2 and RSPO3, but not RSPO1 or RSPO4, are capable of potentiating WNT/β-catenin signaling in cells and mice lacking LGRs, albeit with lower potency and efficacy than in cells containing LGRs (Lebensohn & Rohatgi, 2018; Park et al., 2018; Szenker-Ravi et al., 2018). Furthermore, full length RSPO3 containing inactivating mutations in the LGR-interacting FU2 domain could still promote WNT/β-catenin signaling in haploid human cells, again with lower potency and efficacy than the wild type (WT) counterpart (Lebensohn & Rohatgi, 2018). Similarly, RSPO2 and RSPO3 constructs lacking the BR domain and containing inactivating mutations in the LGR-interacting FU2 domain could also potentiate WNT/β-catenin signaling in HEK293 cells (Szenker-Ravi et al., 2018). These experiments, in which potentiation of WNT/β-catenin signaling was partially retained following perturbations of either the LGR receptors or the LGR-binding FU2 domain on the RSPO ligands themselves, conclusively demonstrated the capacity of RSPOs to signal independently of LGRs. This begged an urgent question: is there an alternative engagement receptor for RSPOs?

To answer that question, we mapped the domains in RSPO3 required for signaling in the absence of LGRs through mutagenesis, domain deletion and domain swapping experiments (Lebensohn & Rohatgi, 2018). The ZNRF3/RNF43-interacting FU1 domain, and the HS-interacting TSP and/or BR domains of RSPO3 were required (constructs lacking either the TSP or the BR domain, but not both, could support signaling) (Figs. 4 and 5C). Furthermore, the precise amino acid composition of the TSP/BR domains is not a critical determinant for signaling in the absence of LGRs: replacing the TSP/BR domains of RSPO3 with those of RSPO1, which cannot signal without LGRs, did not impair signaling (Lebensohn & Rohatgi, 2018) even though the TSP/BR domains of RSPO3 and RSPO1 exhibit very low sequence conservation (21% identical amino acids in the human proteins (de Lau et al., 2012)). These results suggested that LGR-independent signaling may be mediated by electrostatic interactions between the TSP and/or BR domains and the HS chains of HSPGs. Modeling of the TSP/BR domains of RSPO3 predicted two positively charged grooves lined by basic Lys and Arg residues that could potentially dock HS chains like those present in HSPGs (Dubey et al., 2020). Indeed, signaling by RSPO3 in cells lacking all LGRs was nearly completely abolished by three different manipulations that disrupted the interaction between the TSP/BR domains and the HS chains of HSPGs: 1. mutation of some of the Lys/Arg residues in the TSP/BR domains to charge-reversing Glu residues; 2. addition of heparin, which competes for binding to the HS chains of HSPGs; and 3. disruption of the gene encoding EXTL3, a glycosyltransferase specifically required for HSPG biosynthesis but dispensable for the synthesis of other glycosaminoglycans and proteoglycans (Dubey et al., 2020; Lebensohn & Rohatgi, 2018).

The conclusive demonstration that interactions between the TSP and/or BR domains of RSPOs and the HS chains of HSPGs mediate LGR-independent signaling (Fig. 5C) came from ligand engineering experiments (Dubey et al., 2020). A synthetic RSPO3 construct in which the entire TSP and BR domains were replaced with a single-chain variable fragment (scFv) that specifically binds to the HS chains of GPCs potentiated WNT/β-catenin signaling with the same potency and efficacy as WT RSPO3 in cells lacking LGRs. Experiments in which individual or entire families of HSPGs (including all GPCs or all SDCs) were eliminated in haploid human cells demonstrated that RSPO3 can signal in a redundant manner via either GPCs, SDCs or potentially another HSPG by engaging their HS chains rather than their protein cores (Dubey et al., 2020). Furthermore, genome-wide screens in haploid human cells lacking LGR4/5/6 did not reveal additional receptors required for potentiation of WNT signaling by RSPO3, making HSPGs the most likely engagement receptors for RSPOs in the absence of LGRs (Dubey et al., 2020).

The ZNRF3/RNF43-binding FU1 domain is also required for LGR-independent signaling (Fig. 5C) (Lebensohn & Rohatgi, 2018; Park et al., 2018; Szenker-Ravi et al., 2018). In fact, it is the FU1 domain, rather than the HSPG-interacting TSP/BR domains, that determines whether a given RSPO family protein can signal in the absence of LGRs. This was demonstrated by domain-swapping experiments in which the FU1 domain of RSPO3 conferred on RSPO1 the ability to signal without LGRs, and conversely an RSPO3 chimera containing the FU1 domain of RSPO1 lost its ability to signal without LGRs (Lebensohn & Rohatgi, 2018). Because the affinities of the FU1 domains from RSPO3 ($K_D \sim 60\,nM$) and from RSPO1 ($K_D \sim 6.8\,\mu M$) towards ZNRF3 are markedly different (Zebisch et al., 2013), this difference may determine the requirement for LGRs (Lebensohn & Rohatgi, 2018). We speculate that in the presence of a high-affinity interaction between the FU1 domain of RSPO2 ($K_D \sim 25\,nM$) or RSPO3 ($K_D \sim 60\,nM$) and ZNRF3 (Zebisch et al., 2013), the interaction between the FU2 domain and LGRs can be functionally replaced by the interaction between the TSP/BR domains and HSPGs. However, the lower-affinity interaction between the FU1 domain of RSPO1 ($K_D \sim 6.8\,\mu M$) or RSPO4 ($K_D \sim 300\,\mu M$) and ZNRF3 (Zebisch et al., 2013) would require the high-affinity interaction between the FU2 domain and LGRs ($K_D \sim 2$–$3\,nM$) (de Lau et al., 2011; Glinka et al., 2011; Zebisch et al., 2013) in order to signal.

While the TSP/BR domains are not required for signaling in the presence of LGRs, they substantially increase the potency of signaling by RSPOs in cells and small intestinal organoids (Dubey et al., 2020; Kim et al., 2008; Lebensohn & Rohatgi, 2018). In fact, at lower concentrations of RSPO3, the interaction of the FU2 domain with LGRs is not sufficient to drive efficient endocytosis of RNF43, and HSPG binding mediated by the TSP/BR domains is also required even in the presence of LGRs (Dubey et al., 2020). Furthermore, at limiting concentrations, RSPO3 was significantly more potent than RSPO1 in supporting the growth of intestinal organoids (Greicius et al., 2018), consistent with the ability of RSPO3 but not RSPO1 to signal through both LGR-dependent and LGR-independent mechanisms (Dubey et al., 2020; Kim et al., 2008; Lebensohn & Rohatgi, 2018). Although none of the structural studies discussed earlier included the TSP/BR domains of RSPOs, one of the structural models of the LGR5-RSPO1-RNF43 ternary complex suggested that the TSP/BR domains would extend into an open space not occupied by other polypeptides (Fig. 5B) (Chen et al., 2013), and would therefore be available to interact with other molecules such as HSPGs. This would allow RSPOs to bind two engagement receptors—LGRs and HSPGs—and an effector receptor—ZNRF3 or RNF43—simultaneously (Figs. 4 and 5E), consistent with the ability of HSPGs to potentiate LGR-dependent signaling (Dubey et al., 2020). Therefore, HSPGs may enhance the potency of RSPO signaling by trapping RSPOs near the cell surface, increasing their local concentration and promoting binding to LGRs. In support of this model, depletion of HS chains or removal of the TSP/BR domains reduces binding of RSPOs to the cell surface, while depletion of LGR4 does not (Ren et al., 2018).

We and others initially referred to the modality of RSPO signaling that takes place in the absence of LGRs as "LGR-independent" (Lebensohn & Rohatgi, 2018; Park et al., 2018; Szenker-Ravi et al., 2018), but LGR-independent signaling has since been shown to happen in more than one way (see Section 3.2.3). Therefore, in the context of the WNT/β-catenin pathway, where in the absence of LGRs RSPO signaling is mediated by HSPGs (Dubey et al., 2020; Lebensohn & Rohatgi, 2018), we will henceforth refer to this modality as "HSPG-dependent" RSPO signaling.

### *3.2.3 BMPR1A-dependent, ZNRF3/RNF43-mediated inhibition of BMP signaling by RSPO2/3*

A third mechanism of signaling by RSPOs, also independent of LGRs but mediated by ZNRF3/RNF43, has recently been described (Lee et al., 2020). In this case, RSPO2 and RSPO3, but not RSPO1 or RSPO4, antagonize BMP signaling in a process that is independent of WNT/β-catenin and WNT/PCP signaling. RSPO2 and RSPO3 interact directly with ZNRF3 and the type I BMP receptor BMPR1A (Figs. 4 and 5D), triggering internalization and lysosomal degradation of BMPR1A. This results in decreased sensitivity of target cells to BMP ligands.

Domain analysis revealed that the FU1 and TSP domains of RSPO2 are required to antagonize BMP signaling (Fig. 5D) (Lee et al., 2020). RSPO2 interacts with the BMPR1A ECD with high affinity ($K_D \sim 4.8\,nM$), comparable to that of the FU2-mediated RSPO-LGR interaction ($K_D$ ~2–3 nM). The TSP domain of RSPO2 and RSPO3, but not the FU1, FU2 or BR domains, is required for binding to the BMPR1A ECD. Furthermore, domain-swapping experiments revealed that the capacity to downregulate BMP receptor levels resides in the TSP domain: while WT RSPO1 did not antagonize BMP signaling, an RSPO1 chimera containing the TSP domain of RSPO2 bound to BMPR1A and antagonized BMP signaling. siRNA-mediated knock-down of ZNRF3/RNF43 or overexpression of a dominant negative ZNRF3 lacking the RING domain prevented RSPO2-induced destabilization of BMPR1A and inhibition of BMP signaling. On the other hand, siRNA-mediated knock-down of LGR4/5 did not affect inhibition of BMP signaling by RSPO2. These results suggest that BMP antagonism by RSPO2 requires ZNRF3/RNF43 but not LGRs. Consistent with these requirements, the ZNRF3/RNF43-binding FU1 domain of RSPO2, but not the LGR-binding FU2 domain, was required to antagonize BMP receptor signaling (Lee et al., 2020).

RSPO2 triggers BMPR1A clearance from the cell surface by acting as a cross-linking ligand between BMPR1A and ZNRF3 (Fig. 5D) (Lee et al., 2020). *In vitro* binding assays and co-localization experiments demonstrated that ZNRF3 interacted with BMPR1A in the presence of RSPO2, and formation of a ZNRF3-RSPO2-BMPR1A ternary complex depended on the FU1 and TSP domains of RSPO2. In cells that produce RSPO2, BMPR1A was absent from the plasma membrane but colocalized with ZNRF3 in cytoplasmic vesicles, as well as with the early endosome marker EEA1 and the lysosome marker Lamp1. Knock-down of RSPO2 abolished endosomal and lysosomal localization, and resulted in accumulation of BMPR1A at the plasma membrane. Therefore, RSPO2 bridges ZNRF3 and BMPR1A, and routes the ternary complex for lysosomal degradation, antagonizing BMP signaling. The authors proposed that a similar mechanism applies to RSPO3, but not RSPO1 or RSPO4 (Lee et al., 2020).

### 3.2.4 Comparing different modalities of RSPO-ZNRF3/RNF43 signaling

The three different modalities of ZNRF3/RNF43-mediated RSPO signaling described so far, LGR-dependent potentiation of WNT/β-catenin signaling, HSPG-dependent potentiation of WNT/β-catenin signaling, and BMPR1A-dependent antagonism of BMP signaling, illustrate the versatile modularity of the RSPO-ZNRF3/RNF43 signaling system (Table 1). These signaling modes are defined by a "combinatorial code" in which the FU1, FU2, TSP and/or BR domains of RSPOs interact with different combinations of engagement, effector and target receptors to modulate the WNT/β-catenin or BMP pathways (Figs. 4 and 5 and Table 1). Furthermore, differences in the extent to which individual domains of distinct RSPO ligands interact with these receptors, presumably determined by their binding affinities, dictates the modalities through which each RSPO

**Table 1** Summary of the three RSPO signaling modalities regulated by ZNRF3/RNF43.

| Signaling modality | RSPO ligands | Required domains in RSPO | Engagement receptors | Effector receptors | Target receptors | Adapter | Direct/indirect effect of RSPO binding on target receptor levels | Up/down-regulation of target receptor by RSPO |
|---|---|---|---|---|---|---|---|---|
| LGR-dependent potentiation of WNT/β-catenin signaling | RSPO1–4 | FU1, FU2 | LGR4–6 | ZNRF3/ RNF43 | FZD, LRP6 | DVL | Indirect | Up-regulation |
| HSPG-dependent potentiation of WNT/β-catenin signaling | RSPO2/3 | FU1, TSP/BR | HSPGs (GPCs, SDCs) | ZNRF3/ RNF43 | FZD, LRP6? | DVL? | Indirect | Up-regulation |
| BMPR1A-dependent inhibition of BMP signaling | RSPO2/3 | FU1, TSP | BMPR1A | ZNRF3/ RNF43 | BMPR1A | ? | Direct | Down-regulation |

See main text for description.

ligand can signal (Table 1). Finally, depending on whether RSPOs engage target receptors and directly promote their membrane clearance, as in the case of BMPR1A, or indirectly effect changes in target receptor levels by modulating ZNRF3/RNF43 internalization and potentially ubiquitin ligase activity, as in the case of the WNT receptors FZD and LRP6, the functional outcome is either down- or up-regulation of the signaling pathway, respectively (Table 1).

In accordance with the opposite ways in which ZNRF3/RNF43 function during regulation of the WNT receptors FZD and LRP6 versus regulation of the BMP receptor BMPR1A (Table 1), the molecular mechanisms leading to internalization and lysosomal degradation of target receptors are different between the two pathways. Furthermore, the molecular mechanisms leading to ZNRF3/RNF43 membrane clearance and lysosomal degradation, triggered by binding of RSPOs to ZNRF3/RNF43 and to the different engagement receptors, have not been fully elucidated. In Sections 3.2.5 and 3.2.6, we discuss potential molecular mechanisms controlling ZNRF3/RNF43-dependent membrane clearance and degradation of target receptors in the WNT and BMP pathways, as well as those controlling RSPO-dependent membrane clearance of ZNRF3/RNF43 and engagement receptors. In the case of BMPR1A-dependent signaling, these mechanisms are one and the same.

### 3.2.5 Molecular mechanisms controlling internalization and degradation of target receptors in the WNT pathway

In the context of WNT signaling, ZNRF3/RNF43-dependent ubiquitylation of the target receptor FZD on Lys residues within the cytoplasmic loops of the 7TM domain and/or the C-terminal tail targets FZD to RAB5+ early endosomes and CD63+ lysosomes (Hao et al., 2012; Koo et al., 2012). This results in FZD internalization and lysosomal degradation, leading to decreased sensitivity to WNT ligands. ZNRF3 and RNF43 are most likely co-internalized with FZD—RNF43 co-localized with FZD5 in internal vesicles (Koo et al., 2012)—and this endocytic process is regulated by phosphorylation/dephosphorylation of a conserved 4Tyr motif within the DIR of ZNRF3 (Chang et al., 2020; Kim et al., 2021).

Several lines of evidence demonstrate that the ubiquitin ligase activity of ZNRF3/RNF43 is required for FZD ubiquitylation, internalization and degradation. Overexpression of WT ZNRF3 or RNF43 increased ubiquitylation of FZD, decreased cell surface FZD levels and reduced WNT-induced pathway activity (Hao et al., 2012; Koo et al., 2012).

Conversely, overexpression of ZNRF3 or RNF43 mutants containing inactivating point mutations in, or altogether lacking the catalytic RING domain suppressed ubiquitylation, increased the plasma membrane expression and extended the half-life of FZD, abolishing the inhibitory effect of ZNRF3/RNF43 on WNT signaling (Hao et al., 2012; Koo et al., 2012). Inactivating mutations in, or deletion of the RING domain of ZNRF3/RNF43 also enhanced WNT-induced pathway activity by acting in a dominant-negative fashion (Hao et al., 2012; Koo et al., 2012). Furthermore, FZD variants in which all conserved cytoplasmic Lys residues were mutated to Arg were not internalized upon expression of RNF43 (Koo et al., 2012), and the membrane levels of these FZD mutants did not increase upon depletion of ZNRF3 (Hao et al., 2012). Ubiquitylation of FZD was reduced in cells lacking the WNT pathway scaffold protein DVL, which as discussed earlier may serve as an adaptor that targets ZNRF3/RNF43 to FZD (Jiang et al., 2015). Finally, ZNRF3 and RNF43 could be co-immunoprecipitated with FZD (Hao et al., 2012; Koo et al., 2012). This compilation of experiments strongly supports a model in which ZNRF3/RNF43 directly ubiquitylate FZD, but we note that FZD ubiquitylation by ZNRF3/RNF43 has not been reconstituted *in vitro* with purified components.

A recently described "phospho-switch" also modulates the ability of ZNRF3/RNF43 to regulate WNT receptor levels (Tsukiyama et al., 2020). Phosphorylation by casein kinase 1 of 3 Ser residues located in the SRR of RNF43 (also conserved in ZNRF3) was required for down-regulation of cell surface FZD and for suppression of WNT/β-catenin signaling. Phosphorylation of RNF43 at these residues promoted ubiquitylation of FZD, and in turn its endocytosis and lysosomal degradation. The precise mechanism underlying regulation of FZD ubiquitylation by this phospho-switch remains unknown, but does not appear to involve changes in the protein-protein interactions (including binding to the E2 UbcH5C), oligomerization state or subcellular localization of RNF43 (Tsukiyama et al., 2020).

How do ZNRF3/RNF43 recognize FZD for ubiquitylation? As discussed earlier, it has been proposed that DVL, which binds both the DIR of ZNRF3 and the ICD of FZD (Fig. 4), acts as a substrate adaptor that targets ZNRF3/RNF43 to FZD (Jiang et al., 2015). Furthermore, direct binding of the ZNRF3/RNF43 PA domain to the CRD of FZD (Fig. 4) has also been proposed as a recognition mechanism

(Tsukiyama et al., 2015), but this subject is still debated (Radaszkiewicz & Bryja, 2020 and reviewed by Tsukiyama, Koo, & Hatakeyama, 2021). One study detected an interaction between the RNF43 PA domain and the FZD CRD (Tsukiyama et al., 2015), while others did not (Jiang et al., 2015; Peng et al., 2013). Several studies showed that deletion or replacement of the PA domain prevented ZNRF3/RNF43 from promoting FZD internalization and suppressing WNT/β-catenin signaling (Koo et al., 2012; Moffat et al., 2014; Spit et al., 2020; Tsukiyama et al., 2015), while another study found that deletion of the PA domain had none of these effects (Radaszkiewicz & Bryja, 2020). AlphaFold modeling suggests that the FZD1 CRD is well positioned to interact with the PA domain of RNF43, which could drive contacts between the TM helix of RNF43 and the 7TM of FZD, and orient the RING domain for ubiquitin transfer (Fig. 5F). So while the question of how FZD is recognized as a substrate by ZNRF3/RNF43 is still unresolved, one possibility is that extracellular contacts between the ZNRF3/RNF43 PA domain and the FZD CRD, intramembrane packing of ZNRF3/RNF43 and FZD TM helices, and intracellular interactions mediated by DVL all play a role in substrate recognition.

LRP6 internalization and degradation is also regulated by ZNRF3/RNF43 (Chang et al., 2020; Giebel et al., 2021; Hao et al., 2012; Kim et al., 2021) and while ZNRF3 could be co-immunoprecipitated with LRP6 (Hao et al., 2012), no single domain or motif in LRP6 has been identified as a target of ubiquitylation or regulation by ZNRF3/RNF43. Therefore the mechanism of LRP6 receptor regulation by the RSPO-ZNRF3/RNF43 system has not been determined. Some possibilities include direct ubiquitylation of LRP6 by ZNRF3/RNF43—although this has not been demonstrated experimentally—or co-internalization of LRP6 with FZD, mediated by WNT ligands or other mutual binding partners.

### *3.2.6 Molecular mechanisms controlling membrane clearance of ZNRF3/RNF43 and engagement receptors*

Importantly, ubiquitylation and internalization of WNT receptors is not regulated directly by interactions between RSPOs and these target receptors, but is instead prevented indirectly as a result of RSPOs binding to and downregulating ZNRF3/RNF43 through LGR-dependent and/or HSPG-dependent mechanisms (Table 1). On the other hand, downregulation of BMPR1A is the direct result of RSPOs interacting with and promoting the internalization of ZNRF3/RNF43 (Table 1). Therefore, the mechanisms controlling membrane clearance of ZNRF3/RNF43 are crucial to the regulation of target receptors in both the WNT and BMP pathways.

During LGR-dependent signaling, binding of RSPOs to both ZNRF3/RNF43 and LGRs is required for internalization of the ternary complex, since mutation of key residues in the ZNRF3/RNF43-interacting FU1 domain or the LGR-interacting FU2 domain of RSPOs abolishes potentiation of WNT/β-catenin signaling (Peng, de Lau, Forneris, et al., 2013; Xie et al., 2013; Zebisch et al., 2013). However, the precise molecular mechanism whereby formation of this ternary complex drives its internalization is not fully understood. One model is that RSPO acts as a cross-linking ligand that couples ZNRF3/RNF43 to LGRs, and since LGR5 undergoes constitutive clathrin-mediated endocytosis (Snyder, Rochelle, Lyerly, Caron, & Barak, 2013; Snyder et al., 2017), mere coupling could result in the co-internalization of ZNRF3/RNF43. This is consistent with the finding that RSPO-dependent potentiation of WNT/β-catenin signaling requires clathrin-mediated endocytosis (Glinka et al., 2011). Further support of this model comes from the fact that synthetic RSPO ligands that cross-link ZNRF3/RNF43 to constitutively endocytosed receptors can promote ZNRF3/RNF43 internalization and upregulate WNT signaling. Engineered "surrogate RSPO" bispecific ligands comprising a ZNRF3- or RNF43-specific scFv fused to the immune cytokine IL-2, which binds to the constitutively internalized IL-2 receptor CD25, leads to co-internalization of ZNRF3 and stimulation of WNT signaling in CD25+ cells (Luca et al., 2020). Additionally, synthetic RSPO2 ligands retaining only the ability to bind ZNRF3/RNF43 through the FU1 domain and fused to scFvs targeting them to the liver-specific asialoglycoprotein receptor (ASGR), which is predominantly expressed on hepatocytes and undergoes rapid endocytosis, increased cell surface FZD and enhanced WNT signaling specifically in cells that express ASGRs (Zhang et al., 2020). Similar results were obtained when these synthetic RSPO2 ligands were fused to scFvs targeting them to a ubiquitously expressed cell surface receptor, transferrin receptor 1 (TFR1), which undergoes continuous endocytosis (Zhang et al., 2020). Finally, the need for RSPOs can be bypassed altogether as long as their cross-linking functionality is provided: appending DmrA and DmrC heterodimerization domains to the C-termini of ZNRF3 and LGR4, respectively, enabled the membrane clearance of ZNRF3 in response to addition of an A/C dimerizer (Hao et al., 2012).

These disparate systems demonstrate that cross-linking ZNRF3/RNF43 to a constitutively endocytosed cell surface receptor, whether it be through

RSPOs themselves or other artificial cross-linkers, can clear ZNRF3/RNF43 from the plasma membrane and promote upregulation of WNT receptors. However, there is evidence that cross-linking of ZNRF3/RNF43 to the engagement receptors is not sufficient in all physiological contexts, and ubiquitylation/deubiquitylation of ZNRF3/RNF43, or potentially engagement receptors, is also involved in regulating their internalization. First, the RING domain of ZNRF3 and RNF43 is required for RSPO1 (or for the A/C dimerizer discussed earlier) to reduce the membrane level of ZNRF3 (Hao et al., 2012), suggesting that membrane clearance requires the ubiquitin ligase activity of ZNRF3/RNF43. Furthermore, the intracellular portion of ZNRF3 and the full-length protein purified by immunoprecipitation exhibit RING domain-dependent auto-ubiquitylation in *in vitro* ubiquitylation assays (Chang et al., 2020; Hao et al., 2012). Therefore, one possibility is that auto-ubiquitylation of ZNRF3 is required for internalization, although this mechanism has not been directly demonstrated. Second, deubiquitylation of ZNRF3/RNF43 by the DUB USP42 stabilizes ZNRF3/RNF43 at the plasma membrane and "stalls" the LGR4-RSPO-ZNRF3/RNF43 ternary complex, preventing its clearance from the cell surface (Giebel et al., 2021). In this way USP42 antagonizes RSPOs by protecting ZNRF3/RNF43 from RSPO- and ubiquitin-dependent internalization, thereby increasing the ubiquitylation and turnover of FZD and LRP6 receptors, and inhibiting WNT signaling. Since ubiquitylation of membrane proteins can drive their internalization (reviewed in MacGurn, Hsu, & Emr, 2012), auto-ubiquitylation of ZNRF3/RNF43 in response to RSPOs may therefore be a second mechanism promoting membrane clearance of ZNRF3/RNF43.

Alternatively or in addition to auto-ubiquitylation, ubiquitylation of another substrate by ZNRF3/RNF43, for instance the engagement receptors themselves, may promote endocytosis of the receptors and associated ZNRF3/RNF43 molecules. This hypothesis is supported by the fact that bringing RNF43 in close proximity to transmembrane proteins, including a synthetic GFP-TM-NanoLuc construct as well as the endogenous immune checkpoint protein programmed death-ligand 1 (PD-L1), can promote their internalization and lysosomal degradation (Cotton, Nguyen, Gramespacher, Seiple, & Wells, 2021). In the case of PD-L1, a synthetic bispecific IgG, or "abTAC," that bound to the ECDs of both RNF43 and PD-L1 was used to recruit RNF43 to PD-L1. Since neither of these two proteins are internalized or degraded constitutively, this experiment showed that recruitment of RNF43 and a target TM protein in close proximity is sufficient to induce internalization and lysosomal degradation of the target protein, independently of RSPOs. Therefore, simultaneous binding of RSPO ligands to ZNRF3/RNF43 and engagement receptors, which would bring them in close proximity, may be sufficient to promote ubiquitylation of the engagement receptors and internalization of the ternary complex. However, whether RSPOs actively regulate the ubiquitin ligase activity of ZNRF3/RNF43, and therefore affect the endocytic efficiency of this process, remains an unanswered question.

In the case of HSPG-dependent signaling, we surmise that RSPO2/3-mediated cross-linking of ZNRF3/RNF43 and HSPGs (Fig. 5C) promotes ternary complex co-internalization driven by endocytosis of HSPGs. HSPGs are autonomous endocytosis receptors that can mediate the internalization of growth factors and morphogens among other ligands (reviewed in Christianson & Belting, 2014). They can undergo constitutive or ligand-induced endocytosis, followed in some cases by lysosomal degradation (Burbach, Friedl, Mundhenke, & Rapraeger, 2003; Fuki et al., 1997; Fuki, Meyer, & Williams, 2000; Wittrup et al., 2009). During HSPG-dependent potentiation of WNT/β-catenin signaling by RSPO3, RNF43 is internalized in a process that requires the interaction of the TSP/BR domains with HSPGs (Dubey et al., 2020). Since GPCs are tethered to the plasma membrane through a GPI anchor (Fig. 4) and do not have a cytoplasmic domain that can be ubiquitylated by ZNRF3/RNF43, in this signaling modality ternary complex internalization cannot be driven by ubiquitylation of the engagement receptor.

In contrast to the indirect regulation of WNT receptor internalization by RSPOs, BMPR1A clearance from the plasma membrane is driven by direct binding of RSPOs to both BMPR1A and ZNRF3/RNF43 (Table 1 and Fig. 5D), which promotes internalization and lysosomal degradation of the ternary complex (Lee et al., 2020). The molecular mechanism through which the ZNRF3/RNF43-RSPO2/3-BMPR1A complex is internalized has not been defined. In this case, internalization of BMPR1A is the step being regulated rather than being a constitutive process like the endocytosis of LGRs or HSPGs. Therefore, the mere cross-linking of BMPR1A and ZNRF3/RNF43 by RSPOs would not be sufficient to drive internalization of either receptor. We surmise that ubiquitylation of either ZNRF3/RNF43 or BMPR1A, induced by binding of RSPO2 or RSPO3, is likely the main mechanism driving internalization of the ternary complex.

In summary, we described two molecular mechanisms that could drive ZNRF3/RNF43 internalization and lysosomal degradation: (1) co-internalization of ZNRF3/RNF43 promoted by RSPO-mediated cross-linking to a constitutively endocytosed engagement receptor, and (2) endocytosis driven by ubiquitylation of ZNRF3/RNF43, engagement receptors or both, promoted by RSPO-mediated ternary complex formation. The latter could be driven by regulated auto-ubiquitylation of ZNRF3/RNF43 or by trans-ubiquitylation of the engagement receptors.

## 3.3 Physiological, pathological and therapeutic implications of distinct RSPO-ZNRF3/RNF43 signaling modalities

The RSPO-ZNRF3/RNF43 system has important functions during embryonic development and in adult tissue homeostasis. Aberrant regulation caused by mutations in ZNRF3/RNF43 or by RSPO fusions that cause elevated expression can lead to cancer. We refer the reader to some excellent reviews on the physiology and pathology of the RSPO-ZNRF3/RNF43 system (Bugter, Fenderico, & Maurice, 2021; de Lau et al., 2012; Hao et al., 2016; Jin & Yoon, 2012; Nagano, 2019; Raslan & Yoon, 2019; Ter Steege, ter Steege, & Bakke, 2021). Here we describe the principal phenotypes caused by disruption of different components of the system, and discuss how the discovery of the three RSPO signaling modalities presented earlier compels us to re-interpret some of these phenotypes. We also posit that the modular nature of RSPO proteins presents a unique opportunity to manipulate the RSPO-ZNRF3/RNF43 system for therapeutic benefit.

To the best of our knowledge, the comprehensive phenotype of the ubiquitous *Znrf3/Rnf43* double KO mouse has not been published, but would be predicted to result in early embryonic lethality. However, conditional *Znrf3/Rnf43* double KO in the intestinal epithelium (driven by *Cyp1a1-cre* or *Villin-creERT2*) resulted in marked expansion of the proliferative compartment (with hyperproliferative cells containing high levels of β-catenin), upregulation of WNT target genes, and increased numbers of intestinal stem and Paneth cells (Koo et al., 2012). Clonal deletion of *Znrf3/Rnf43* in the intestinal epithelium or in intestinal stem cells (driven by *Lgr5-creERT2*) resulted in adenoma formation, with continuous expansion of stem cells and generation of Paneth cells but no other differentiated cell types (Koo et al., 2012). Intestinal organoids derived from *Znrf3/Rnf43* double KO mice grew faster than controls and lost the dependence on RSPO1 supplementation, but not on secreted WNT3A, consistent with the role of ZNRF3/RNF43 in mediating RSPO-dependent potentiation of WNT signaling (Koo et al., 2012). The WNT/β-catenin pathway is also a major regulator of liver metabolic zonation, development and regeneration (reviewed in Hu & Monga, 2021; Monga, 2014; Yang et al., 2014). Inducible, systemic combined deletion of *Znrf3* and *Rnf43* (driven by *Rosa26-creERT2*) in mice induced hepatocyte proliferation and extended metabolic zonation, measured as a marked increase in the expression, as well as zonal expansion, of the liver-specific WNT/β-catenin target proteins GS and CYP2E1 (Planas-Paz et al., 2016). Furthermore, deletion of *Znrf3* and *Rnf43* specifically in hepatocytes (driven by Ad5cre virus) led to the formation of multiple liver tumors, primarily classified as hepatocellular carcinomas (Sun et al., 2021). Thus, ZNRF3/RNF43 control the hepatic WNT/β-catenin signaling gradient and metabolic liver zonation, and prevent liver tumor formation. Simultaneous disruption of *znrf3* and *rnf43* by injection of TALENs into two-cell stage *Xenopus tropicalis* embryos resulted in development of ectopic limbs, ranging from diplopodia (duplication of digits) to complete polymelia (presence of supernumerary limbs), including quadruplication of forelimbs in extreme cases (Szenker-Ravi et al., 2018). These three examples, and many others not discussed here, demonstrate that loss of ZNRF3 and/or RNF43 results in elevated WNT/β-catenin signaling in various tissues. However, given that ZNRF3/RNF43 regulate all three RSPO signaling modalities discussed earlier, the phenotypes caused by their disruption do not distinguish the specific physiological functions of LGR-dependent, HSPG-dependent and BMPR1A-dependent RSPO signaling.

The expression patterns of the four RSPOs in mice are distinct (Nam, Turcotte, & Yoon, 2007) and, not surprisingly, so are the phenotypes associated with their disruption, illustrating the pleiotropic roles of RSPOs during embryogenesis (reviewed in de Lau et al., 2012; Jin & Yoon, 2012; Nagano, 2019). Mutations in *RSPO1* cause a rare human syndrome characterized by XX male sex reversal, palmoplantar hyperkeratosis (abnormal thickening of the palms and soles) and predisposition to squamous cell carcinoma of the skin (Parma et al., 2006). Loss of *Rspo1* in mice confirmed that the absence of RSPO1 at the gonadal differentiation stage causes partial sex reversal (Tomizuka et al., 2008). Human mutations in *RSPO2* cause tetra-amelia with lung hypo/aplasia syndrome (TETAMS), a severe condition characterized by amelia (the complete absence of limbs), lung hypo/aplasia, cleft lip-palate, and labioscrotal fold aplasia

(Szenker-Ravi et al., 2018). Consistently, loss of *Rspo2* in mice causes limb malformations or amelia, severe malformations of laryngeal-tracheal cartilages, lung hypoplasia, and palate malformations (Aoki, Kiyonari, Nakamura, & Okamoto, 2008; Bell et al., 2008; Nam et al., 2007; Yamada et al., 2009). Loss of *Rspo3* in mice results in severe placental vascular defects, causing death of the mutant mice around embryonic day (E)10 (Aoki et al., 2007; Kazanskaya et al., 2008). Mutations in human *RSPO4* were found in individuals affected with anonychia, a rare autosomal recessive congenital syndrome characterized by partial or complete absence of fingernails and toenails (Bergmann et al., 2006; Blaydon et al., 2006; Brüchle et al., 2008; Ishii et al., 2008).

While previously it may have been tempting to attribute these different phenotypes to the distinct expression patterns of RSPO1–4, an additional explanation must be considered in light of the multiple RSPO signaling modalities discussed in this chapter. Differences in *RSPO* KO phenotypes may also be explained by the capacity of different RSPOs to signal through LGR-dependent, HSPG-dependent and BMPR1A-dependent mechanisms. This possibility is supported by the finding that ubiquitous *Lgr4/5/6* triple KO mice do not exhibit many of the phenotypes observed in *Rspo2* KO or *Rspo3* KO mice (Szenker-Ravi et al., 2018). *Lgr4/5/6* triple KO mice die around E14.5–18.5, but the embryos undergo normal development of the limbs and lungs, as well as normal placental vascularization, suggesting RSPO2 and RSPO3 signaling is largely unaffected. From these experiments it can be surmised that certain developmental processes governed by RSPO2 and RSPO3 occur independently of LGRs. However, other processes that are also regulated by RSPO2, such as palate and tongue development, rely on LGRs, since both *Rspo2* KO and *Lgr4/5/6* triple KO mice exhibit cleft palate and ankyloglossia (tongue-tie).

Comparison of the phenotypes caused by *Lgr4/5/6* KO and by *RSPO1–4* loss-of-function mutations in humans and mice can help distinguish between LGR-dependent and LGR-independent effects of RSPOs, but whether these effects are driven by potentiation of WNT signaling through HSPGs or by downregulation of BMP signaling through BMPR1A (or by yet other pathways regulated by RSPOs) is less clear. Little is known about the physiological contexts in which HSPG-dependent and BMPR1A-dependent RSPO signaling operate. During nephrogenesis, strong RSPO-dependent activation of WNT/β-catenin signaling is essential for nephron progenitors to differentiate and undergo mesenchymal to epithelial transition, and this process occurs largely in an LGR-independent manner (Vidal et al., 2020), suggesting the possibility that nephrogenesis is driven in part by HSPG-dependent RSPO signaling. In multiple myeloma cells, RSPO binds to SDC1 in a HS-dependent manner, and this event is required for optimal stimulation of WNT/β-catenin signaling (Ren et al., 2018). In *Xenopus*, RSPO2 cooperates with Spemann organizer effectors to inhibit BMP signaling during embryonic axis formation (Lee et al., 2020), and BMP signaling inhibition by RSPO2 maintains autocrine self-renewal in acute myeloid leukemia (AML) (Sun et al., 2021). Elucidating the complete repertoire of biological and pathological processes controlled by RSPOs through different signaling modalities will require a combination of approaches. Disrupting entire families of engagement receptors, as was done with the LGRs in mice (Szenker-Ravi et al., 2018), could provide additional insights, but this is a challenging prospect for HSPGs, since RSPO3 (and presumably RSPO2) can signal redundantly through GPCs, of which there are six members in mammals, and SDCs, of which there are an additional four (Dubey et al., 2020; Lebensohn & Rohatgi, 2018).

Given the modular structure of RSPO proteins (Fig. 4) and the different domain requirements for distinct signaling modalities (Table 1), "modality-specific" engineered RSPO ligands—ligands that can signal exclusively through a single signaling modality—could yield further insights. This concept was demonstrated by experiments with RSPO chimeras in which domains from RSPOs capable of signaling through distinct modalities were swapped, rendering the chimeras competent or incompetent to signal through a different modality (Lebensohn & Rohatgi, 2018; Lee et al., 2020). In other experiments, a domain required for one signaling modality was mutated, deleted or replaced by synthetic scFvs, leaving only the domains necessary to target RSPOs to engagement receptors that mediate a different signaling modality (Dubey et al., 2020; Lebensohn & Rohatgi, 2018; Lee et al., 2020). Furthermore, the modularity of the RSPO-ZNRF3/RNF43 signaling system should enable the use of such modality-specific ligands for therapeutic or regenerative applications. The ability to target RSPOs to desired tissues through engagement of tissue-specific receptors has been demonstrated (Luca et al., 2020; Zhang et al., 2020) and could potentially be combined with modality-specific mutations to selectively target a single signaling pathway specifically in an affected tissue.

# 4. Regulation of Hedgehog and melanocortin receptor abundance by the membrane-recruited E3 MGRN1

MGRN1 and its paralog RNF157 (Fig. 1C) belong to a unique class of E3s that do not contain a TM domain but are recruited to the plasma membrane through interactions with other single-pass TM proteins. The TM partners of MGRN1 regulate its substrate specificity, much like substrate adaptors do in multi-subunit RING class E3s (Metzger et al., 2014). We describe two such systems in which association with two related single-pass TM proteins directs MGRN1 to ubiquitylate different GPCRs. We predict that the recruitment of cytoplasmic E3s to the plasma membrane is a mechanism used more broadly to regulate the cell surface abundance of membrane receptors, and consequently to regulate signaling sensitivity.

## 4.1 The MGRN1-MEGF8-MOSMO complex, an attenuator of Hedgehog signaling strength

*Mgrn1* encodes an eponymous RING family E3 (also known as RNF156, Figs. 1C and 6A), and was identified as the mutated gene at the *mahoganoid* locus in mice (He et al., 2003; Phan, Lin, LeDuc, Chung, & Leibel, 2002). *Mgrn1* has been studied because of its effects on coat color determination and spongiform neurodegeneration. However, ~25% of *Mgrn1*$^{-/-}$ embryos die during gestation with heterotaxy (defects in left-right patterning of organs) and complex cardiac anomalies, suggesting an additional role for *Mgrn1* during development (Cota et al., 2006).

*MEGF8*, the gene encoding multiple epidermal growth factor-like (EGFL) domains protein 8 (MEGF8, Fig. 6B–D) was among those compiled by a bioinformatics screen for genes that encode uncommonly large (>1000 amino acids) membrane-embedded proteins that contain multiple EGFL domains (Nakayama et al., 1998). This effort was motivated by the observation that these characteristics—large size, a TM domain and multiple EGFL domains—were seen in proteins that play important roles in cell-cell or cell-extracellular matrix interactions, such as AGRIN or receptors and ligands of the NOTCH family. MEGF8 was subsequently noted to have homology to a single-pass TM protein commonly referred to as attractin (ATRN, Fig. 7A), which was identified by positional cloning of a mouse gene from the *mahogany* locus implicated in both body weight and coat color (discussed in Section 4.2) (Gunn et al., 1999; Nagle et al., 1999).

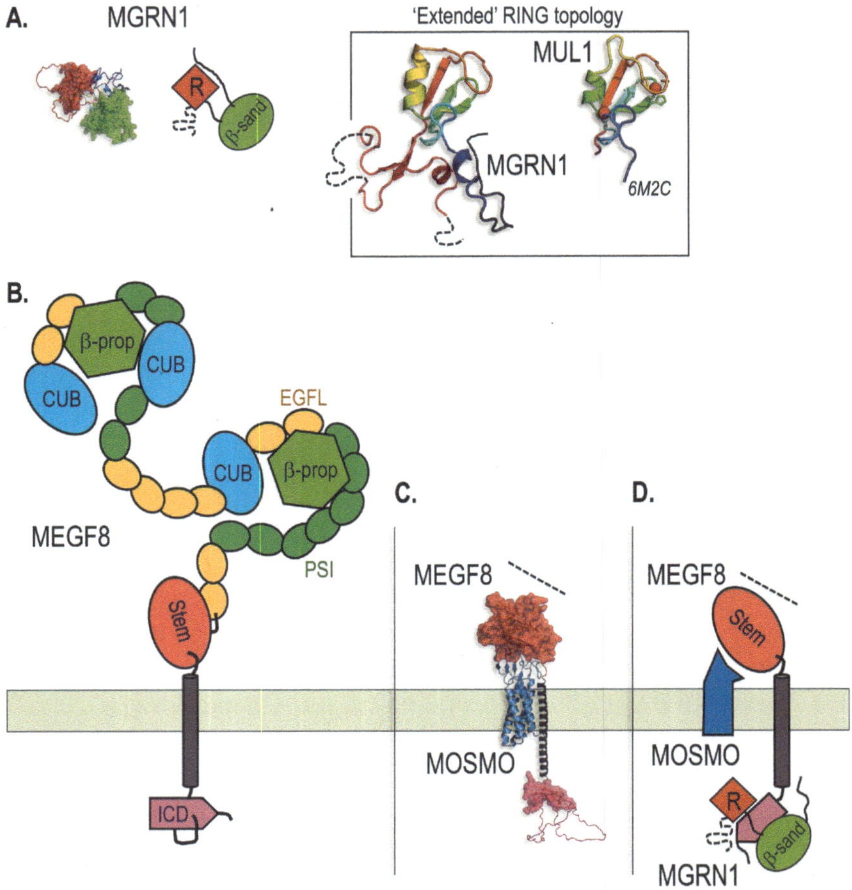


**Fig. 6** Architecture of the MMM complex, an attenuator of Hh signaling. (A) MGRN1 is a cytoplasmic E3 containing a RING domain and a β-sandwich domain. An AlphaFold structural model is shown on the left and a cartoon representation on the right. The box shows the AlphaFold prediction of the MGRN1 "extended" RING domain, which most closely resembles the structure of the RING domain from MUL1 (PDB ID 6M2C). (B) MEGF8 contains a massive ECD with a pseudo-repeat architecture. A central spine composed of multiple EGFL and PSI domains is decorated with two 6-blade β-propellers and three complement C1r/C1s, uegf, bmp1 (CUB) domains. The extracellular domain is perched on a juxta-membrane, extracellular Stem domain, followed by a TM helix that extends into the cytoplasm and connects to an ICD. (C) AlphaFold model of the 4-pass TM protein MOSMO (related to the Claudin family of 4TM proteins) complexed with a fragment of MEGF8 containing the Stem, TM and ICD. The Stem stacks on top of the extracellular β-sheet of MOSMO, promoting the "zippering" of the 4TM bundle of MOSMO with the single TM helix of MEGF8. (D) Cartoon depicting the assembly of the MGRN1-MEGF8-MOSMO complex, excluding the large pseudo-repeat ECD of MEGF8 (shown in (B)).

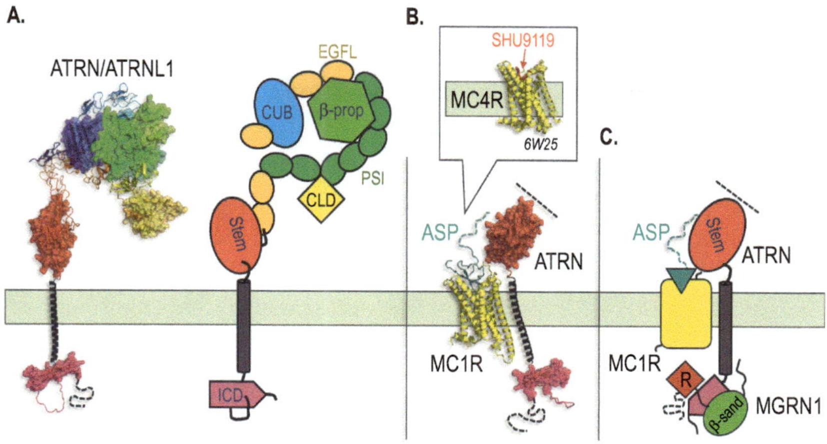


**Fig. 7** The MGRN1-ATRN complex, an attenuator of melanocortin receptor signaling. (A) AlphaFold model and cartoon representation of ATRN/ATRNL1 (ATRN and ATRNL1 are two closely related proteins, so only one of them is shown). Note the similarities between ATRN and MEGF8 (Fig. 6B). The ECD of ATRN has only one of the two repeats present in MEGF8. ATRN has a cyclophilin-like domain (CLD) not found in MEGF8, but shares a 6-blade β-propeller and a CUB domain. The domain coloring in the cartoon is matched to the structural model. (B) and (C) AlphaFold model (B) and cartoon representation (C) of how the ligand ASP could cross-link MC1R to ATRN. The C-terminus of ASP forms a β-hairpin that occupies a putative ligand binding site in MC1R. The box in (B) shows the solved structure of MC4R in complex with the antagonist SHU9119 (PDB ID 6W25) (Yu et al., 2020), which occupies the same site predicted to interact with ASP by AlphaFold. The N-terminus of ASP (dotted aqua line) is well positioned to interact with the Stem domain of ATRN.

Insights into the biological function of MEGF8 came from mouse embryos carrying loss-of-function mutations in *Megf8*, as well as from human patients with a recessively inherited birth defect syndrome called Carpenter's Syndrome (Engelhard et al., 2013; Twigg et al., 2012; Zhang et al., 2009). In both cases, loss of MEGF8 resulted in heterotaxy, severe congenital heart defects, pre-axial polydactyly, and skeletal and craniofacial defects. While *Megf8* clearly plays a widespread role in the development of multiple tissues, the underlying mechanisms remained unclear. MEGF8 was proposed to be a modifier of BMP and nodal signaling due to its role in left-right patterning and peripheral axon guidance (Engelhard et al., 2013; Zhang et al., 2009). In addition, a possible connection to Hedgehog (Hh) signaling was suggested by the observations that (1) Carpenter's Syndrome can also be caused by mutations in *RAB23*, a negative regulator of Hh signaling and that (2) Carpenter's Syndrome phenotypes resembled those of a related syndrome caused by mutations in the Hh transcription factor GLI3 (Eggenschwiler, Espinoza, & Anderson, 2001; Twigg et al., 2012). However, the molecular mechanism of action of both MGRN1 and MEGF8 remained unknown.

MGRN1, MEGF8 and MOSMO, a completely uncharacterized 4-pass TM protein of the Claudin family (Fig. 6C–D), had never been linked to each other until they were all identified in a genome-wide, loss-of-function CRISPR screen designed to uncover attenuators of Hh signaling strength (Pusapati et al., 2018). This screen was performed by exposing cells to a sub-saturating concentration of the Hh ligand Sonic Hedgehog (SHH) and selecting cells containing mutations that enhanced the activity of a Hh transcriptional reporter. Biochemical analyses demonstrated that all three proteins form a complex, which we named the MMM complex. MGRN1 stably associates with the C-terminal, cytoplasmic tail of MEGF8, and MOSMO associates with this MEGF8-MGRN1 subcomplex (Kong et al., 2020, 2021) (Fig. 6D). The MMM complex represents a new architecture for E3 complexes: a TM protein with a massive extracellular domain that stably associates with a cytoplasmic RING E3 through a short cytoplasmic tail.

#### *4.1.1 Biochemical and cellular functions of the MMM complex*

While there is little homology between the MMM proteins and ZNRF3/RNF43 (except for the presence of a RING domain in MGRN1), there are conceptual similarities in their mechanisms of action. The MMM complex regulates the abundance of the Hh signal transducer smoothened (SMO) at the cell surface by ubiquitylation (Kong et al., 2020), similar to the way in which ZNRF3/RNF43 regulate the abundance of FZD receptors. SMO transmits the Hh signal across the membrane and is the closest relative of the FZD receptors in the Class F family of GPCRs. Loss of MGRN1, MEGF8 or MOSMO markedly reduced SMO endocytosis and degradation, leading to SMO accumulation on the cell surface and the membrane of the primary cilium (Kong et al., 2020, 2021; Pusapati et al., 2018). Primary cilia function as compartments for Hh signaling, and increased SMO accumulation in primary cilia enhances Hh signaling strength (Huangfu & Anderson, 2005; Huangfu et al., 2003). Consequently, the concentration of SHH required to induce target genes is reduced by nearly an order of magnitude when any of the MMM proteins are lost (Pusapati et al., 2018). In summary, just as FZDs are target receptors for ZNRF3/RNF43, SMO is the target receptor for the MMM complex.

Both the integrity of the RING domain of MGRN1 and the physical interaction between MGRN1 and the MEGF8 ICD are required for ubiquitylation of SMO by the MMM complex (Kong et al., 2020). The MGRN1-MEGF8 interaction requires the amino acid sequence MASRPFA, a motif that is conserved in proteins of the MEGF8/ATRN family across evolution (Gunn et al., 1999; Nagle et al., 1999). Direct binding of SMO to MMM complex components has not been reported, so the mechanism of SMO recognition by the MMM complex remains to be elucidated. The large ECD of MEGF8 is dispensable for SMO recognition. A truncated protein containing only the TM domain (TMD) and ICD of MEGF8 (which also cannot bind to MOSMO) is nonetheless sufficient to mediate SMO ubiquitylation, reduce its levels at the cell surface and dampen Hh signaling (Kong et al., 2020). However, it is not known whether this TMD-ICD segment of MEGF8 recognizes SMO directly or through an adaptor protein. Importantly, mere recruitment of MGRN1 to the plasma membrane is not sufficient to drive SMO ubiquitylation. The MEGF8 TMD, or the precise arrangement of the TMD relative to the ICD, is likely to be important because replacement of the MEGF8 TMD with the TMD of a different single-pass TM protein abolished SMO ubiquitylation, despite the fact that MGRN1 was still recruited to the membrane and retained ubiquitin ligase activity (Kong et al., 2020). Therefore, MEGF8 likely functions as a substrate adaptor that recruits SMO, and perhaps other substrates, for ubiquitylation by MGRN1.

The function of the 4TM protein MOSMO remains to be fully elucidated. MOSMO is a Claudin family protein whose distinctive extracellular loop structure, which folds into a compact, disulfide-locked β-sheet, forms a cell surface interaction platform for a previously uncharacterized extracellular juxtamembrane domain in MEGF8, which we named the Stem domain (Fig. 6C–D) (Kong et al., 2021). Loss of MOSMO partially reduces MEGF8 levels at the cell surface, suggesting a role in trafficking (Kong et al., 2021). While MOSMO is not absolutely required for SMO ubiquitylation by the MEGF8-MGRN1 subcomplex in an overexpression system, the severe phenotypes of *Mosmo*$^{-/-}$ mice, which are similar to those of *Megf8*$^{-/-}$ mice, suggest that the association of MOSMO with MEGF8 and MGRN1 may in fact play a critical role in activation of the MMM complex.

A key unanswered question in MMM complex function is the role of the large MEGF8 ECD (Fig. 6B). Just like ZNRF3/RNF43 are regulated by RSPO ligands, it is likely that the MMM complex is also regulated by interactions of the MEGF8 ECD with a soluble extracellular protein, a component of the extracellular matrix or another membrane protein on the same or an adjacent cell. The unique architecture of the MMM complex suggests a regulatory mechanism whereby the TM topology of the MEGF8 substrate adaptor allows extracellular signals to directly regulate the selection and ubiquitylation of specific substrates by MGRN1 in the cytoplasm.

In both cultured cells and mouse embryos, loss of MMM complex components in target cells receiving Hh signals leads to elevated sensitivity to Hh ligands (Kong et al., 2020, 2021). The pattern of elevated Hh signaling in embryos lacking MMM components is distinct in a very specific way from what is observed in embryos lacking patched 1 (PTCH1) or suppressor of fused (SUFU), two negative regulators of Hh signaling. Loss of PTCH1 or SUFU leads to the ectopic activation of Hh signaling in multiple tissues, showing that these proteins suppress basal signaling activity even in the absence of Hh ligands (Cooper et al., 2005; Goodrich, Milenkovic, Higgins, & Scott, 1997; Svard et al., 2006). In contrast, Hh signaling in MMM mutant mice remains dependent on Hh ligands: Hh target gene expression is localized correctly in the embryo, but the strength of signaling is elevated (Kong et al., 2021). Thus, like ZNRF3 and RNF43, the MMM complex proteins are best characterized as "attenuators" of signaling rather than negative regulators, because their effects remain dependent on the presence of WNT or Hh ligands, respectively. In summary, the MMM complex forms a signaling module that calibrates how the Hh ligand gradient is decoded by target cells.

### 4.1.2 Developmental roles of the MMM complex

The common function of the three proteins in the MMM complex that is suggested by biochemical analyses is also supported by mouse genetic studies (Cota et al., 2006; Kong et al., 2020, 2021; Zhang et al., 2009). *Mosmo*$^{-/-}$ and *Megf8*$^{-/-}$ mouse embryos exhibit similar developmental defects, also shared by Carpenter's Syndrome patients: heterotaxy, severe congenital heart defects, pre-axial digit duplication, skeletal defects, craniofacial defects and neurodevelopmental abnormalities. In addition, SMO abundance in the ciliary membrane is markedly elevated in nearly all *Mosmo*$^{-/-}$ and *Megf8*$^{-/-}$ embryonic tissues, consistent with observations in cultured cells (Kong et al., 2020, 2021). Some of these embryonic phenotypes were initially not observed in *Mgrn1*$^{-/-}$ embryos because MGRN1 is partially redundant with RNF157 (Cota et al., 2006; Kong et al., 2020). However, *Mgrn1*$^{-/-}$; *Rnf157*$^{-/-}$ embryos exhibit a constellation of defects very similar to those seen in *Mosmo*$^{-/-}$ and *Megf8*$^{-/-}$ embryos. Beyond common phenotypes, strong genetic interactions and gene dosage effects between mutant alleles of *Megf8*, *Mgrn1* and *Rnf157* support the conclusion that MGRN1/RNF157 and MEGF8 function together to regulate a common set of developmental processes (Kong et al., 2020).

The widespread nature of the defects observed in mouse embryos carrying mutations in the MMM genes points to a central role of the MMM complex in the regulation of cell-cell communication. MMM complex components have been linked to nodal, BMP and Hh signaling, but a direct role in regulation of signaling components has only been established for SMO in the Hh pathway (Engelhard et al., 2013; Pusapati et al., 2018). The expression patterns of *Megf8*, *Mgrn1* and *Mosmo* do not provide many clues to the developmental functions of the MMM complex: they are ubiquitously expressed in all three germ layers and in all major cardiac populations (Cota et al., 2006; Kong et al., 2021; Wang et al., 2020; Zhang et al., 2009). Thus, control of MMM complex activity, perhaps by a ligand or by post-translational modifications, is likely to be the key regulated step that explains its tissue-specific roles.

An unresolved question is whether all of the embryonic defects seen in MMM mutant mice are caused by upregulation of SMO and elevated Hh signaling, or whether some are caused by regulation of other unidentified substrates that function in other pathways. The pre-axial digit duplication seen in MMM mutant embryos is likely to be driven by elevated Hh signaling, since this phenotype can be completely reversed by administration of vismodegib, a placenta-permeable small molecule inhibitor of SMO (Kong et al., 2020, 2021). However, vismodegib only partially rescues the congenital heart defects observed in MMM mutant embryos, which could be due to a suboptimal schedule of *in utero* vismodegib administration, or could also indicate that other pathways are regulated by the MMM complex. Further research will be necessary to comprehensively identify substrates of the MMM complex other than SMO and to test for possible roles of the MMM complex in other developmental signaling pathways.

Heterotaxy is a prominent feature of MMM mutant mouse embryos and Carpenter's Syndrome patients, and may be the root cause of the severe congenital heart defects observed in both (Cota et al., 2006; Kong et al., 2020, 2021; Zhang et al., 2009; Twigg et al., 2012). The MMM complex regulates left-right patterning at an early stage in embryogenesis. In both *Megf8*$^{-/-}$ and *Mgrn1*$^{-/-}$ embryos, abnormal expression of all three canonical left-expressed genes (*Nodal*, *Lefty*, *Pitx2*) was observed (Cota et al., 2006; Zhang et al., 2009). Conditional disruption of *Megf8* in various cardiac lineages using a panel of cre drivers (*cTnt-cre*, *Wt1-cre*, *Tie2-cre*, *Wnt1-cre*, *Mesp1-cre*) did not reproduce the heart defects seen in the global *Megf8* KO. Timed global deletion of *Megf8* at E7.5 also did not reproduce the cardiac defects (Wang et al., 2020). These data suggest that *Megf8* is required for cardiac development at a time point earlier than cardiac organogenesis and supports the hypothesis that the heart defects seen in MMM mutant mice are a consequence of disrupted left-right patterning early in development.

Current models suggest that Hh signaling plays a permissive role in left-right patterning of the mouse embryo: reduced Hh signaling caused by loss of SMO leads to a midline heart tube that fails to loop and an embryo that fails to turn (Levin, Johnson, Stern, Kuehn, & Tabin, 1995; Tsiairis & McMahon, 2009; Zhang, Ramalho-Santos, & McMahon, 2001). While this model does not readily explain how the elevated Hh signaling seen in MMM mutant embryos leads to left-right patterning defects, loss-of-function mutations in the Hh negative regulator *Sufu* do cause abnormalities in embryo turning, heart looping and expression of the left-expressed gene *Pitx2* (Cooper et al., 2005). Thus, normal left-right patterning may depend on a just-right, or "goldilocks," level of Hh signaling strength calibrated by the MMM complex. Alternatively, the MMM complex may regulate another pathway involved in left-right patterning, such as nodal or BMP signaling. Elucidating how the MMM complex regulates left-right patterning is critical to understanding its developmental roles and, consequently, the etiology of birth defects caused by mutations in MMM genes.

## 4.2 The MGRN1-ATRN system regulates melanocortin receptors

Regulation of Hh signaling by the MMM complex shares many themes with another membrane-tethered E3 complex, formed by MGRN1 and the MEGF8-related protein ATRN, that functions in a paracrine fashion to regulate melanocortin receptor signaling (reviewed in He, Eldridge, et al., 2003). The four melanocortin receptors, MC1R, MC2R, MC3R and MC4R, are GPCRs that bind to peptide agonists, including α-MSH and ACTH, and regulate diverse physiological processes in vertebrates. We focus here on the regulation of mouse coat color, which serves as a useful paradigm for paracrine cell–cell interactions that orchestrate both tissue patterning during development and tissue homeostasis in adults.

Melanocortin receptor activity is controlled by agonists, like α-MSH, and inverse agonists. Agouti signaling protein (ASP) and agouti-related protein (AGRP) are both inverse agonists of the MC1R receptors in the skin and the MC4R receptors in the hypothalamus. They reduce basal and α-MSH-stimulated receptor activity (Lu et al., 1994).

Hair follicle melanocytes switch between producing the pigments eumelanin (dark) and pheomelanin (light). In mice, the presence of a subapical light band in an otherwise dark hair leads to the agouti coat color. The dark eumelanic hair pigmentation is driven by MC1R signaling in melanocytes. The light pheomelanic band in each hair is generated by a paracrine signaling interaction between dermal cells at the base of each hair follicle and neighboring melanocytes. Transient secretion of ASP from the dermal cells inhibits MC1R signaling, causing a switch to pheomelanin synthesis and the generation of the light band on each hair. Ubiquitous and constitutive expression of ASP results in a light coat as well as hyperphagia and obesity, caused by inhibition of MC4R in the hypothalamus. This short-range dermal-melanocyte signaling circuit is similar to how RSPOs secreted by stromal cells influence WNT signaling in epithelial stem cells of the intestinal crypts (Greicius et al., 2018).

*Mgrn1* and *Atrn* were identified as genes required for the inhibitory effect of ASP on melanocortin receptor signaling (Gunn et al., 1999; He, Eldridge, et al., 2003; He, Lu, et al., 2003; Nagle et al., 1999; Phan et al., 2002). The pigmentation changes caused by ectopic ASP expression can be suppressed by mutations in *Mgrn1* and *Atrn*. Epistasis analyses have placed *Mgrn1* and *Atrn* downstream of ASP but upstream of MC1R. Strikingly, genetic analyses show that *Mgrn1* and *Atrn* are required for ASP signaling, despite the fact that purified ASP alone is a high-affinity antagonist of MC1R in biochemical assays (Willard et al., 1995). These genetic studies suggested that MGRN1 and ATRN are required for the inhibitory effects of ASP on MC1R in target melanocytes.

### *4.2.1 Cellular and biochemical models for the regulation of melanocortin receptors by ASP, MGRN1 and ATRN*

ATRN is a single pass TM protein related to MEGF8. The massive extracellular domain of MEGF8 is composed of two tandem repeats (Fig. 6B). ATRN is more compact and has only one of these repeats, but shares the single TM helix and short intracellular tail found in MEGF8 (Fig. 7A). While the C-terminal domain of ASP binds to MC1R with high affinity

($K_D \sim 1$ nM), the N-terminal chain of ASP binds to the ECD of ATRN with ~500-fold lower affinity (Fig. 7B-C) (He et al., 2001; Willard et al., 1995). These genetic and biochemical studies are most consistent with a model in which the binding of ASP to ATRN transmits a signal to MGRN1 in the cytoplasm that ultimately leads to the downregulation of melanocortin receptors.

There are striking similarities between the ASP-MGRN1-ATRN module and the MMM complex: ATRN and MEGF8 are related proteins, MGRN1 is shared, and both systems regulate GPCRs. The simplest model that emerges from this comparison is that the MGRN1-ATRN complex ubiquitylates and downregulates MC1R and MC4R at the cell surface in response to binding of ASP. In this model, ASP acts as a ligand that cross-links melanocortin receptors to ATRN (Fig. 7B-C), analogous to how RSPO2 and RSPO3 cross-link BMPR1A to ZNRF3/RNF43 (Fig. 5D). AlphaFold modeling suggests that ASP can simultaneously bind to MC1R through its C-terminus and to the Stem domain in ATRN through its N-terminus, thereby positioning the RING domain of ATRN for Ub transfer to the cytoplasmic surface of MC1R (Fig. 7B-C).

There is some evidence to support this model. ASP promotes trafficking of MC4R to the lysosome and its subsequent degradation in a manner that depends on both MGRN1 and ATRN (Kim, Olzmann, Barsh, Chin, & Li, 2007; Overton & Leibel, 2011). Loss of ATRN leads to elevated MC4R levels at the cell surface (Overton & Leibel, 2011), analogous to how loss of MEGF8 leads to elevated SMO levels at the cell surface (Pusapati et al., 2018). However, a physical interaction between the C-terminal tail of ATRN and MGRN1, analogous to that between MEGF8 and MGRN1, has not been demonstrated, and neither has a role for such an interaction in melanocortin receptor downregulation. Notably, the MASRPFA motif in the MEGF8 cytoplasmic tail that is required for binding to MGRN1 is conserved in ATRN (Kong et al., 2020; Nagle et al., 1999), and this motif in the *Drosophila* ATRN ortholog is required for association with *Drosophila* MGRN1 (Nawaratne, Kudumala, Kakad, & Godenschwege, 2021). Ubiquitylation of MC1R by an MGRN1-ATRN complex also remains to be established. MGRN1 was shown to ubiquitylate MC2R, but it is not clear whether this reaction required ATRN (Cooray, Guasti, & Clark, 2011). MGRN1 also ubiquitylates tumor suppressor gene 101 (TSG101), a component of the endosomal sorting complex required for transport-I (ESCORT-I) complex that mediates the trafficking of ubiquitylated cell surface receptors from the plasma membrane to the lysosome for degradation (Jiao et al., 2009; Kim et al., 2007). Based on this finding, a different model has been proposed in which the effect of MGRN1 on melanocortin receptor trafficking is an indirect consequence of its regulation of TSG101. Further experiments will therefore be required to elucidate the biochemical function of the MGRN1-ATRN complex.

### 4.3 Evolutionary insights into MGRN1 and the MEGF8/ATRN protein family

MEGF8 and ATRN family proteins are conserved across metazoans and in their closest living relatives, the choanoflagellates (Pusapati et al., 2018). The *Drosophila* ortholog of *Megf8* is also required for early embryonic development (Lloyd, Toegel, Fulga, & Wilkie, 2018). However, MGRN1 is more widely distributed throughout evolution, found in all major eukaryotic lineages (Pusapati et al., 2018). The *Arabidopsis thaliana* ortholog of MGRN1, called LOG2, can be functionally replaced by human MGRN1 (Guerra, Pratelli, Kraft, Callis, & Pilot, 2013). LOG2 is recruited to the plasma membrane by binding to the cytoplasmic tail of the single-pass TM protein glutamine dumper-1 (GDU1) (Guerra et al., 2017). The LOG2-GDU1 complex has been implicated in amino acid transport. These observations suggest that members of the MGRN1 family of RING E3s have evolved to associate with multiple single-pass TM proteins across eukaryotes to regulate the activity of diverse membrane receptors and transporters. In this scheme, MGRN1 provides the ubiquitin ligase function while the single-pass TM protein functions as a substrate adaptor to select targets for ubiquitylation.

This ancient membrane-tethered E3 system has likely been adapted to regulate signaling pathways at multiple times during evolution. While MGRN1 is found in all eukaryotes, Hh signaling is only present in a subset of the metazoan lineages where MEGF8 is found. The MGRN1-ATRN system seems to have been co-opted to regulate melanocortin receptor signaling much later in evolution, since ASP and melanocortin receptors are only found in vertebrates.

## 5. Conclusions

Membrane-tethered E3s can tune the sensitivity of cells to ligands by promoting the internalization and degradation of specific signaling receptors. They can target substrates constitutively, like ZNRF3/RNF43 target FZDs, or they can be regulated by secreted ligands that direct them to specific substrates, like RSPOs direct ZNRF3/RNF43 to BMPR1A or ASP directs MGRN1-ATRN to melanocortin receptors. Thus, membrane-tethered E3s enable extracellular ligands to directly control the ubiquitylation of substrates in the cytoplasm. We propose that regulation of membrane receptors by membrane-tethered E3s plays a widespread and understudied role in tuning cell sensitivity to paracrine signals that control embryonic development and tissue homeostasis, as exemplified by the two conceptually analogous systems described in this chapter: (1) the RSPO-ZNRF3/RNF43 module that regulates WNT and BMP receptors, and (2) the MMM and MGRN1-ATRN modules that regulate Hh and melanocortin receptors, respectively.

Compared to cytoplasmic or nuclear E3s, membrane-tethered E3s face unique challenges in substrate recognition and modification because of a physical barrier, the plasma membrane, that creates three discrete zones for protein-protein interactions: the intracellular and extracellular spaces, and the plane of the membrane. These three regions likely create composite binding surfaces that promote the assembly of protein complexes in which the RING domain of the membrane-tethered E3 is optimally positioned in the cytoplasm to transfer Ub to the substrate. In this respect, there are architectural and mechanistic parallels between the assembly of membrane-tethered E3-substrate complexes and of cytokine-nucleated signaling receptor assemblies, which are driven by the "zippering" together of the full complex, from binding of ligands to receptors, to ECD and TM contacts between receptors, and finally cytoplasmic domain interactions that may also recruit downstream signaling proteins (Spangler, Moraga, Mendoza, & Christopher Garcia, 2015). These aggregate contacts contribute to the stability, lifetime and signaling strength of the receptor complex, and can be sites for therapeutic modulation or for engineering of tunable receptors (Rosenbaum, Clemmensen, Bredt, Bettler, & Strømgaard, 2020).

For cytoplasmic and nuclear E3s, proteolysis-targeting chimeras (PROTACs) have emerged as a therapeutic modality that enables target degradation driven by small molecules (Bond & Crews, 2021; Schneider et al., 2021). PROTACs function as bivalent linkers that direct E3s to specific substrates for ubiquitylation and degradation. Recently, a strategy conceptually analogous to PROTACs has been applied to induce the degradation of PD-L1, a TM immune checkpoint ligand, by RNF43 (Cotton et al., 2021). Targeting membrane-tethered E3s to heterologous signaling receptors using small molecules, surrogate ligands or bivalent nanobody or antibody constructs is a promising strategy to modulate signaling strength for therapeutic purposes.

Molecular models and protein-protein interaction predictions enabled by AlphaFold and RoseTTaFold, like those presented in this chapter, will enable the rapid generation of hypotheses about how membrane-tethered E3s fold, assemble and recognize substrates. The ability of these new deep learning-based algorithms to predict protein folds and interactions relies in good part on their capacity to tease out faint signals of *co*-evolutionary linkage between amino acid positions on the same and on partner polypeptide chains by mining sequence databases. These algorithmic approaches should work well with the evolutionary pairings that underlie membrane-tethered E3-substrate systems. We expect that the characterization of other membrane-tethered E3-receptor systems will unravel new regulatory layers in many signaling pathways.

## Note added in proof

While the manuscript was being processed for publication, two groups reported on the regulation of additional cell surface proteins by ZNRF3/RNF43. Zhu and colleagues described the regulation of Hulula (Hwa), a determinant of the Spemann organizer and dorsal body axis formation in *Xenopus laevis*, by ZNRF3 (Zhu et al., 2021). ZNRF3 binds and ubiquitylates Hwa, thereby regulating its lysosomal trafficking and degradation. Radaszkiewicz and colleagues described the regulation of β-catenin-independent, WNT5A-induced signaling by RNF43 in normal physiology and during melanoma invasion (Radaszkiewicz et al., 2021). RNF43 interacts with components of the WNT5A signaling pathway, including the receptors ROR1 and ROR2, and the signal transducers VANGL1 and VANGL2. RNF43 induces VANGL2 ubiquitylation and proteasomal degradation, promotes ROR1 internalization, and inhibits ROR2 activation. We presume that some of the mechanisms described in this chapter may apply to the regulation of ZNRF3/RNF43 in these contexts.

## Acknowledgments

A.M.L. is supported by the Intramural Research Program of the National Institutes of Health, National Cancer Institute, Center for Cancer Research. R.R. is supported by grants from the NIH (GM118082, HL157103, HD101980). We thank Feng Cong, Christof Niehrs and Sara Konopelski Snavely for feedback on the manuscript.

## References

Acebron, S. P., Karaulanov, E., Berger, B. S., Huang, Y.-L., & Niehrs, C. (2014). Mitotic Wnt signaling promotes protein stabilization and regulates cell size. *Molecular Cell*, *54*(4), 663–674.

Aoki, M., Kiyonari, H., Nakamura, H., & Okamoto, H. (2008). R-spondin2 expression in the apical ectodermal ridge is essential for outgrowth and patterning in mouse limb development. *Development, Growth & Differentiation*, *50*(2), 85–95.

Aoki, M., Mieda, M., Ikeda, T., Hamada, Y., Nakamura, H., & Okamoto, H. (2007). R-spondin3 is required for mouse placental development. *Developmental Biology*, *301*(1), 218–226.

Ayadi, L. (2008). Molecular modelling of the TSR domain of R-spondin 4. *Bioinformation*. https://doi.org/10.6026/97320630003119.

Baek, M., DiMaio, F., Anishchenko, I., Dauparas, J., Ovchinnikov, S., Lee, G. R., et al. (2021). Accurate prediction of protein structures and interactions using a three-track neural network. *Science*, *373*(6557), 871–876.

Barker, N., Tan, S., & Clevers, H. (2013). Lgr proteins in epithelial stem cell biology. *Development*, *140*(12), 2484–2494.

Bauer, J., Bakke, O., & Preben Morth, J. (2017). Overview of the membrane-associated RING-CH (MARCH) E3 ligase family. *New Biotechnology*, *38*(Pt. A), 7–15.

Bell, S. M., Schreiner, C. M., Wert, S. E., Mucenski, M. L., Scott, W. J., & Whitsett, J. A. (2008). R-Spondin 2 is required for Normal laryngeal-tracheal, lung and limb morphogenesis. *Development*, *135*(6), 1049–1058.

Bergmann, C., Senderek, J., Anhuf, D., Thiel, C. T., Ekici, A. B., Poblete-Gutierrez, P., et al. (2006). Mutations in the gene encoding the Wnt-signaling component R-spondin 4 (RSPO4) cause autosomal recessive anonychia. *American Journal of Human Genetics*, *79*(6), 1105–1109.

Bhanot, P., Brink, M., Samos, C. H., Hsieh, J. C., Wang, Y., Macke, J. P., et al. (1996). A new member of the frizzled family from Drosophila functions as a wingless receptor. *Nature*, *382*(6588), 225–230.

Blaydon, D. C., Ishii, Y., O'Toole, E. A., Unsworth, H. C., Teh, M.-T., Rüschendorf, F., et al. (2006). The gene encoding R-spondin 4 (RSPO4), a secreted protein implicated in Wnt signaling, is mutated in inherited anonychia. *Nature Genetics*, *38*(11), 1245–1247.

Bond, M. J., & Crews, C. M. (2021). Proteolysis targeting chimeras (PROTACs) come of age: Entering the third decade of targeted protein degradation. *RSC Chemical Biology*, *2*(3), 725–742.

Bourhis, E., Tam, C., Franke, Y., Bazan, J. F., & Ernst, J. (2010). Reconstitution of a Frizzled8·Wnt3a·LRP6 signaling complex reveals multiple Wnt and Dkk1 binding sites on LRP6. *Journal of Biological*. https://www.jbc.org/article/S0021-9258(20)87312-8/abstract.

Brüchle, N. O., Frank, J., Frank, V., Senderek, J., Akar, A., Koc, E., et al. (2008). RSPO4 is the major gene in autosomal-recessive anonychia and mutations cluster in the Furin-like cysteine-rich domains of the Wnt signaling ligand R-spondin 4. *The Journal of Investigative Dermatology*, *128*(4), 791–796.

Bryant, P., Pozzati, G., & Elofsson, A. (2021). Improved prediction of protein-protein interactions using AlphaFold2. *bioRxiv*. https://doi.org/10.1101/2021.09.15.460468.

Bugter, J. M., Fenderico, N., & Maurice, M. M. (2021). Mutations and mechanisms of WNT pathway tumour suppressors in cancer. *Nature Reviews. Cancer*, *21*(1), 5–21.

Burbach, B. J., Friedl, A., Mundhenke, C., & Rapraeger, A. C. (2003). Syndecan-1 accumulates in lysosomes of poorly differentiated breast carcinoma cells. *Matrix Biology*. https://doi.org/10.1016/s0945-053x(03)00009-x.

Cappadocia, L., & Lima, C. D. (2018). Ubiquitin-like protein conjugation: Structures, chemistry, and mechanism. *Chemical Reviews*, *118*(3), 889–918.

Carmon, K. S., Gong, X., Lin, Q., Thomas, A., & Liu, Q. (2011). R-Spondins function as ligands of the orphan receptors LGR4 and LGR5 to regulate Wnt/β-catenin signaling. *Proceedings of the National Academy of Sciences of the United States of America, 108*(28), 11452–11457.

Carmon, K. S., Gong, X., Yi, J., Thomas, A., & Liu, Q. (2014). RSPO-LGR4 functions via IQGAP1 to potentiate Wnt signaling. *Proceedings of the National Academy of Sciences of the United States of America, 111*(13), E1221–E1229.

Chang, C.-F., Hsu, L.-S., Weng, C.-Y., Chen, C.-K., Wang, S.-Y., Chou, Y.-H., et al. (2016). N-glycosylation of human R-spondin 1 is required for efficient secretion and stability but not for its heparin binding ability. *International Journal of Molecular Sciences, 17*(6). https://doi.org/10.3390/ijms17060937.

Chang, L.-S., Kim, M., Glinka, A., Reinhard, C., & Niehrs, C. (2020). The tumor suppressor PTPRK promotes ZNRF3 internalization and is required for Wnt inhibition in the Spemann organizer. *eLife, 9*. https://doi.org/10.7554/eLife.51248.

Chen, P.-H., Chen, X., Lin, Z., Fang, D., & He, X. (2013). The structural basis of R-spondin recognition by LGR5 and RNF43. *Genes & Development, 27*(12), 1345–1350.

Chen, J.-Z., Wang, S., Tang, R., Yang, Q.-S., Zhao, E., Chao, Y., et al. (2002). Cloning and identification of a cDNA that encodes a novel human protein with thrombospondin type I repeat domain, hPWTSR. *Molecular Biology Reports, 29*(3), 287–292.

Christianson, H. C., & Belting, M. (2014). Heparan sulfate proteoglycan as a cell-surface endocytosis receptor. *Matrix Biology: Journal of the International Society for Matrix Biology, 35*, 51–55.

Chu, M. L.-H., Ahn, V. E., Choi, H.-J., Daniels, D. L., Nusse, R., & Weis, W. I. (2013). Structural studies of Wnts and identification of an LRP6 binding site. *Structure, 21*(7), 1235–1242.

Clague, M. J., Urbé, S., & Komander, D. (2019). Breaking the chains: Deubiquitylating enzyme specificity begets function. *Nature Reviews. Molecular Cell Biology, 20*(6), 338–352.

Consonni, S. V., Maurice, M. M., & Bos, J. L. (2014). DEP domains: Structurally similar but functionally different. *Nature Reviews. Molecular Cell Biology, 15*(5), 357–362.

Cooper, A. F., Yu, K. P., Brueckner, M., Brailey, L. L., Johnson, L., McGrath, J. M., et al. (2005). Cardiac and CNS defects in a mouse with targeted disruption of suppressor of fused. *Development, 132*(19), 4407–4417.

Cooray, S. N., Guasti, L., & Clark, A. J. L. (2011). The E3 ubiquitin ligase Mahogunin ubiquitinates the melanocortin 2 receptor. *Endocrinology, 152*(11), 4224–4231.

Cota, C. D., Bagher, P., Piotr Pelc, C., Smith, O., Bodner, C. R., & Gunn, T. M. (2006). Mice with mutations in Mahogunin Ring Finger-1 (Mgrn1) exhibit abnormal patterning of the left-right Axis. *Developmental Dynamics: An Official Publication of the American Association of Anatomists, 235*(12), 3438–3447.

Cotton, A. D., Nguyen, D. P., Gramespacher, J. A., Seiple, I. B., & Wells, J. A. (2021). Development of antibody-based PROTACs for the degradation of the cell-surface immune checkpoint protein PD-L1. *Journal of the American Chemical Society, 143*(2), 593–598.

de Lau, W., de Lau, W., Barker, N., Low, T. Y., Koo, B.-K., Li, V. S. W., et al. (2011). Lgr5 homologues associate with Wnt receptors and mediate R-spondin signalling. *Nature*. https://doi.org/10.1038/nature10337.

de Lau, W., Peng, W. C., Gros, P., & Clevers, H. (2014). The R-spondin/Lgr5/Rnf43 module: Regulator of Wnt signal strength. *Genes & Development, 28*(4), 305–316.

de Lau, W. B. M., Snel, B., & Clevers, H. C. (2012). The R-spondin protein family. *Genome Biology, 13*(3), 242.

Derynck, R., & Miyazono, K. (2017). *The biology of the TGF-[beta] family*. Cold Spring Harbor Laboratory Press.

Deshaies, R. J., & Joazeiro, C. A. P. (2009). RING domain E3 ubiquitin ligases. *Annual Review of Biochemistry, 78*, 399–434.

Dubey, R., van Kerkhof, P., Jordens, I., Malinauskas, T., Pusapati, G. V., McKenna, J. K., et al. (2020). R-Spondins engage heparan sulfate proteoglycans to potentiate WNT signaling. *eLife, 9*. https://doi.org/10.7554/eLife.54469.

Eggenschwiler, J. T., Espinoza, E., & Anderson, K. V. (2001). Rab23 is an essential negative regulator of the mouse sonic hedgehog signalling pathway. *Nature, 412*(6843), 194–198.

Engelhard, C., Sarsfield, S., Merte, J., Wang, Q., Li, P., Beppu, H., et al. (2013). MEGF8 is a modifier of BMP signaling in trigeminal sensory neurons. *eLife, 2*, e01160.

Evans, R., O'Neill, M., Pritzel, A., Antropova, N., Senior, A., Green, T., et al. (2021). Protein complex prediction with alphafold-multimer. *bioRxiv*. https://doi.org/10.1101/2021.10.04.463034.

Fenech, E. J., Lari, F., Charles, P. D., Fischer, R., Laétitia-Thézénas, M., Bagola, K., et al. (2020). Interaction mapping of endoplasmic reticulum ubiquitin ligases identifies modulators of innate immune signalling. *eLife, 9*. https://doi.org/10.7554/eLife.57306.

Fiedler, M., Mendoza-Topaz, C., Rutherford, T. J., Mieszczanek, J., & Bienz, M. (2011). Dishevelled interacts with the DIX domain polymerization interface of axin to interfere with its function in down-regulating β-catenin. *Proceedings of the National Academy of Sciences*. https://doi.org/10.1073/pnas.1017063108.

Foot, N., Henshall, T., & Kumar, S. (2017). Ubiquitination and the regulation of membrane proteins. *Physiological Reviews, 97*(1), 253–281.

Fuki, I. V., Kuhn, K. M., Lomazov, I. R., Rothman, V. L., Tuszynski, G. P., Iozzo, R. V., et al. (1997). The syndecan family of proteoglycans. Novel receptors mediating internalization of atherogenic lipoproteins in vitro. *Journal of Clinical Investigation*. https://doi.org/10.1172/jci119685.

Fuki, I. V., Meyer, M. E., & Williams, K. J. (2000). Transmembrane and cytoplasmic domains of syndecan mediate a multi-step endocytic pathway involving detergent-insoluble membrane rafts. *Biochemical Journal*. https://doi.org/10.1042/0264-6021:3510607.

Gammons, M., & Bienz, M. (2018). Multiprotein complexes governing Wnt signal transduction. *Current Opinion in Cell Biology, 51*, 42–49.

Gao, C., & Chen, Y.-G. (2010). Dishevelled: The hub of Wnt signaling. *Cellular Signalling, 22*(5), 717–727.

Geng, A., Wu, T., Cai, C., Song, W., Wang, J., Yu, Q. C., et al. (2020). A novel function of R-spondin1 in regulating estrogen receptor expression independent of Wnt/β-catenin signaling. *eLife, 9*. https://doi.org/10.7554/eLife.56434.

Giebel, N., de Jaime-Soguero, A., Del Arco, A. G., Landry, J. J. M., Tietje, M., Villacorta, L., et al. (2021). USP42 protects ZNRF3/RNF43 from R-spondin-dependent clearance and inhibits Wnt signalling. *EMBO Reports, 22*(5), e51415.

Glaeser, K., Urban, M., Fenech, E., Voloshanenko, O., Kranz, D., Lari, F., et al. (2018). ERAD-dependent control of the Wnt secretory factor Evi. *The EMBO Journal, 37*(4). https://doi.org/10.15252/embj.201797311.

Glinka, A., Dolde, C., Kirsch, N., Huang, Y.-L., Kazanskaya, O., Ingelfinger, D., et al. (2011). LGR4 and LGR5 are R-spondin receptors mediating Wnt/β-catenin and Wnt/PCP signalling. *EMBO Reports, 12*(10), 1055–1061.

Goodrich, L. V., Milenkovic, L., Higgins, K. M., & Scott, M. P. (1997). Altered neural cell fates and medulloblastoma in mouse patched mutants. *Science, 277*(5329), 1109–1113.

Greicius, G., Kabiri, Z., Sigmundsson, K., Liang, C., Bunte, R., Singh, M. K., et al. (2018). PDGFRα+ pericryptal stromal cells are the critical source of Wnts and RSPO3 for murine intestinal stem cells in vivo. *Proceedings of the National Academy of Sciences of the United States of America, 115*(14), E3173–E3181.

Guerra, D., Chapiro, S. M., Pratelli, R., Shi, Y., Jia, W., Leary, J., et al. (2017). Control of amino acid homeostasis by a ubiquitin ligase-coactivator protein complex. *The Journal of Biological Chemistry*, *292*(9), 3827–3840.

Guerra, D. D., Pratelli, R., Kraft, E., Callis, J., & Pilot, G. (2013). Functional conservation between mammalian MGRN1 and plant LOG2 ubiquitin ligases. *FEBS Letters*, *587*(21), 3400–3405.

Gunn, T. M., Miller, K. A., He, L., Hyman, R. W., Davis, R. W., Azarani, A., et al. (1999). The mouse mahogany locus encodes a transmembrane form of human attractin. *Nature*, *398*(6723), 152–156.

Hao, H.-X., Jiang, X., & Cong, F. (2016). Control of Wnt receptor turnover by R-spondin-ZNRF3/RNF43 signaling module and its dysregulation in cancer. *Cancers*, *8*(6). https://doi.org/10.3390/cancers8060054.

Hao, H.-X., Xie, Y., Zhang, Y., Charlat, O., Oster, E., Avello, M., et al. (2012). ZNRF3 promotes Wnt receptor turnover in an R-spondin-sensitive manner. *Nature*, *485*(7397), 195–200.

Harper, J. W., & Schulman, B. A. (2021). Cullin-RING ubiquitin ligase regulatory circuits: A quarter century beyond the F-box hypothesis. *Annual Review of Biochemistry*, *90*, 403–429.

He, L., Eldridge, A. G., Jackson, P. K., Gunn, T. M., & Barsh, G. S. (2003). Accessory proteins for melanocortin signaling: Attractin and Mahogunin. *Annals of the New York Academy of Sciences*, *994*, 288–298.

He, L., Gunn, T. M., Bouley, D. M., Lu, X. Y., Watson, S. J., Schlossman, S. F., et al. (2001). A biochemical function for attractin in Agouti-induced pigmentation and obesity. *Nature Genetics*, *27*(1), 40–47.

He, L., Lu, X.-Y., Jolly, A. F., Eldridge, A. G., Watson, S. J., Jackson, P. K., et al. (2003). Spongiform degeneration in mahoganoid mutant mice. *Science*, *299*(5607), 710–712.

He, X., Semenov, M., Tamai, K., & Zeng, X. (2004). LDL receptor-related proteins 5 and 6 in Wnt/beta-catenin signaling: Arrows point the way. *Development*, *131*(8), 1663–1677.

Heldin, C.-H., & Moustakas, A. (2016). Signaling receptors for TGF-β family members. *Cold Spring Harbor Perspectives in Biology*, *8*(8). https://doi.org/10.1101/cshperspect.a022053.

Hershko, A., Ciechanover, A., & Varshavsky, A. (2000). Basic medical research award. The ubiquitin system. *Nature Medicine*, *6*(10), 1073–1081.

Hirai, H., Matoba, K., Mihara, E., Arimori, T., & Takagi, J. (2019). Crystal structure of a mammalian Wnt–frizzled complex. *Nature Structural & Molecular Biology*, *26*(5), 372–379.

Hoppler, S. P., & Moon, R. T. (2014). *Wnt signaling in development and disease: Molecular mechanisms and biological functions.* John Wiley & Sons.

Hu, S., & Monga, S. P. (2021). Wnt/-catenin signaling and liver regeneration: Circuit, biology, and opportunities. *Gene Expression*. https://doi.org/10.3727/105221621x16111780348794.

Huang, H.-C., & Klein, P. S. (2004). The frizzled family: Receptors for multiple signal transduction pathways. *Genome Biology*, *5*(7), 234.

Huangfu, D., & Anderson, K. V. (2005). Cilia and hedgehog responsiveness in the mouse. *Proceedings of the National Academy of Sciences of the United States of America*, *102*(32), 11325–11330.

Huangfu, D., Liu, A., Rakeman, A. S., Murcia, N. S., Niswander, L., & Anderson, K. V. (2003). Hedgehog signalling in the mouse requires intraflagellar transport proteins. *Nature*, *426*(6962), 83–87.

Ishii, Y., Wajid, M., Bazzi, H., Fantauzzo, K. A., Barber, A. G., Blaydon, D. C., et al. (2008). Mutations in R-Spondin 4 (RSPO4) underlie inherited anonychia. *The Journal of Investigative Dermatology*, *128*(4), 867–870.

Janda, C. Y., Waghray, D., Levin, A. M., Thomas, C., & Christopher Garcia, K. (2012). Structural basis of Wnt recognition by frizzled. *Science*, *337*(6090), 59–64.

Jiang, X., Charlat, O., Zamponi, R., Yang, Y., & Cong, F. (2015). Dishevelled promotes Wnt receptor degradation through recruitment of ZNRF3/RNF43 E3 ubiquitin ligases. *Molecular Cell, 58*(3), 522–533.

Jiao, J., Sun, K., Walker, W. P., Bagher, P., Cota, C. D., & Gunn, T. M. (2009). Abnormal regulation of TSG101 in mice with spongiform neurodegeneration. *Biochimica et Biophysica Acta, 1792*(10), 1027–1035.

Jin, Y.-R., & Yoon, J. K. (2012). The R-Spondin family of proteins: Emerging regulators of WNT signaling. *The International Journal of Biochemistry & Cell Biology, 44*(12), 2278–2287.

Jumper, J., Evans, R., Pritzel, A., Green, T., Figurnov, M., Ronneberger, O., et al. (2021). Highly accurate protein structure prediction with alphafold. *Nature, 596*(7873), 583–589.

Kamata, T., Katsube, K.-I., Michikawa, M., Yamada, M., Takada, S., & Mizusawa, H. (2004). R-Spondin, a novel gene with thrombospondin type 1 domain, was expressed in the dorsal neural tube and affected in Wnts mutants. *Biochimica et Biophysica Acta, 1676*(1), 51–62.

Kazanskaya, O., Glinka, A., del Barco, I., Barrantes, P. S., Niehrs, C., & Wei, W. (2004). R-Spondin2 is a secreted activator of Wnt/beta-catenin signaling and is required for xenopus myogenesis. *Developmental Cell,* 7(4), 525–534.

Kazanskaya, O., Ohkawara, B., Heroult, M., Wei, W., Maltry, N., Augustin, H. G., et al. (2008). The Wnt signaling regulator R-spondin 3 promotes angioblast and vascular development. *Development, 135*(22), 3655–3664.

Kim, K.-A., Kakitani, M., Zhao, J., Oshima, T., Tang, T., Binnerts, M., et al. (2005). Mitogenic influence of human R-spondin1 on the intestinal epithelium. *Science, 309*(5738), 1256–1259.

Kim, B. Y., Olzmann, J. A., Barsh, G. S., Chin, L.-S., & Li, L. (2007). Spongiform neurodegeneration-associated E3 ligase Mahogunin ubiquitylates TSG101 and regulates endosomal trafficking. *Molecular Biology of the Cell, 18*(4), 1129–1142.

Kim, M., Reinhard, C., & Niehrs, C. (2021). A MET-PTPRK kinase-phosphatase rheostat controls ZNRF3 and Wnt signaling. *eLife, 10*, e70885.

Kim, K.-A., Wagle, M., Tran, K., Zhan, X., Dixon, M. A., Liu, S., et al. (2008). R-Spondin family members regulate the Wnt pathway by a common mechanism. *Molecular Biology of the Cell, 19*(6), 2588–2596.

Kinzel, B., Pikiolek, M., Orsini, V., Sprunger, J., Isken, A., Zietzling, S., et al. (2014). Functional roles of Lgr4 and Lgr5 in embryonic gut, kidney and skin development in mice. *Developmental Biology, 390*(2), 181–190.

Kishida, S., Yamamoto, H., Hino, S.-I., Ikeda, S., Kishida, M., & Kikuchi, A. (1999). DIX domains of Dvl and Axin are necessary for protein interactions and their ability to regulate β-catenin stability. *Molecular and Cellular Biology*. https://doi.org/10.1128/mcb.19.6.4414.

Komander, D., & Rape, M. (2012). The ubiquitin code. *Annual Review of Biochemistry, 81*, 203–229. https://doi.org/10.1146/annurev-biochem-060310-170328. Epub 2012 Apr 10.

Kong, J. H., Young, C. B., Pusapati, G. V., Hernán Espinoza, F., Patel, C. B., Beckert, F., et al. (2021). Gene-teratogen interactions influence the penetrance of birth defects by altering hedgehog signaling strength. *Development*. https://doi.org/10.1242/dev.199867.

Kong, J. H., Young, C. B., Pusapati, G. V., Patel, C. B., Ho, S., Krishnan, A., et al. (2020). A membrane-tethered ubiquitination pathway regulates hedgehog signaling and heart development. *Developmental Cell, 55*(4). 432–49.e12.

Koo, B.-K., & Clevers, H. (2014). Stem cells marked by the R-Spondin receptor LGR5. *Gastroenterology, 147*(2), 289–302.

Koo, B.-K., Spit, M., Jordens, I., Low, T. Y., Stange, D. E., van de Wetering, M., et al. (2012). Tumour suppressor RNF43 is a stem-cell E3 ligase that induces endocytosis of Wnt receptors. *Nature*, *488*(7413), 665–669.

Kwon, Y. T., & Ciechanover, A. (2017). The ubiquitin code in the ubiquitin-proteasome system and autophagy. *Trends in Biochemical Sciences*, *42*(11), 873–886.

Lebensohn, A. M., Dubey, R., Neitzel, L. R., Tacchelly-Benites, O., Yang, E., Marceau, C. D., et al. (2016). Comparative genetic screens in human cells reveal new regulatory mechanisms in WNT signaling. *eLife*, *5*. https://doi.org/10.7554/eLife.21459.

Lebensohn, A. M., & Rohatgi, R. (2018). R-Spondins can potentiate WNT signaling without LGRs. *eLife*, 7(February). https://doi.org/10.7554/eLife.33126.

Lee, H., Seidl, C., Sun, R., Glinka, A., & Niehrs, C. (2020). R-Spondins are BMP receptor antagonists in xenopus early embryonic development. *Nature Communications*, *11*(1), 5570.

Leung, C., Tan, S. H., & Barker, N. (2018). Recent advances in Lgr5 stem cell research. *Trends in Cell Biology*. https://doi.org/10.1016/j.tcb.2018.01.010.

Levin, M., Johnson, R. L., Stern, C. D., Kuehn, M., & Tabin, C. (1995). A molecular pathway determining left-right asymmetry in chick embryogenesis. *Cell*, *82*(5), 803–814.

Li, W., Bengtson, M. H., Ulbrich, A., Matsuda, A., Reddy, V. A., Orth, A., et al. (2008). Genome-wide and functional annotation of human E3 ubiquitin ligases identifies MULAN, a mitochondrial E3 that regulates the organelle's dynamics and signaling. *PLoS One*, *3*(1), e1487.

Li, S.-J., Yen, T.-Y., Endo, Y., Klauzinska, M., Baljinnyam, B., Macher, B., et al. (2009). Loss-of-function point mutations and two-furin domain derivatives provide insights about R-spondin2 structure and function. *Cellular Signalling*, *21*(6), 916–925.

Lian, F.-M., Xie, S., & Qian, C. (2016). Crystal structure and SUMO binding of Slx1-Slx4 complex. *Scientific Reports*, *6*(January), 19331.

Lineberry, N., Leon, S., Soares, L., & Garrison Fathman, C. (2008). The single subunit transmembrane E3 ligase gene related to anergy in lymphocytes (GRAIL) captures and then ubiquitinates transmembrane proteins across the cell membrane. *The Journal of Biological Chemistry*, *283*(42), 28497–28505.

Lloyd, D. L., Toegel, M., Fulga, T. A., & Wilkie, A. O. M. (2018). The drosophila homologue of MEGF8 is essential for early development. *Scientific Reports*, *8*(1), 8790.

Loregger, A., Grandl, M., Mejías-Luque, R., Allgäuer, M., Degenhart, K., Haselmann, V., et al. (2015). The E3 ligase RNF43 inhibits Wnt signaling downstream of mutated β-catenin by sequestering TCF4 to the nuclear membrane. *Science Signaling*, *8*(393), ra90.

Lu, D., Willard, D., Patel, I. R., Kadwell, S., Overton, L., Kost, T., et al. (1994). Agouti protein is an antagonist of the melanocyte-stimulating-hormone receptor. *Nature*, *371*(6500), 799–802.

Luca, V. C., Miao, Y., Li, X., Hollander, M. J., Kuo, C. J., & Christopher Garcia, K. (2020). Surrogate R-Spondins for tissue-specific potentiation of Wnt signaling. *PLoS One*. https://doi.org/10.1371/journal.pone.0226928.

MacDonald, B. T., & He, X. (2012). Frizzled and LRP5/6 receptors for Wnt/β-catenin signaling. *Cold Spring Harbor Perspectives in Biology*, *4*(12). https://doi.org/10.1101/cshperspect.a007880.

MacGurn, J. A., Hsu, P.-C., & Emr, S. D. (2012). Ubiquitin and membrane protein turnover: From cradle to grave. *Annual Review of Biochemistry*. https://doi.org/10.1146/annurev-biochem-060210-093619.

Malbon, C. C. (2004). Frizzleds: New members of the superfamily of G-protein-coupled receptors. *Frontiers in Bioscience: A Journal and Virtual Library*, *9*(May), 1048–1058.

Metzger, M. B., Pruneda, J. N., Klevit, R. E., & Weissman, A. M. (2014). RING-type E3 ligases: Master manipulators of E2 ubiquitin-conjugating enzymes and ubiquitination. *Biochimica et Biophysica Acta*, *1843*(1), 47–60.

Middleton, A. J., Zhu, J., & Day, C. L. (2020). The RING domain of RING finger 12 efficiently builds degradative ubiquitin chains. *Journal of Molecular Biology, 432*(13), 3790–3801.
Mlodzik, M. (2016). The dishevelled protein family: Still rather a mystery after over 20 years of molecular studies. *Current Topics in Developmental Biology, 117*(January), 75–91.
Moad, H. E., & Pioszak, A. A. (2013). Reconstitution of R-spondin:LGR4:ZNRF3 adult stem cell growth factor signaling complexes with recombinant proteins produced in *Escherichia coli*. *Biochemistry*. https://doi.org/10.1021/bi401090h.
Moffat, L. L., Robinson, R. E., Bakoulis, A., & Clark, S. G. (2014). The conserved transmembrane RING finger protein PLR-1 downregulates Wnt signaling by reducing frizzled, Ror and Ryk cell-surface levels in *C. elegans*. *Development, 141*(3), 617–628.
Monga, S. P. S. (2014). Role and regulation of β-catenin signaling during physiological liver growth. *Gene Expression*. https://doi.org/10.3727/105221614x13919976902138.
Morreale, F. E., & Walden, H. (2016). Types of ubiquitin ligases. *Cell, 165*(1), 248–248.e1.
Nagano, K. (2019). R-spondin signaling as a pivotal regulator of tissue development and homeostasis. *The Japanese Dental Science Review, 55*(1), 80–87.
Nagle, D. L., McGrail, S. H., Vitale, J., Woolf, E. A., Dussault, B. J., Jr., DiRocco, L., et al. (1999). The mahogany protein is a receptor involved in suppression of obesity. *Nature, 398*(6723), 148–152.
Nakayama, M., Nakajima, D., Nagase, T., Nomura, N., Seki, N., & Ohara, O. (1998). Identification of high-molecular-weight proteins with multiple EGF-like motifs by motif-trap screening. *Genomics, 51*(1), 27–34.
Nam, J.-S., Park, E., Turcotte, T. J., Palencia, S., Zhan, X., Lee, J., et al. (2007). Mouse R-spondin2 is required for apical ectodermal ridge maintenance in the hindlimb. *Developmental Biology, 311*(1), 124–135.
Nam, J.-S., Turcotte, T. J., Smith, P. F., Choi, S., & Yoon, J. K. (2006). Mouse cristin/R-spondin family proteins are novel ligands for the frizzled 8 and LRP6 receptors and activate β-catenin-dependent gene expression. *Journal of Biological Chemistry*. https://doi.org/10.1074/jbc.m508324200.
Nam, J.-S., Turcotte, T. J., & Yoon, J. K. (2007). Dynamic expression of R-spondin family genes in mouse development. *Gene Expression Patterns*. https://doi.org/10.1016/j.modgep.2006.08.006.
Nawaratne, V., Kudumala, S., Kakad, P. P., & Godenschwege, T. A. (2021). The conserved MASRPF motif in the attractin homolog, distracted, is required for association with drosophila E3-ligase Mgrn1. *microPublication Biology, 2021*(July). https://doi.org/10.17912/micropub.biology.000416.
Neutzner, A., Neutzner, M., Benischke, A.-S., Ryu, S.-W., Frank, S., Youle, R. J., et al. (2011). A systematic search for endoplasmic reticulum (ER) membrane-associated RING finger proteins identifies Nixin/ZNRF4 as a regulator of calnexin stability and ER homeostasis. *The Journal of Biological Chemistry, 286*(10), 8633–8643.
Niehrs, C. (2012). The complex world of WNT receptor signalling. *Nature Reviews. Molecular Cell Biology, 13*(12), 767–779.
Ohkawara, B., Glinka, A., & Niehrs, C. (2011). Rspo3 binds syndecan 4 and induces Wnt/PCP signaling via clathrin-mediated endocytosis to promote morphogenesis. *Developmental Cell, 20*(3), 303–314.
Overton, J. D., & Leibel, R. L. (2011). Mahoganoid and mahogany mutations rectify the obesity of the yellow mouse by effects on endosomal traffic of MC4R protein. *The Journal of Biological Chemistry, 286*(21), 18914–18929.
Park, S., Cui, J., Yu, W., Wu, L., Carmon, K. S., & Liu, Q. J. (2018). Differential activities and mechanisms of the four R-Spondins in potentiating Wnt/β-catenin signaling. *The Journal of Biological Chemistry, 293*(25), 9759–9769.

Parma, P., Radi, O., Vidal, V., Chaboissier, M. C., Dellambra, E., Valentini, S., et al. (2006). R-spondin1 is essential in sex determination, skin differentiation and malignancy. *Nature Genetics*. https://doi.org/10.1038/ng1907.

Peng, W. C., de Lau, W., Forneris, F., Granneman, J. C. M., Huch, M., Clevers, H., et al. (2013). Structure of stem cell growth factor R-spondin 1 in complex with the ectodomain of its receptor LGR5. *Cell Reports*, *3*(6), 1885–1892.

Peng, W. C., de Lau, W., Madoori, P. K., Forneris, F., Granneman, J. C. M., Clevers, H., et al. (2013). Structures of Wnt-antagonist ZNRF3 and its complex with R-spondin 1 and implications for signaling. *PLoS One*, *8*(12), e83110.

Phan, L. K., Lin, F., LeDuc, C. A., Chung, W. K., & Leibel, R. L. (2002). The mouse Mahoganoid coat color mutation disrupts a novel C3HC4 RING domain protein. *The Journal of Clinical Investigation*, *110*(10), 1449–1459.

Phillips, B. P., & Miller, E. A. (2021). Membrane protein folding and quality control. *Current Opinion in Structural Biology*, *69*(August), 50–54.

Planas-Paz, L., Orsini, V., Boulter, L., Calabrese, D., Pikiolek, M., Nigsch, F., et al. (2016). The RSPO–LGR4/5–ZNRF3/RNF43 module controls liver zonation and size. *Nature Cell Biology*. https://doi.org/10.1038/ncb3337.

Pusapati, G. V., Kong, J. H., Patel, B. B., Krishnan, A., Sagner, A., Kinnebrew, M., et al. (2018). CRISPR screens uncover genes that regulate target cell sensitivity to the morphogen sonic hedgehog. *Developmental Cell*, *44*(1). 113–29.e8.

Radaszkiewicz, T., & Bryja, V. (2020). Protease associated domain of RNF43 is not necessary for the suppression of Wnt/β-catenin signaling in human cells. *Cell Communication and Signaling: CCS*, *18*(1), 91.

Radaszkiewicz, T., Nosková, M., Gömöryová, K., Blanářová, O. V., Radaszkiewicz, K. A., Picková, M., et al. (2021). RNF43 inhibits WNT5A-driven signaling and suppresses melanoma invasion and resistance to the targeted therapy. *eLife*, *10*(October). https://doi.org/10.7554/eLife.65759.

Raslan, A. A., & Yoon, J. K. (2019). R-Spondins: Multi-mode WNT signaling regulators in adult stem cells. *The International Journal of Biochemistry & Cell Biology*. https://doi.org/10.1016/j.biocel.2018.11.005.

Reis, A. H., & Sokol, S. Y. (2020). Rspo2 antagonizes FGF signaling during vertebrate mesoderm formation and patterning. *Development*, *147*(10). https://doi.org/10.1242/dev.189324.

Reis, A. H., & Sokol, S. Y. (2021). Rspo2 inhibits TCF3 phosphorylation to antagonize Wnt signaling during vertebrate anteroposterior axis specification. *Scientific Reports*, *11*(1), 13433.

Ren, Z., van Andel, H., de Lau, W., Hartholt, R. B., Maurice, M. M., Clevers, H., et al. (2018). Syndecan-1 promotes Wnt/β-catenin signaling in multiple myeloma by presenting Wnts and R-spondins. *Blood*, *131*(9), 982–994.

Rosenbaum, M. I., Clemmensen, L. S., Bredt, D. S., Bettler, B., & Strømgaard, K. (2020). Targeting receptor complexes: A new dimension in drug discovery. *Nature Reviews Drug Discovery*. https://doi.org/10.1038/s41573-020-0086-4.

Ruffner, H., Sprunger, J., Charlat, O., Leighton-Davies, J., Grosshans, B., Salathe, A., et al. (2012). R-spondin potentiates Wnt/β-catenin signaling through orphan receptors LGR4 and LGR5. *PLoS One*, 7(7), e40976.

Sardana, R., & Emr, S. D. (2021). Membrane protein quality control mechanisms in the endo-lysosome system. *Trends in Cell Biology*, *31*(4), 269–283.

Sarrazin, S., Lamanna, W. C., & Esko, J. D. (2011). Heparan sulfate proteoglycans. *Cold Spring Harbor Perspectives in Biology*, *3*(7). https://doi.org/10.1101/cshperspect.a004952.

Schneider, M., Radoux, C. J., Hercules, A., Ochoa, D., Dunham, I., Zalmas, L.-P., et al. (2021). The PROTACtable genome. *Nature Reviews. Drug Discovery*, *20*(10), 789–797.

Schulte, G., & Bryja, V. (2007). The frizzled family of unconventional G-protein-coupled receptors. *Trends in Pharmacological Sciences*, *28*(10), 518–525.

Schulte, G., & Wright, S. C. (2018). Frizzleds as GPCRs—More conventional than we thought! *Trends in Pharmacological Sciences*, *39*(9), 828–842.

Schwarz-Romond, T., Fiedler, M., Naoki, S., Jonathan, P., Butler, G., Kikuchi, A., et al. (2007). The DIX domain of dishevelled confers Wnt signaling by dynamic polymerization. *Nature Structural & Molecular Biology*. https://doi.org/10.1038/nsmb1247.

Schwarz-Romond, T., Metcalfe, C., & Bienz, M. (2007). Dynamic recruitment of axin by dishevelled protein assemblies. *Journal of Cell Science*. https://doi.org/10.1242/jcs.002956.

Sharma, M., Castro-Piedras, I., Simmons, G. E., Jr., & Pruitt, K. (2018). Dishevelled: A masterful conductor of complex Wnt signals. *Cellular Signalling*, *47*(July), 52–64.

Simons, M., Gault, W. J., Gotthardt, D., Rohatgi, R., Klein, T. J., Shao, Y., et al. (2009). Electrochemical cues regulate assembly of the frizzled/Dishevelled complex at the plasma membrane during planar epithelial polarization. *Nature Cell Biology*, *11*(3), 286–294.

Snyder, J. C., Rochelle, L. K., Lyerly, H. K., Caron, M. G., & Barak, L. S. (2013). Constitutive internalization of the leucine-rich G protein-coupled receptor-5 (LGR5) to the trans-Golgi network. *The Journal of Biological Chemistry*, *288*(15), 10286–10297.

Snyder, J. C., Rochelle, L. K., Ray, C., Pack, T. F., Bock, C. B., Veronica, L., et al. (2017). Inhibiting clathrin-mediated endocytosis of the leucine-rich G protein-coupled receptor-5 diminishes cell fitness. *The Journal of Biological Chemistry*, *292*(17), 7208–7222.

Spangler, J. B., Moraga, I., Mendoza, J. L., & Christopher Garcia, K. (2015). Insights into cytokine–receptor interactions from cytokine engineering. *Annual Review of Immunology*. https://doi.org/10.1146/annurev-immunol-032713-120211.

Spit, M., Fenderico, N., Jordens, I., Radaszkiewicz, T., Lindeboom, R. G. H., Bugter, J. M., et al. (2020). RNF 43 truncations Trap CK 1 to drive niche-independent self-renewal in cancer. *The EMBO Journal*, *39*(18). https://doi.org/10.15252/embj.2019103932.

Stevens, P. D., & Williams, B. O. (2021). LGR4: Not just for Wnt anymore? *Cancer Research*. https://doi.org/10.1158/0008-5472.can-21-2266.

Su, L. L., Iwai, H., Lin, J. T., & Garrison Fathman, C. (2009). The transmembrane E3 ligase GRAIL ubiquitinates and degrades CD83 on CD4 T cells. *Journal of Immunology*, *183*(1), 438–444.

Sun, T., Annunziato, S., Bergling, S., Sheng, C., Orsini, V., Forcella, P., et al. (2021). ZNRF3 and RNF43 cooperate to safeguard metabolic liver zonation and hepatocyte proliferation. *Cell Stem Cell*, *28*(10). 1822–37.e10.

Sun, R., He, L., Lee, H., Glinka, A., Andresen, C., Hübschmann, D., et al. (2021). RSPO2 inhibits BMP signaling to promote self-renewal in acute myeloid leukemia. *Cell Reports*, *36*(7), 109559.

Svard, J., Heby-Henricson, K., Persson-Lek, M., Rozell, B., Lauth, M., Bergstrom, A., et al. (2006). Genetic elimination of suppressor of fused reveals an essential repressor function in the mammalian hedgehog signaling pathway. *Developmental Cell*, *10*(2), 187–197.

Szenker-Ravi, E., Altunoglu, U., Leushacke, M., Bosso-Lefèvre, C., Khatoo, M., Tran, H. T., et al. (2018). RSPO2 inhibition of RNF43 and ZNRF3 governs limb development independently of LGR4/5/6. *Nature*, *557*(7706), 564–569.

Tamai, K., Semenov, M., Kato, Y., Spokony, R., Liu, C., Katsuyama, Y., et al. (2000). LDL-receptor-related proteins in Wnt signal transduction. *Nature*, *407*(6803), 530–535.

Tauriello, D. V. F., Jordens, I., Kirchner, K., Slootstra, J. W., Kruitwagen, T., Bouwman, B. A. M., et al. (2012). Wnt/β-catenin signaling requires interaction of the Dishevelled DEP domain and C terminus with a discontinuous motif in frizzled. *Proceedings of the National Academy of Sciences of the United States of America*, *109*(14), E812–E820.

Ter Steege, E. J., ter Steege, E. J., & Bakker, E. R. M. (2021). The role of R-spondin proteins in cancer biology. *Oncogene*. https://doi.org/10.1038/s41388-021-02059-y.

Tomizuka, K., Horikoshi, K., Kitada, R., Sugawara, Y., Iba, Y., Kojima, A., et al. (2008). R-spondin1 plays an essential role in ovarian development through positively regulating Wnt-4 signaling. *Human Molecular Genetics*, *17*(9), 1278–1291.

Trenker, R., Wu, X., Nguyen, J. V., Wilcox, S., Rubin, A. F., Call, M. E., et al. (2021). Human and viral membrane-associated E3 ubiquitin ligases MARCH1 and MIR2 recognize different features of CD86 to downregulate surface expression. *The Journal of Biological Chemistry*, *297*(1), 100900.

Tsiairis, C. D., & McMahon, A. P. (2009). An Hh-dependent pathway in lateral plate mesoderm enables the generation of left/right asymmetry. *Current Biology: CB*, *19*(22), 1912–1917.

Tsukiyama, T., Fukui, A., Terai, S., Fujioka, Y., Shinada, K., Takahashi, H., et al. (2015). Molecular role of RNF43 in canonical and noncanonical Wnt signaling. *Molecular and Cellular Biology*, *35*(11), 2007–2023.

Tsukiyama, T., Koo, B.-K., & Hatakeyama, S. (2021). Post-translational Wnt receptor regulation: Is the fog slowly clearing? The molecular mechanism of RNF43/ZNRF3 ubiquitin ligases is not yet fully elucidated and still controversial. *BioEssays: News and Reviews in Molecular, Cellular and Developmental Biology*, *43*(4), e2000297.

Tsukiyama, T., Zou, J., Kim, J., Ogamino, S., Shino, Y., Masuda, T., et al. (2020). A Phospho-switch controls RNF43-mediated degradation of Wnt receptors to suppress tumorigenesis. *Nature Communications*, *11*(1), 4586.

Tunyasuvunakool, K., Adler, J., Wu, Z., Green, T., Zielinski, M., Žídek, A., et al. (2021). Highly accurate protein structure prediction for the human proteome. *Nature*, *596*(7873), 590–596.

Twigg, S. R., Lloyd, D., Jenkins, D., Elcioglu, N. E., Cooper, C. D., Al-Sannaa, N., et al. (2012). Mutations in multidomain protein MEGF8 identify a carpenter syndrome subtype associated with defective lateralization. *American Journal of Human Genetics*, *91*(5), 897–905.

Umbhauer, M., Djiane, A., Goisset, C., Penzo-Méndez, A., Riou, J. F., Boucaut, J. C., et al. (2000). The C-terminal cytoplasmic Lys-Thr-X-X-X-Trp motif in frizzled receptors mediates Wnt/beta-catenin Signalling. *The EMBO Journal*, *19*(18), 4944–4954.

Vidal, V. P., Jian-Motamedi, F., Rekima, S., Gregoire, E. P., Szenker-Ravi, E., Leushacke, M., et al. (2020). R-spondin signalling is essential for the maintenance and differentiation of mouse nephron progenitors. *eLife*, *9*(May). https://doi.org/10.7554/eLife.53895.

Wang, Y., Chang, H., Rattner, A., & Nathans, J. (2016). Frizzled receptors in development and disease. *Current Topics in Developmental Biology*, *117*(January), 113–139.

Wang, D., Huang, B., Zhang, S., Yu, X., Wu, W., & Wang, X. (2013). Structural basis for R-spondin recognition by LGR4/5/6 receptors. *Genes & Development*. https://doi.org/10.1101/gad.219360.113.

Wang, W., Zheng, X., Song, H., Yang, J., Liu, X., Wang, Y., et al. (2020). Spatial and temporal deletion reveals a latent effect of Megf8 on the left-right patterning and heart development. *Differentiation; Research in Biological Diversity*, *113*(May), 19–25.

Whiting, C. C., Su, L. L., Lin, J. T., & Garrison Fathman, C. (2011). GRAIL: A unique mediator of CD4 T-lymphocyte unresponsiveness. *The FEBS Journal*, *278*(1), 47–58.

Willard, D. H., Bodnar, W., Harris, C., Kiefer, L., Nichols, J. S., Blanchard, S., et al. (1995). Agouti structure and function: Characterization of a potent alpha-melanocyte stimulating hormone receptor antagonist. *Biochemistry*, *34*(38), 12341–12346.

Wittrup, A., Zhang, S.-H., ten Dam, G. B., van Kuppevelt, T. H., Bengtson, P., Johansson, M., et al. (2009). ScFv antibody-induced translocation of cell-surface heparan sulfate proteoglycan to endocytic vesicles. *Journal of Biological Chemistry*. https://doi.org/10.1074/jbc.m109.036129.

Wong, H.-C., Bourdelas, A., Krauss, A., Lee, H.-J., Shao, Y., Wu, D., et al. (2003). Direct binding of the PDZ domain of Dishevelled to a conserved internal sequence in the C-terminal region of frizzled. *Molecular Cell*. https://doi.org/10.1016/s1097-2765(03)00427-1.

Wong, H. C., Mao, J., Nguyen, J. T., Srinivas, S., Zhang, W., Liu, B., et al. (2000). Structural basis of the recognition of the Dishevelled DEP domain in the Wnt signaling pathway. *Nature Structural Biology*, 7(12), 1178–1184.

Xie, Y., Zamponi, R., Charlat, O., Ramones, M., Swalley, S., Jiang, X., et al. (2013). Interaction with both ZNRF3 and LGR4 is required for the signalling activity of R-spondin. *EMBO Reports*, *14*(12), 1120–1126.

Xu, K., Xu, Y., Rajashankar, K. R., Robev, D., & Nikolov, D. B. (2013). Crystal structures of Lgr4 and its complex with R-spondin1. *Structure*, *21*(9), 1683–1689.

Yamada, W., Nagao, K., Horikoshi, K., Fujikura, A., Ikeda, E., Inagaki, Y., et al. (2009). Craniofacial malformation in R-spondin2 knockout mice. *Biochemical and Biophysical Research Communications*, *381*(3), 453–458.

Yamazaki, Y., Schönherr, C., Varshney, G. K., Dogru, M., Hallberg, B., & Palmer, R. H. (2013). Goliath family E3 ligases regulate the recycling endosome pathway via VAMP3 ubiquitylation. *The EMBO Journal*, *32*(4), 524–537.

Yang, J., Mowry, L. E., Nejak-Bowen, K. N., Okabe, H., Diegel, C. R., Lang, R. A., et al. (2014). Beta-catenin signaling in murine liver zonation and regeneration: A Wnt-Wnt situation! *Hepatology*. https://doi.org/10.1002/hep.27082.

Yu, J., Gimenez, L. E., Hernandez, C. C., Wu, Y., Wein, A. H., Han, G. W., et al. (2020). Determination of the melanocortin-4 receptor structure identifies $Ca^{2+}$ as a cofactor for ligand binding. *Science*, *368*(6489), 428–433.

Yue, F., Jiang, W., Ku, A. T., Young, A. I. J., Zhang, W., Souto, E. P., et al. (2021). A Wnt-independent LGR4-EGFR signaling axis in cancer metastasis. *Cancer Research*.

Zebisch, M., Xu, Y., Krastev, C., MacDonald, B. T., Chen, M., Gilbert, R. J. C., et al. (2013). Structural and molecular basis of ZNRF3/RNF43 transmembrane ubiquitin ligase inhibition by the Wnt agonist R-spondin. *Nature Communications*, *4*, 2787.

Zebisch, M., & Yvonne Jones, E. (2015a). Crystal structure of R-spondin 2 in complex with the ectodomains of its receptors LGR5 and ZNRF3. *Journal of Structural Biology*, *191*(2), 149–155.

Zebisch, M., & Yvonne Jones, E. (2015b). ZNRF3/RNF43- -A direct linkage of extracellular recognition and E3 ligase activity to modulate cell surface Signalling. *Progress in Biophysics and Molecular Biology*, *118*(3), 112–118.

Zhang, Z., Alpert, D., Francis, R., Chatterjee, B., Yu, Q., Tansey, T., et al. (2009). Massively parallel sequencing identifies the gene Megf8 with ENU-induced mutation causing heterotaxy. *Proceedings of the National Academy of Sciences of the United States of America*, *106*(9), 3219–3224.

Zhang, Z., Broderick, C., Nishimoto, M., Yamaguchi, T., Lee, S.-J., Zhang, H., et al. (2020). Tissue-targeted R-spondin mimetics for liver regeneration. *Scientific Reports*, *10*(1), 13951.

Zhang, X. M., Ramalho-Santos, M., & McMahon, A. P. (2001). Smoothened mutants reveal redundant roles for shh and Ihh signaling including regulation of L/R symmetry by the mouse node. *Cell*, *106*(2), 781–792.

Zhang, M., Zhang, P., Liu, Y., Lv, L., Zhang, X., Liu, H., et al. (2017). RSPO3-LGR4 regulates osteogenic differentiation of human adipose-derived stem cells via ERK/FGF Signalling. *Scientific Reports*, 7(February), 42841.

Zheng, N., & Shabek, N. (2017). Ubiquitin ligases: Structure, function, and regulation. *Annual Review of Biochemistry*, *86*(June), 129–157.

Zheng, C., Zhou, F., Shi, L. L., Xu, G. F., Zhang, B., Wang, L., et al. (2020). R-spondin2 suppresses the progression of hepatocellular carcinoma via MAPK signaling pathway. *Molecular Cancer Research: MCR, 18*(10), 1491–1499.

Zhou, X., Geng, L., Wang, D., Yi, H., Talmon, G., & Wang, J. (2017). R-Spondin1/LGR5 activates TGFβ signaling and suppresses colon cancer metastasis. *Cancer Research*, 77(23), 6589–6602.

Zhu, X., Wang, P., Wei, J., Li, Y., Zhai, J., Zheng, T., et al. (2021). Lysosomal degradation of the maternal dorsal determinant Hwa safeguards dorsal body axis formation. *EMBO Reports*, e53185.

CHAPTER THREE

# An itch for things remote: The journey of Wnts

**Lorenz Mittermeier and David M. Virshup***
Program in Cancer and Stem Cell Biology, Duke-NUS Medical School, Singapore, Singapore
*Corresponding author: e-mail address: david.virshup@duke-nus.edu.sg

## Contents



## Abstract

Wnts are a family of secreted, lipid-modified signaling glycoproteins that regulate a multiplicity of fundamental biological processes. Wnt signaling is essential for embryonic development, controlling body axis patterning, cell proliferation, cell migration and cell fate specification needed for proper tissue and organ formation. In adulthood, Wnt signaling controls tissue regeneration in a wide range of organs, and disturbance of this system is common in cancer and other diseases. A key feature of Wnt signaling is that it is a local process. Wnts signal via paracrine, cell-to-cell communication. Upon synthesis and transport to the plasma membrane in the "sending" cell, Wnts travel to nearby "receiving" cells. At the plasma membrane of these receiving cells, they interact with a variety of cell-surface receptors. This interaction triggers a diversity of different downstream signaling events, including the stabilization of β-catenin and tissue-specific changes in gene expression. Wnt signaling is a local event because as an indispensable step in their maturation, Wnts are palmitoleated immediately after synthesis. This lipid modification is essential for Wnts to be transported and biologically active, but it also renders them highly hydrophobic. This makes all Wnts highly dependent on carrier

*Current Topics in Developmental Biology*, Volume 150
ISSN 0070-2153
https://doi.org/10.1016/bs.ctdb.2022.03.007

proteins and specialized cellular structures both for intra- and inter-cellular movement. How this complex machinery acts in concert to deliver its highly important payload from the place of synthesis to the correct site of delivery is under active investigation. Here, we review the current understanding of how lipid-modified Wnts are processed, transported, and guided to their place of action.

## 1. Introduction to Wnt signaling

Wnt signaling is found in all animals with a patterned body axis, making it one of the most primordial intercellular signal transduction pathways in biology (Bazan, Janda, & Garcia, 2012; Clevers & Nusse, 2012). Wnts control a myriad of biological processes including proliferation, differentiation, and morphogenetic movements in multiple tissues during both development and adult homeostasis. Moreover, abnormal Wnt signaling is linked to a variety of human diseases, including cancer, congenital malformations, fibrosis, and inflammation, making it a potential drug target (Clevers & Nusse, 2012).

After synthesis in a "sending" cell (Fig. 1) (Clevers & Nusse, 2012) Wnts require post-translational acylation (mono-palmitoleation) on a single conserved serine residue, catalyzed by Porcupine (PORCN), an endoplasmic reticulum (ER) resident membrane-bound O-acyltransferase enzyme (Fig. 1) (Kadowaki et al., 1996; Takada et al., 2006). Acylated Wnts then bind to the Wnt carrier protein, the eight-pass transmembrane protein Wntless (WLS) (Coombs et al., 2010). WLS carries Wnts from the ER through the Golgi to the plasma membrane in secretory vesicles (Fig. 1) (Banziger et al., 2006; Bartscherer, Pelte, Ingelfinger, & Boutros, 2006; Clevers & Nusse, 2012; Yu et al., 2014).

Downstream Wnt signaling can be usefully divided into the Wnt/β-catenin (also called the canonical) pathway and the β-catenin-independent Wnt-regulated events such as morphogenetic movements and planar cell polarity (PCP). Both pathways require Wnts to bind to a Frizzled (Fzd) family receptor on the Wnt "receiving" cell together with various co-receptors that provide specificity to the downstream events. In the Wnt/β-catenin pathway, low-density lipoprotein receptor-related protein-5/6 (LRP-5/6) is an essential co-receptor (Fig. 1). Activation of the canonical Wnt pathway results in an accumulation of β-catenin in the cytoplasm via disruption of its destruction complex (Clevers & Nusse, 2012; Nusse & Clevers, 2017). β-catenin also translocates into the nucleus and interacts with TCF/LEF and other transcription factors to drive expression of a broad array of genes (Clevers & Nusse, 2012; Doumpas et al., 2019).

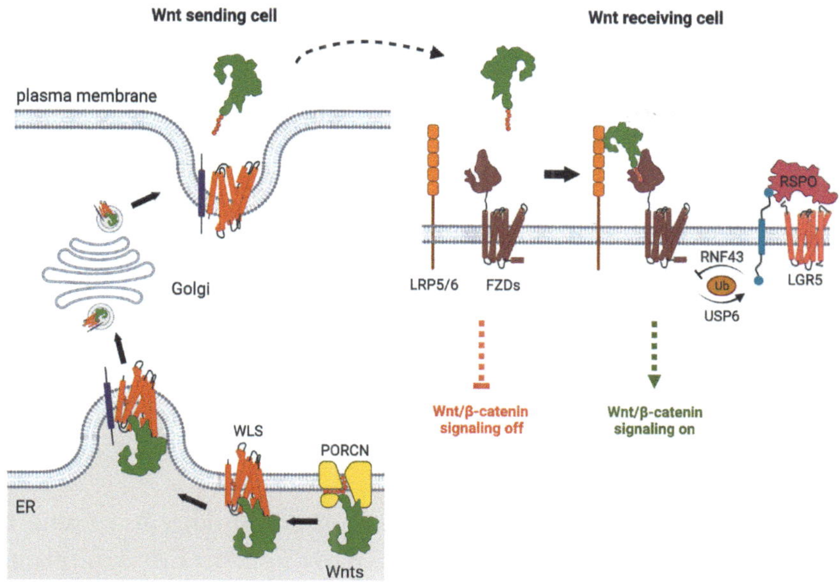


**Fig. 1** The journey from "sending" to "receiving" cells. "Wnt sending" cell: Wnt proteins are palmitoleated by Porcupine (PORCN) in the endoplasmic reticulum (ER). After acylated Wnt is transferred to Wntless (WLS), the Wnt-WLS complex is transported to the Golgi and then the plasma membrane, supported by accessory proteins (purple) such as TMEM132. At the plasma membrane, Wnt is somehow released from WLS and transported to the "Wnt receiving" cell. Binding of Wnt to Frizzled receptors (FZDs) and low-density lipoprotein receptor-related proteins 5 and 6 (LRP5/6) activates downstream β-catenin signaling. In parallel, R-spondins (RSPO) interact with LGR5 and ubiquitin ligase RNF43 or its paralog ZNRF3, preventing the ubiquitylation of FZD and LRP. This, and deubiquitylation by USP6, inhibits the endocytosis and degradation of the Wnt receptors, enhancing their abundance on the cell surface, leading to increased cellular sensitivity to Wnt ligands. *All figures were created with biorender.com.*

The sensitivity of cells to Wnt ligands depends on the abundance of Fzds and other coreceptors on the cell surface. This is regulated by two paralogous ubiquitin ligases, ring finger protein 43 (RNF43) or zinc and ring finger 3 (ZNRF3), that ubiquitylate the Wnt receptors Fzd and LRP, thereby driving their endocytosis and degradation (Fig. 1). The action of these ubiquitin ligases is opposed by a deubiquitylase, USP6 (Madan et al., 2016). RNF43 and ZNRF3 are in turn tightly regulated by the secreted R-spondin (RSPO) proteins (Fig. 1) (Binnerts et al., 2007; Hao et al., 2012; Kazanskaya et al., 2004). Several of the RSPOs are frequently co-expressed with Wnts and thus are also released from "sending" cells to regulate the Wnt sensitivity of a "receiving" cell (Greicius et al., 2018; Kabiri et al., 2014). RSPOs interact with leucine-rich repeat-containing G-protein coupled receptor-5 (LGR5) and with RNF43/ZNRF3. Formation of this complex prevents

RNF43/ZNRF3 from ubiquitylation of the Wnt receptors Fzd and LRP, thereby inhibiting their endocytosis and degradation (Fig. 1) (Binnerts et al., 2007; Hao et al., 2012). Accordingly, the RSPO-mediated increased abundance of these Wnt receptors on the cell membrane sensitizes cells to Wnts (de Lau et al., 2014; Xie et al., 2013).

The β-catenin-independent Wnt pathways (Yang & Mlodzik, 2015) do not use LRP-5/6 as a co-receptor. Instead, alternative membrane proteins, most notably Protein tyrosine kinase-7 (PTK7) (Peradziryi et al., 2011) and Receptor tyrosine kinase-like Orphan Receptors 1 and 2 (ROR1 and ROR2) (Oishi et al., 2003) act as Wnt coreceptors. Wnt binding to these receptors drives a cascade of β-catenin independent signaling activities, eventually leading to regulation of the cytoskeleton organization by actin polymerization and cell polarity (Grumolato et al., 2010; Ho et al., 2012; Oishi et al., 2003).

Interestingly, despite the importance of Wnt signaling for organismal regulation, how highly hydrophobic Wnts move from the ER to the plasma membrane of the "sending" cell and then in a site-directed manner to the receptors on a "receiving" cell is still controversial (Langton, Kakugawa, & Vincent, 2016; Routledge & Scholpp, 2019; Stanganello & Scholpp, 2016). Several different models have been proposed, and they are not mutually exclusive. Direct diffusion, carrier proteins, exosomes and cytonemes each can explain some aspects of how palmitoleated Wnts travel from the "sending" to the "receiving" cells. In this review, we will discuss what is known about the intracellular and extracellular transport mechanisms facilitating the journey of highly hydrophobic Wnt proteins through hydrophilic spaces.

# 2. Biogenesis and transport to the plasma membrane

## 2.1 Wnt ligand processing

The Wnt pathway begins with production of one of the ~40 kDa Wnt proteins in a "sending" cell (Clevers & Nusse, 2012). Newly synthesized Wnts have an N-terminal signal peptide that targets them to the lumen of the endoplasmic reticulum (ER). In the ER, the signal peptide is cleaved and Wnts fold into a unique finger-thumb structure with a core saposin homology domain and two extended β-hairpins, all stabilized by (in humans) 11 or 12 intramolecular disulfide bonds (Bazan et al., 2012; MacDonald et al., 2014; Torpe et al., 2019). Hairpin 2 then undergoes a unique mono-palmitoleation (PAM) on a totally conserved serine residue (MacDonald et al., 2014; Takada et al., 2006). This acylation is catalyzed by Porcupine (PORCN) (Fig. 1), an ER resident O-acyltransferase enzyme (Kadowaki et al., 1996; Lum & Clevers, 2012; Rios-Esteves & Resh, 2013; van den Heuvel, Harryman-Samos, Klingensmith, Perrimon, & Nusse, 1993) with structural homology to Hedgehog acyl transfers, HHAT and other membrane bound O-acyl transferase (Kadowaki et al., 1996; Yu et al., 2021). The palmitoleic acid residue makes Wnts very hydrophobic, explaining why they are so difficult to purify.

Wnt acylation plays several essential roles, and multiple studies have shown that interference with palmitoleation severely affects Wnt signaling. Deletion or inhibition of *Porcn* was followed by the retention of Wnt in the ER and consequently a block in global Wnt secretion and signaling (Barrott et al., 2011; Biechele et al., 2011; Najdi et al., 2012). A similar result is seen for individual Wnts when the acylated serine is replaced by alanine (S209A in Wnt3a) preventing PORCN-mediated acylation (Coombs et al., 2010; Takada et al., 2006). In addition to ER exit, the lipid modification of Wnts is also indispensable for Wnt signaling activity. The solved crystal structures of xWnt8 and Wnt3 bound to the FZD cysteine-rich domain (CRD) both show the palmitoleate in a hydrophobic groove of the CRD, stabilizing the Wnt:FZD interaction (Hirai et al., 2019; Janda et al., 2012). One possible function of this palmitoleate-CRD interaction is to mediate FZD dimerization as part of the signal transmission to Wnt receiving cells (Nile et al., 2017).

Wnts are subsequently post-translationally modified via glycosylation in the ER and Golgi apparatus (Komekado et al., 2007; K. Willert et al., 2003). Wnt glycosylation is heterogeneous; different Wnts have different predicted N-glycosylation sites, Wnt glycosylation might serve to target Wnt to different exit sites at the plasma membrane within the Wnt producing cell. In cultured polarized epithelial cells, N-glycosylation influences if Wnt secretion takes place at the apical or the basolateral side of the cell (Yamamoto et al., 2013). Whether this diversity in Wnt glycosylation is an essential factor for the intracellular distribution of Wnt proteins bringing them to the plasma membrane in a spatially specific manner is currently not well known.

In conclusion, diverse posttranslational modifications of Wnts regulate their proper transport, targeting and signaling activity. In particular, palmitoleation renders Wnt proteins highly hydrophobic, making them completely dependent on specialized cellular transport mechanisms for both intra- and intercellular movement (Hannoush, 2015).

## 2.2 Intracellular trafficking

The most important protein involved in the transport of Wnt is the eight-pass transmembrane protein Wntless (WLS), a highly evolutionarily conserved protein present in metazoans from cnidarians to humans (Banziger et al., 2006; Bartscherer et al., 2006; Goodman et al., 2006; Guder et al., 2006). After Wnts are acylated in the ER, they require WLS for transport through the Golgi and secretory vesicles to the plasma membrane (Fig. 1) (Banziger et al., 2006; Bartscherer et al., 2006; Clevers & Nusse, 2012; Yu et al., 2014). A range of studies indicate that palmitoleation of Wnts is critical for their transport from the ER to the plasma membrane. Wnt acylation is required for it to bind to WLS in the ER (Coombs et al., 2010; Herr & Basler, 2012; Najdi et al., 2012). Downregulation of WLS expression results in retention of Wnts in the ER and, consequently, blocks their secretion (Moti et al., 2019; Yu et al., 2014). Global knockout (KO) of *Wls* in mice causes early lethality during embryogenesis (Fu et al., 2009) just as loss-of-function mutations in the X-linked *PORCN* gene are lethal in male mice (Grzeschik et al., 2007; X. Wang et al., 2007). Tissue-specific KO of WLS results in multiple Wnt-related phenotypes (Carpenter et al., 2010; Z. Zhong et al., 2012) and humans with WLS mutations and a severe Wnt-related phenotype have recently been described (Chai et al., 2021). The requirement for Wnt to be palmitoleated to interact with its transport protein WLS explains the retention of Wnts in the ER after either deletion or inhibition of PORCN or mutations of the Wnt palmitoleation site (Barrott et al., 2011; Biechele et al., 2011; Takada et al., 2006). Altogether, this highlights the indispensable functions of both PORCN and WLS in the modification and secretion of Wnt molecules.

Our understanding of how WLS carries Wnts from PORCN in the ER to the plasma membrane has been advanced by two recent cryo-electron microscopy structures of palmitoleated Wnt in complex with WLS (Nygaard et al., 2021; Q. Zhong et al., 2021). WLS has two distinct domains. Its luminal domain shows structural homology to seipin, an ER-resident integral membrane protein that interacts with triacylglycerols during lipid droplet formation, as well as to NPC2, a protein involved in cholesterol transport (Szymanski et al., 2007). The remaining WLS membrane domain shows structural homology to G protein-coupled receptors (GPCRs) including Frizzled and the hedgehog pathway regulator Smoothened. While phospholipid binding sites are not a general feature of GPCRs, several structures of lipid-binding GPCRs have been reported (Audet & Stevens, 2019). The thumb-finger structure of Wnt grasps the WLS luminal domain and inserts the thumb of hairpin 2 deep into the GPCR domain, where the PAM slides between WLS membrane helices 4 and 5 and into the lipid bilayer. WLS is open to the lipid bilayer on the opposite side, between helices 6 and 7, suggesting a route for Wnt loading from PORCN and Wnt unloading, perhaps to FZD or other intermediate carriers. The C-terminal 45 amino acids, essential for WLS export from and back into the ER (Yu et al., 2014), are not visible in the structure. The transport of WNT from PORCN to WLS in the ER may be stabilized by transmembrane protein 132A (TMEM132A) (Li & Niswander, 2020). TMEM132A interacts physically with WLS enhancing the WLS–Wnt interaction (Fig. 1) and, consequently, activates Wnt signaling activity. Thereafter, WLS leaves the ER in coat protein II (COPII) vesicles. Interaction between WLS and the guanine nucleotide-exchange factor activator SEC12 is necessary for WLS incorporation into the COPII vesicles. Binding of palmitoleated Wnt to WLS enhanced WLS-SEC12 complex formation, and, consequently, promoted the incorporation of the complex into the COPII vesicles. This ER export machinery regulates the formation of Wnt secretory vesicles and initiates the first step of Wnt's journey (Sun et al., 2017). Before getting to the plasma membrane, the Wnt-WLS complex passes through the Golgi apparatus where further glycosylation can occur, potentially guiding Wnt to the final site of secretion as described above (Fig. 1) (Komekado et al., 2007; Yamamoto et al., 2013). From the Golgi, the modified Wnts bound to WLS travel further to the plasma membrane in small vesicles (Fig. 1) (Banziger et al., 2006; Bartscherer et al., 2006; Herr & Basler, 2012; Moti et al., 2019). Live imaging of HeLa cells revealed that WNT3A vesicles pause below the plasma membrane for a couple of minutes before fusion (Moti et al., 2019). WLS itself undergoes recycling and is transported back from the plasma membrane to the ER where it can bind and transport additional Wnts (Belenkaya et al., 2008; Franch-Marro et al., 2005; Yu et al., 2014).

In conclusion, WLS is key for the intracellular transport of Wnts from the ER via the Golgi to the plasma membrane where the Wnts are eventually released to Wnt receiving cells (Fig. 1). However, it's not known if WLS cycles constitutively to the PM, or if this process might be triggered by the binding to WNT. Moreover, if WNT-containing vesicles follow the bulk vesicular flow or if they are guided via a specific mechanism needs still to be defined.

## 3. Modes of Wnt secretion

Despite the enormous importance of Wnt signals in regulating cell fate and decisions, it is still controversial how the highly lipophilic Wnts are secreted and travel through the aqueous extracellular space. Importantly, since a malfunction of Wnt signaling can lead to disturbances of cellular pathways and functions, a precise and targeted transport mechanism is needed. While the Wnt-sensitizing RSPOs are soluble and can diffuse, several competing models try to explain how palmitoleated Wnts can travel from the "sending" to the "receiving" cell.

The simplest model is diffusion in a gradient dependent manner (Fig. 2). However, since Wnts are modified with a hydrophobic palmitoleate moiety, they are not freely diffusible. Indeed, purified Wnt proteins aggregate in solution followed by a complete loss of signaling activity unless stabilized by detergents or serum (Fuerer et al., 2010; K. Willert et al., 2003; K. H. Willert, 2008). Free extracellular spreading of a Wnt orthologue was recently suggested, but not directly demonstrated, in *Caenorhabditis elegans* (Pani & Goldstein, 2018). How diffusion of the highly hydrophobic Wnt molecules might occur is unclear. It might be achieved by mechanisms such as binding to carrier proteins or components of the extracellular matrix (ECM) (Fig. 2). However, free diffusion of the lipid-modified hydrophobic Wnt proteins over long distances in the aqueous extracellular space to fulfill paracrine signaling activity in neighboring cells seems inefficient and unreliable if Wnt signaling is to be timely and spatially targeted. Thus, Wnt signaling without any assistance is most likely restricted to autocrine or maybe signaling between adjacent cells (Farin et al., 2016). Further travel requires additional mechanisms that may include extracellular vesicles, cytonemes, soluble or membrane bound proteins, and proteoglycans (Franch-Marro et al., 2005; Port & Basler, 2010; Routledge & Scholpp, 2019; Stanganello & Scholpp, 2016).

### 3.1 Wnt carrier proteins

The transport of a hydrophobic Wnt protein presents unique challenges. After dissociation from WLS at the plasma membrane, the extracellular

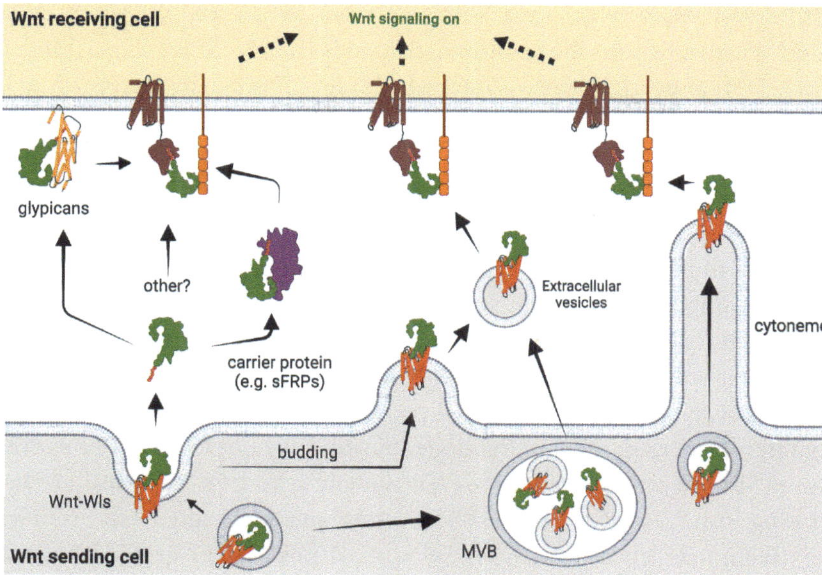


**Fig. 2** Mechanisms of Wnt transport. After Wnt is released at the plasma membrane of the "Wnt sending" cell, the glycosylated Wnt travels to the "Wnt receiving" cell. This process might be facilitated by glypicans or carrier proteins, such as sFRPs or other extracellular lipid-binding proteins such as SWIM or afamin (not shown). The lipid modified Wnts can also be shuttled to the receiving cell via extracellular vesicles, either derived by budding from the plasma membrane (microvesicles) or via exosomes from a multivesicular body (MVB). Cytonemes that extend from the sending cell allow the highly targeted transport of Wnt ligands to the receiving cell.

diffusion of Wnts might be facilitated by binding to additional Wnt-binding proteins that also accommodate their hydrophobic nature. On the other side, Wnts could also interact with cell surface proteins improving their lateral diffusion (Fig. 2).

Other intensively studied Wnt-binding molecules include the secreted frizzled-related proteins (SFRPs) (Fig. 2), a family of glycoproteins comprising a frizzled-like cysteine-rich domain but no membrane domain. This enables them to bind to extracellular Wnts, modulating their signaling activity (Galli et al., 2006; Hoang et al., 1996; Uren et al., 2000; van Loon et al., 2021). SFRPs are present in most metazoans except Drosophila, which might explain why flies have elaborate transcytosis mechanisms for Wnt localization (Yamazaki et al., 2016). Interestingly, the exact function of the SFRPs in metazoans is controversial. Initially, SFRPs were thought to be Wnt inhibitors (Leyns et al., 1997). For example, overexpression of

SFRP2 in mesenchymal stem cells decreased Wnt activity (Alfaro et al., 2010). Further studies have shown that SFRP1 and SFRP2 can bind to Wnt3a (Hua et al., 2018; Wawrzak et al., 2007). Overexpression of SFRPs results in a potent inhibition of Wnt activity in enteroendocrine L cells (Galli et al., 2006; Wawrzak et al., 2007), in the neural tube of chick embryos (Galli et al., 2006) and in COS7, a fibroblast cell line derived from monkey kidneys (Morello et al., 2008) affecting accumulation of β-catenin and proliferation. These studies propose that SFRPs can inhibit Wnt signaling in vitro and in vivo presumably as decoy receptors, sequestering Wnt ligands and therefore preventing the binding to FZD receptors in an antagonistic manner. However, in contrast to its potential role as Wnt pathway antagonist when over-expressed, multiple studies have highlighted an agonist role of endogenous SFRPs. Lower levels of expression of SFRP2 activate canonical Wnt signaling both in vitro and in vivo, resulting in increased target gene expression and augmented cell proliferation (Heinosalo et al., 2018; Kress et al., 2009; Mastri et al., 2014). In *Xenopus*, diffusion of Wnt8 and Wnt11 is significantly increased by forming a complex with SFRPs (Mii & Taira, 2009).

There is also genetic evidence that SFRPs enhance Wnt signaling. Mice with mutations that affect Wnt delivery have neural tube defects. Thus, WLS mutations (Chai et al., 2021), loss of TMEM132A (Dickinson et al., 2016) and mutations in the PCP genes Vangl2 and CELSR result in this phenotype (M. Wang et al., 2019). Interestingly, *Sfrp1/2* double knockout mice show caudal neural tube defect as well, suggesting that SFRPs function as Wnt agonists in vivo (Misra & Matise, 2010). Other studies proposed that SFRPs are involved in both inhibition and activation of the Wnt pathway depending on their concentration (Houart et al., 2002; Uren et al., 2000; Xavier et al., 2014). On the one hand, SFRPs at low concentrations might bind to Wnts, support their transport and, this way, activate Wnt signaling. On the other hand, especially at high concentrations, SFRPs can also bind to Wnts and/or Wnt receptors resulting in inhibition of Wnt signaling. Most of the publications showing that SFRPs function as Wnt inhibitors used overexpression, which might lead to supraphysiologic expression levels. In summary, the exact function as well as SFRP's mechanism of action on Wnt signaling remains an open question.

Other soluble Wnt carrier proteins have been identified, with uncertain physiological significance. Several groups have purified acylated Wnts in complex with plasma proteins that have subsequently been identified. These include afamin, high-density lipoproteins (HDL) and from Drosophila,

Swim (Fig. 2). Afamin is an albumin family plasma glycoprotein and vitamin E carrier associated with metabolic syndrome (Dieplinger & Dieplinger, 2015). It forms a stable and soluble 1:1 complex with most of the known 19 human Wnt proteins when co-expressed in HEK293 cells, markedly improving their solubility, especially in serum-free media (Mihara et al., 2016). The PAM has been modeled to lie in a lipophilic pocket in the afamin structure (Naschberger et al., 2017). However, only the afamin-Wnt3a and 5a complexes have been shown to have significant biological activity. It is unlikely that afamin is a physiologically important Wnt carrier protein, since it is expressed primarily in the liver, and is not present in invertebrates (Baker, 1998). Nevertheless, afamin is a useful tool to generate soluble Wnt protein for in vitro studies.

Lipoproteins were also found to bind to Wnts in various systems. Lipoproteins are a class of extracellular proteins typically used for transporting lipids, especially cholesterol, in the circulation. It was shown that Wg colocalizes with lipoprotein particles in *Drosophila* (Panakova et al., 2005). In another publication, Wnt3a was found to be associated with lipoproteins in the media of in vitro mouse fibroblasts. Here, high-density lipoproteins (HDLs) lead to a release of WNT3A in the media, indicating that WNT3As bind to exogenous HDLs (Neumann et al., 2009). In a recent publication, WNT5A was found to be secreted from the choroid plexus to the cerebrospinal fluid (CSF) on lipoprotein particles rather than exosomes, in both mouse and human developing hindbrain. Deletion of WNT5A in the choroid plexus resulted in impaired cerebellar morphogenesis of hindbrain progenitor cells presumably due to the lack of WNT5A in the CSF. This suggests that lipoproteins are a crucial mechanism for long-range transport of WNT5A in the central nervous system (Kaiser et al., 2019). Another Wg-interacting Molecule (Swim), a lipocalin protein, was proposed to transport Wg over long distances to Wg-receiving cells by increasing its solubility in *Drosophila* (Mulligan et al., 2012). However, a homolog of Swim is not identified in vertebrates and few further studies have addressed Swim's function.

Next to Wnt-binding proteins enhancing the solubility and diffusion of Wnt proteins, components of the extracellular matrix (ECM) could also facilitate the transmission of Wnts through the ECM (Fig. 2). Here, glypicans constituting one of the two main heparan sulfate proteoglycans (HSPGs) might play an essential role. Glypicans are known to distribute signaling molecules like morphogens in the ECM, and, by this, modulate binding of the signaling molecules to their receptors at the cell surface influencing diverse cellular processes including proliferation, motility, and adhesion. Since they can be critical for the binding of ligands to their receptors, they are sometimes considered coreceptors (Kirkpatrick & Selleck, 2007). In *Drosophila*, two glypicans, namely the cell surface glypicans Dally and Dally-like (Dlp), have been identified. Both Dally and Dlp contain a protein core element with attached glycosaminoglycan chains facilitating the stabilization and lateral diffusion of Wnt proteins. Overexpression of Dlp led to an accumulation, whereas ablation of glycosaminoglycan biosynthesis (resulting in sugar-deficient HSPGs) decreased the abundance of extracellular wingless (Wg), suggesting that HSPGs are critically involved in the extracellular distribution of Wg (Baeg et al., 2001). Studies in wing imaginal discs have highlighted that Dlp can both retain the availability of Wg on the surface of the producing cell and promote transport of Wg to receiving cells, whereas Dally acts as a Wg co-receptor in this process presenting Wg to the signaling receptors (Baeg et al., 2004; Franch-Marro et al., 2005; Han et al., 2005; Yan et al., 2009).

In vertebrates, several lines of evidence support the involvement of glypicans in Wnt signaling. In vitro, both gain- and loss-of-function experiments indicate that glypican 4 (GPC4) is involved in activation of both canonical and noncanonical Wnt signaling pathways, in dependence on different membrane microdomains (Sakane et al., 2012). Mutations in zebrafish GPC4 impair gastrulation movements of convergent extension associated with abnormal cell polarity. The authors showed that GPC4 potentiates Wnt11 signaling that mediates convergent extension (Topczewski et al., 2001). Analysis of Xenopus GPC4 revealed that its depletion ablated Wnt signaling, and thereby interrupts cell movements during gastrulation via inhibition of the noncanonical Wnt pathway. The authors also showed binding of GPC4 to Wnt11 to activate the non-canonical Wnt pathway (Ohkawara et al., 2003). In mouse embryos lacking glypican 3 *(Gpc3)*, non-canonical Wnt signaling pathway was reduced. The authors concluded that at least in some cell types GPC3 might be a positive regulator of non-canonical Wnt signaling (Song et al., 2005). In zebrafish, mutation of *gpc4, wnt5b* or *wnt9a* results in severely delayed endochondral ossification during the onset of osteogenic differentiation, another Wnt phenotype. This highlights the importance of the non-canonical Wnt pathway components and *gpc4* in regulating coordinated ossification (Ling et al., 2017). As an alternative model to the direct transfer of Wnts by glypicans, HSPGs may also form a transport machinery regulating the transmission of larger Wnt-containing complexes. Several lines of evidence support the hypothesis that HSPGs serve as cell-surface receptors for the endocytosis of different cargo structures and thus support the transmission to and uptake of exosomes and other structures in target cells (Christianson & Belting, 2014; Christianson, Svensson, van Kuppevelt, Li, & Belting, 2013; Olsson et al., 2001; Stanford et al., 2010; Willert et al., 2003). This vesicle transfer model complements the alternative model that glypicans interact with the palmitoleate moiety of Wnts to directly facilitate their diffusion (McGough et al., 2020).

## 3.2 Exosomes

As noted, another route for Wnt cell-cell transfer is via extracellular vesicles, including exosomes and microvesicles (Fig. 2). Various types of extracellular vesicles surrounded by a lipid bilayer membrane are produced and secreted by a majority of cell types as a fundamental part of intercellular communication (Dai et al., 2020; Maas et al., 2017; Ruivo et al., 2017).

Exosomes develop through the maturation of early endosomes by loading with cargo proteins into so-called multivesicular bodies (MVBs) (Fig. 2), in which they are stored. Via fusion of the MVBs with the plasma membrane, exosomes are secreted from the cell to the extracellular space to transmit its signal peptides to neighboring cells facilitating cell-cell communication (Fig. 2). Microvesicles are small vesicles that are derived by directly budding off of the plasma membrane, releasing them into the extracellular environment (Fig. 2) (Schubert & Boutros, 2021).

Wnt was first identified in extracellular vesicles in *Drosophila*, as Wg was found on vesicles in the wing imaginal discs (Greco et al., 2001). In another study in *Drosophila*, it was shown that Wg can be carried by WLS-containing exosomes across synapses (Korkut et al., 2009). Wg transported across the synaptic cleft via exosomes containing WLS affected synaptic function, growth and plasticity. Wnts were demonstrated to be secreted on exosomes both during *Drosophila* development and in human HEK293 cells (Gross et al., 2012). Together with WLS, Wnts are transported onto exosomes via SNARE-mediated processes requiring Ykt6 (Gross et al., 2012), a longin-type of R-SNARE involved in various trafficking events (McNew et al., 1997; Meiringer et al., 2008). A critical step of maturating MVBs is endosomal acidification, which can be blocked either by weak bases such as $NH_4Cl$ or by the V-ATPase inhibitor bafilomycin A1 (Clague et al., 1994; Coombs et al., 2010). Inhibiting acidification and, subsequently MVB maturation impairs WLS-dependent Wnt transport onto exosomes and, consequently, results in intracellular accumulation of the Wnt-WLS complex (Gross et al., 2012). Both Wnt and WLS co-localized with the MVB markers CD81 and TSG101, indicating that they were incorporated into MVBs. Wnt-WLS containing exosomes were also shown to induce Wnt signaling in target cells, highlighting their functional role (Gross et al., 2012). In the *Drosophila* wing disc, Wg colocalized with Cd63 and Rab4, two exosomal proteins (Gross et al., 2012). Interestingly, extracellular Wg was localized together with Cd63 and Rab4 in a different set of exosomes, suggesting that Wg is incorporated in distinct exosome subpopulations that might have different functions. Colocalization was identified both in intracellular MVBs and outside the producing cell, implying that Wg/WLS-containing exosomes are secreted and can diffuse (Gross et al., 2012).

Wnt secretion on exosomes is seen in vertebrates as well. In a study of axonal regeneration in the injured CNS, fibroblast-derived exosomes induced recruitment of Wnt10b toward lipid rafts. This activated mTOR signaling and promoted axonal regeneration after optic nerve injury, and were strongly reduced in Wnt10b$^{-/-}$ mice (Tassew et al., 2017). A similar mechanism was identified by another study reporting that Cd81-positive exosomes secreted from fibroblasts load Wnt11 into human breast adenocarcinoma MDA-MB-231 cells (Luga et al., 2012). This process increased the protrusive activity and motility of breast cancer cells by activating PCP signaling. Wnt-containing exosomes were also associated with poor prognosis in colorectal cancer (CRC) (Hu et al., 2019). There, fibroblast-derived exosomes induced dedifferentiation of differentiated CRC cells. The cancer cells acquired a cancer stem cell (CSC) phenotype, including increased Wnt pathway activity and the expression of CSC marker genes, a process that promotes drug resistance in CRC. The authors identified exosomal Wnts as the major reprogramming regulator since they induced Wnt activity and drug resistance in differentiated CRC cells. Inhibition of both exosome release and Wnt palmitoleation with a PORCN inhibitor blocked this effect both in vitro and in vivo (Hu et al., 2019). Another study examining the role of macrophage-derived WNT highlighted the importance of extracellular vesicles-packaged WNTs in intestinal repair. Mice carrying a macrophage-restricted ablation of Porcupine, and thus, depleted Wnt production only in these cells, exhibit normal intestinal morphology but were very sensitive to radiation injury. Radiation lethality of Porcn-null mice could be compensated by treating the mice with macrophage-conditioned medium from wildtype but not Porcn-null bone marrow. Interestingly, removement of extracellular vesicles from the conditioned medium abolished its ability to rescue radiation lethality, presumably due to the loss of Wnts localized in these vesicles. In another example, Wnt5a and Wnt7a, secreted by cultured hippocampal central neurons via exosomes, were crucial both for synaptogenesis and synapse maintenance (Dickins & Salinas, 2013; Thery, Ostrowski, & Segura, 2009).

All these data show that Wnt can be released from the cells via exosomes. Wnts are carried to the plasma membrane by WLS, but the role of this carrier in Wnt extracellular vesicle formation is not well established. Interestingly, in *Drosophila* embryos, like human cells, WLS gets endocytosed from the plasma membrane and remains within the producing cell (Pfeiffer et al., 2002) facilitated by a retromer-dependent recycling machinery (Port et al., 2008). The robust WLS recycling machinery suggests that WLS is primarily destined to release Wnts from the sending cell, and that other mechanisms (glypicans, SFRPs) are responsible for the next step in the Wnt journey. However, as noted above, there is some data that at least in flies, Wg is secreted along with Wls. During *Drosophila* imaginal disc development Wls/Evi is found on exosomes, but only 10% of extracellular Wg co-localized with Wls, suggesting that they are not released on the same exosomes (Gross et al., 2012). WLS has also been localized on secreted exosomes in *Drosophila* in the neuromuscular junction (Ataman et al., 2008; Koles et al., 2012) indicating that the function of WLS might vary dependent on the cell type. The presence of WLS on exosomes may be rare. In proteomic surveys of exosomes and extracellular vesicles, Wnts are occasionally detected, but WLS is not seen, e.g., reference (Servage et al., 2020). This suggests that Wnts are more often released from sending-cell WLS during the secretion process. Inhibition of vacuolar acidification was shown to effectively prevent Wnt secretion both in cultured human cells and in vivo and results in accumulation of the WNT3A-WLS complex at the plasma membrane. This led to the conclusion that acidification is needed to release WNT3A from WLS in secretory vesicles, potentially to transfer WNT3A to other proteins, e.g. soluble carrier protein or components of the extracellular matrix (Coombs et al., 2010), indicating that WLS is most likely left behind at the sending cell. In summary, several lines of evidence support the concept that Wnt can be transported on exosomes from Wnt producing to Wnt receiving cells in health and disease, although it is not clearly defined if WLS is an obligate part of this process.

## 3.3 Cytonemes and telocytes

An attractive additional model to deliver acylated Wnts to target cells is via long and thin cytoplasmic cellular projections, so-called cytonemes (Fig. 2). A cytoneme is a specialized subset of filopodia, a dynamic actin-based tubular extension of the plasma membrane, that is capable of transporting signaling molecules between cells. The length of cytonemes is very variable, ranging between 1 and 200 μm whereas the diameter is usually between 0.1 μm and 0.4 μm (Kornberg & Roy, 2014). Cytonemes emanate from signal-producing cells and extend to the signal-receiving cell (Callejo et al., 2011). Signaling molecules can be released from the cellular extensions to the cell surface in discrete domains. The cytoneme model is so attractive since signaling molecules can be secreted specifically at sites where the cytoneme contacts the signal-receiving cell, enabling a temporally and spatially controlled delivery (Kornberg & Roy, 2014). Cytonemes were discovered in *Drosophila* wing imaginal disc (Ramirez-Weber & Kornberg, 1999) and are primarily linked to the transport of a variety of morphogens involved in several central pathways, including but not limited to epidermal growth factor (EGF) (Lidke et al., 2005), fibroblast growth factor (FGF) (Koizumi et al., 2012) and Decapentaplegic (Dpp) (Roy et al., 2011). Cytonemes are also known to be important in propagation of another lipid modified morphogen, Hedgehog (Hh) (Kornberg & Roy, 2014; Ramirez-Weber & Kornberg, 1999; Rojas-Rios, Guerrero, & Gonzalez-Reyes, 2012). More precisely, Hh ligands are dually lipid modified by the long-chain fatty acid palmitate on the N-terminus of the protein and by a C-terminally linked cholesterol moiety (Pepinsky et al., 1998; Porter et al., 1996). Analogous to Wnt signaling, this step is fundamental for Hh to travel from their site of production to target cells to induce appropriate responses (Hall et al., 2019). Several studies localized Hh ligands, the Hh transporter-like protein Dispatched (*Disp*), the co-receptor BOC (BOI in *Drosophila*), CDON (iHOG, in *Drosophila*) as well as the Hh receptor patched (Ptch) on cytonemes in chick and fly systems (Bischoff et al., 2013; Bodeen et al., 2017; Chen et al., 2017; Gradilla et al., 2014; Sanders et al., 2013), consistent with an important role for cytonemes in Hh signal transmission. Despite Hh pathway's importance in embryonic development and adult tissue homeostasis through regulating polarity, cell proliferation, cell fate, and stem cell maintenance (Briscoe & Therond, 2013), the proteins involved in initiation, expansion and maintenance of the Hh distributing cytonemes are not completely understood. In a very recent study, Hall and colleagues revealed that Myosin 10 (MYO 10) is needed for the transport of SHH to the cytoneme tips to trigger a response in signal receiving cells in vitro (Hall et al., 2021). Here, they could identify MYO10 being involved in both cytoneme biogenesis as well as cytoneme-based transport of SHH via its cargo binding domain. Interestingly, MYO10 ablation in mice results in neural tube patterning defects that are consistent with an altered SHH gradient, indicating that MYO10 is also a part of the SHH delivering machinery in vivo. As neural tube defects are seen in Wnt distribution mutants as well, it raises the question of whether MYO10 is relevant to the distribution of lipid modified Wnt signals.

Cytonemes are also involved in Wg signaling in *Drosophila*. Recall that Wg was first identified in fly mutants that lacked wings (Sharma & Chopra, 1976). Interestingly, it was Wg's receptor, FZD, rather than Wg itself, which was found to be present on cytonemes of wing disc myoblasts. Here, FZD-containing cytonemes emanate from the myoblast to the disc, form a complex with Wingless (Wg), and retract again toward the Wg-receiving cell (Huang & Kornberg, 2015). This occurs in vertebrates as well. Murine embryonic stem cells (ESCs) generate specialized Wnt ligand-recognizing cytonemes that can contact trophoblast stem cells (TSCs), leading to ESC–TSC pairing and self-renewal of the ESCs. This connection is needed for the transmission of Wnt signals from the TSCs to ESCs to activate the Wnt/β-catenin pathway (Junyent et al., 2020). Similarly, Fzd7 was detected on cytonemes of epithelial cells of the somites in chicken embryos contacting the ectoderm, indicating a role in signal transmission (Sagar et al., 2015). Whether this very interesting mechanism of a FZD receptor reaching out for Wnt signals via long cellular protrusions is limited to developmental processes, or also functional in adult signaling systems awaits further study.

More commonly, Wnts rather than their receptors are transported by cytonemes to the signal-receiving target cells. WNT2B was found to be transported intracellularly along cytonemes before transfer to a Wnt receiving cell in *Xenopus* (Holzer et al., 2012). In zebrafish embryos, Wnt8A localized to dynamic clusters found on Wnt producing cell's cytonemes. From there, Wnt8A was found to be released and colocalized on signal receiving cells in FZD receptor-containing clusters, indicating an important role of cytonemes in transmission of Wnt signals in vertebrates (Luz et al., 2014). In another study, Wnt8A was transported via Cdc42/N-Wasp-positive actin-based cytonemes to activate Wnt signaling, influencing tissue patterning during formation of the neural plate in zebrafish (Stanganello et al., 2015). Interestingly, the authors could find this cytoneme-based distribution of Wnt signals in a variety of vertebrate cell lines, indicating that this mechanism might be conserved and utilized in a variety of cell types and tissues, not just during early zebrafish development (Stanganello et al., 2015). This demonstrates that cytoneme based transport of Wnt signals could be fundamental in transmission of Wnt signals in vertebrates.

A distinctive type of interstitial cells, so called telocytes (Fig. 3), are characterized by extraordinary long cytonemes commonly referred to as telopodes, and these telopodes contain podoms, which are dilated segments

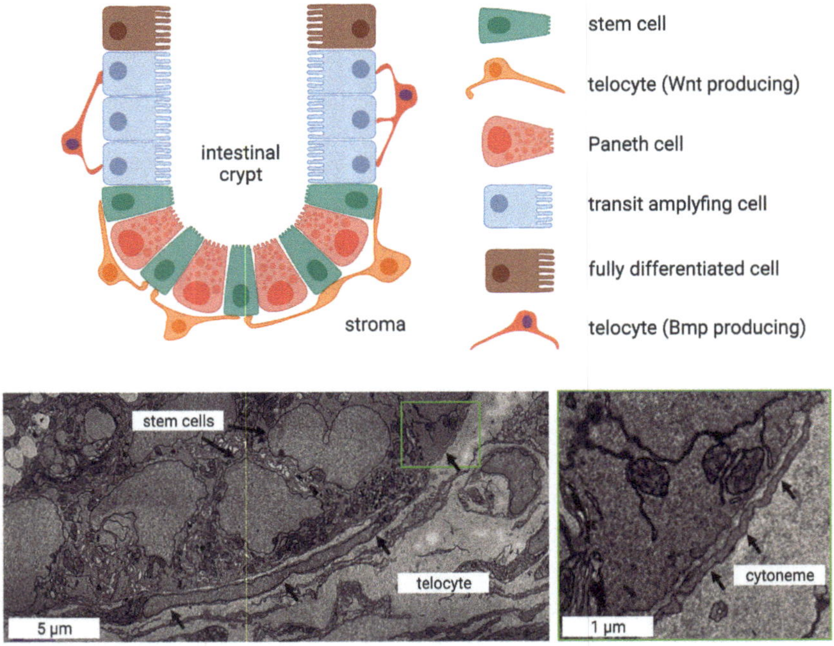


**Fig. 3** Wnt transport in the intestinal stem cell niche. Transit amplifying (TA) cells are produced by division of stem cells interspersed between Paneth cells located in the crypt of Lieberkühn. Differentiation is driven by BMP signals derived from stromal telocytes. TA cells migrate out of the crypt to provide new fully differentiated epithelial cells. The division and renewal of stem cells is regulated by stromal Wnt- and RSPO3-producing telocytes that are located directly beneath the intestinal epithelial cells. Transmission electron microscopy (TEM) illustrates a telocyte spanning the basolateral membrane of a murine colonic intestinal crypt (lower left). Cytonemes emanate from the cell body and connect to the cell surface of the stem cells for the transport of signaling molecules including Wnts (lower right, magnification). TEM images courtesy of David Hildebrand, Diego Bohórquez, and Wei-Chung Lee.

that accommodate cell compartments like mitochondria or rough ER. Telocytes form three-dimensional networks to build contacts with neighboring cells to transmit signaling molecules that can be released both from the tip of cellular extensions and from the podoms (El Maadawi, 2016). Telocytes can readily be seen close to the stem cell niche in a broad range of organs including the heart (Gherghiceanu et al., 2010; Popescu et al., 2010), lung (Popescu et al., 2011; Zheng et al., 2011), intestine (Cantarero Carmona et al., 2011; Snyder et al., 2015; Vannucchi et al., 2013), kidney (L. Li et al., 2014; Qi et al., 2012), and many other organs (Cretoiu et al., 2017) indicating that cytoneme-mediated transport of signaling molecules might be widely prevalent (Fig. 3). In human skin, telocytes were found to form sheath layers surrounding the stem cell niche of hair follicles and directly contacting the dermal stem cells (Ceafalan et al., 2012). In the lung, telocytes connect to type 2 epithelial cells, which function as stem cells (Chung et al., 2018; Nabhan et al., 2018; Sirianni et al., 2003), and form stem cell contacts with their telopodes in prostate tissue (Corradi et al., 2013).

The intestinal stem cell niche provides a well-studied example of the role of telocytes (Fig. 3). Similar to hair follicles, intestinal telocytes form a thin sheath wrapping the epithelial basal membrane (Deane, 1964), regulating stem cell activity in the intestinal crypts (Shoshkes-Carmel et al., 2018). The epithelium of the intestine is continually renewed every 4–5 days (Clevers, 2013; Lipkin, 1965; Potten et al., 1974). This tightly regulated process of self-renewal including rapid proliferation and regulated differentiation is a hallmark of the intestinal stem-cell niche located at the base of the crypts and is indispensable for the proper function of the intestine (Clevers, 2013). This process needs high Wnt activity to regulate the stem cell's fast proliferation, however, the source of these Wnt signals had been controversial for a long time. Paneth cells located in the small intestinal crypt base express several Wnts and have close interactions with the stem cells (Farin et al., 2012; Sasaki et al., 2016; Sato et al., 2011). Moreover, secretion of Wnts by Paneth cells in ex vivo organoids, cultured mini-guts growing from isolated intestinal stem cells, can regulate the proliferation and differentiation of ISCs (Farin et al., 2012). This led to the prediction that Paneth cells served the same function in vivo. However, depletion of Paneth cells or complete block of epithelial Wnt production did not result in any change in intestinal homeostasis or any detectable phenotype (Durand et al., 2012; Kabiri et al., 2014). This indicated that the production of Wnts by epithelial cells (including Paneth cells) is completely dispensable and other nearby sources of Wnts must exist. Recently, several independent studies demonstrated this Wnt source was telocytes in the underlying intestinal stroma that express multiple Wnts (Fig. 3) (Farin et al., 2012; Gregorieff et al., 2005; Greicius et al., 2018; Kabiri et al., 2014; Karpus et al., 2019; Shoshkes-Carmel et al., 2018). Co-culture experiments showed that cells expressing these stromal Wnts can effectively support the formation of intestinal organoids, including when the organoids cannot make their own Wnts (Farin et al., 2012; Kabiri et al., 2014). The stromal cells were also identified as the key source of RSPOs (Greicius et al., 2018; Kabiri et al., 2014). Due to the fundamental importance of proper spatial and temporal delivery of stromal-produced Wnt signals, telocytes and their cytonemes appear to play a pivotal role in the proper regulation of the intestinal stem cell Wnt niche.

How are the Wnt-transporting cytonemes formed and regulated? Recent studies indicate that Wnt signaling itself plays a pivotal role in these processes. Activation of the β-catenin-independent Wnt-planar cell polarity (PCP) pathway drives formation and extension of cytonemes (Schlessinger et al., 2009) by downstream activation of actin polymerization driving cytoskeletal regulators including small GTPases Rac1, Rho, and Cdc42 (Spiering & Hodgson, 2011). The receptor-tyrosine kinase-like orphan receptor 2 (Ror2) is a co-receptor of the Wnt PCP pathway (Oishi et al., 2003). Studies from the Scholpp lab highlighted that the biogenesis of Wnt-carrying cytonemes can be driven directly by Wnt ligands (Mattes et al., 2018). Autocrine-released WNT8A bound to the ROR2 co-receptor triggering the PCP pathway. This activates the small GTPase Cdc42 resulting in actin polymerization and de novo formation of cytonemes. Afterwards, WNT8A is loaded onto the outgrowing cytonemes and transmitted to the signal-receiving cells. Here, WNT8A binds to the β-catenin-specific co-receptor LRP6 in a paracrine fashion activating the β-catenin pathway leading to target-gene induction in HEK293T human embryonic kidney cells and PAC2 fish fibroblasts. This is especially remarkable, since WNT8A can control its own dissemination via this mechanism - activating the PCP-cytoneme extension pathway by binding to ROR2, and activation of the β-catenin-dependent canonical Wnt pathway by interaction with Lrp6. This mechanism was also present in human gastric cancer cells and in the Wnt-dependent regulation of intestinal stem cells in an ex vivo intestinal organoid model. Here, ROR2/PCP-induced cytonemes are essential for the maintenance of murine intestinal crypt organoids (Mattes et al., 2018).

More recently, this model was refined by the discovery that the distribution of Wnt molecules by cytonemes was regulated by the cell polarity protein van gogh-like (Vangl2) (Brunt et al., 2021). In several cellular systems, including zebrafish epiblast cells, human gastric cancer cells, and intestinal murine stromal cells, activation of Vangl2 resulted in the formation of very long cytonemes transporting Wnt molecules at their tips to surrounding cells (Brunt et al., 2021). The authors highlighted that activation of the PCP pathway by autocrine Vangl2-dependent signaling induces the generation of cytonemes to transmit Wnt ligands to the surrounding cells. The activation of cytoneme-dependent mobilization of Wnt signals by Vangl2 appears widespread in vertebrates and may be essential for both development and tissue homeostasis.

Taken together, a growing body of evidence indicates that cytonemes comprise a crucial transport mechanism for Wnts in a wide variety of organisms, cell types and organs. Due to their flexibility, they are capable of delivering Wnts and other morphogens to target cells in a highly temporally and spatially precise manner.

## 4. Wnt receptors and their regulation

Once Wnts reach the nearby signal-receiving cells, they interact in a palmitoleation-dependent way with a variety of cell-surface receptors including FZDs, co-receptors such as LRP5/6, and ROR2 to activate downstream signaling (Langton et al., 2016; Niehrs, 2012; Routledge & Scholpp, 2019; Yu et al., 2014). Whereas FZDs serve as receptors for most or all Wnt molecules, LRP5/6 are thought to be coreceptors solely for the β-catenin dependent canonical Wnt pathway and other coreceptors like ROR2 activate the noncanonical PCP pathway (MacDonald & He, 2012; Oishi et al., 2003). A cell's Wnt responsiveness is controlled by the level of FZD protein at the plasma membrane. Here, RSPO proteins, a family of secreted Wnt agonists, enhance FZD abundance at the plasma membrane. The RSPOs do this by inhibiting two transmembrane ubiquitin ligases, namely ZNRF3 and RNF43, via direct binding facilitated by co-receptor leucine-rich repeat-containing G protein-coupled receptors (LGR) (Cruciat & Niehrs, 2013; de Lau, Peng, Gros, & Clevers, 2014; Hao, Jiang, & Cong, 2016; Jin & Yoon, 2012). Recent studies including single cell transcriptome studies identified that Wnt producing cell populations can also be the source for RSPO signals (Greicius et al., 2018; Karpus et al., 2019; McCarthy et al., 2020) indicating that

Wnt-producing cells might also control the responsiveness of the target cell. This system is very tightly controlled to keep Wnt signals "just right", as in the Wnt-receiving cell, ZNRF3 and RNF43 are robust Wnt target genes. Activation of Wnt/β-catenin signaling promotes ZNRF3 and RNF43 expression, which in turn inhibit Wnt signaling via ubiquitination and degradation of FZD and LRP6, creating a potent negative feedback loop (Hao et al., 2012; Koo et al., 2012).

When acylated Wnts reach the target cell, they bind to the cysteine-rich extracellular domain (CRD) of FZD. The crystal structures of palmitoleate or acylated Wnts bound to FZD CRDs have been reported and provided some provocative insights (Hirai et al., 2019; Janda et al., 2012; Nile et al., 2017). The palmitoleate moiety of Wnts lies in a hydrophobic groove in the CRD. However, it may be more interesting, as two further studies showed that the palmitoleate may serve to mediate FZD receptor dimerization by bridging two CRD monomers (Nile et al., 2017). The Melcher lab reported the crystal structure of palmitoleic acid in complex with the $FZD_4$ CRD, forming a CRD tetramer (DeBruine et al., 2017). One tetramer consists of two cross-braced CRD dimers which are each stabilized via interactions with the palmitoleic acid. These studies suggest that the palmitoleated Wnts stimulate CRD-CRD interaction and, dimer and higher order oligomer formation, potentially accelerating the formation of the signalosome. This FZD multimerization may explain how disparate Wnts can act together synergically at the plasma membrane to drive downstream signaling (Alok et al., 2017).

## 5. The last mile problem

Several models of Wnt delivery to target cells, including carrier molecules, exosomes and cytonemes, have been described in the literature and reviewed above. As noted, the various mechanisms for secreting Wnts at the plasma membrane to traverse the intercellular space, swinging the palmitoleate from one membrane to bind to the Fzd CRD at the signal receiving cell, remains unresolved. Even the cytoneme model that brings Wnt in very close contact to the receiving cells, does not explain how Wnt can go the very last step. Most commonly, it's thought that WLS releases Wnt once the vesicles carrying the complex fuse with the membrane (Fig. 2). Then, Wnts interact with glypicans or other lipid binding proteins to diffuse toward the recipient cell (Langton, Kakugawa, & Vincent, 2016; McGough et al., 2020; Routledge & Scholpp, 2019; Stanganello & Scholpp, 2016).

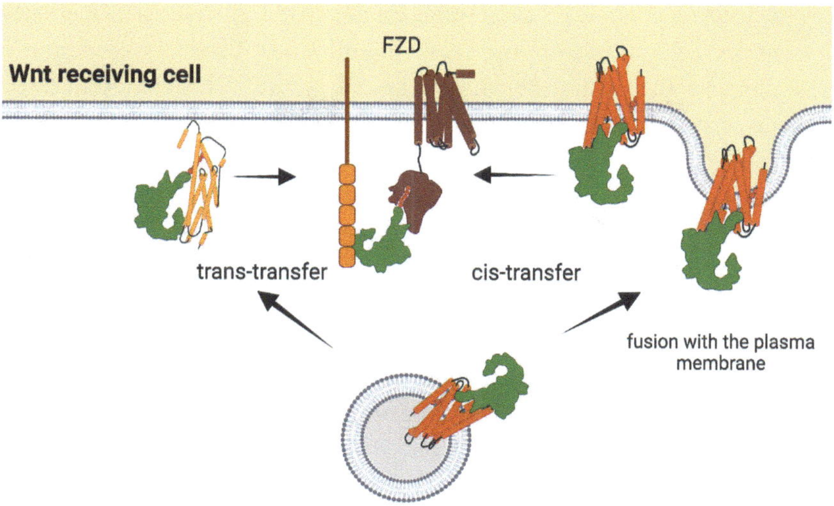


**Fig. 4** Speculative model of Wnt transfer from Wntless to Frizzled. In the trans-transfer model, Wnt is released from WLS on the sending cell and binds to carrier molecules like glypicans that facilitate the transport to Frizzled on the receiving cell. In the cis-transfer model, the vesicle carrying the Wnt-WLS complex fuses with the membrane of the receiving cell. Afterwards, Wnt can be transferred in cis from WLS to FZD. This mechanism may also occur in autocrine Wnt signaling.

This model is called the *trans*-transfer model (Fig. 4) (Nygaard et al., 2021). However, when bound to WLS, the palmitoleate residue is buried deep into the membrane, resulting in a very tight interaction (Coombs et al., 2010). The *trans*-transfer model does not explain how Wnt might be pulled out of the membrane, although conformational shifts of WLS following vesicular acidification have been proposed. Moreover, even supported by glypican or other components of the extracellular matrix (Fig. 4), the actual mode of delivery would be diffusion, albeit on a very local scale. However, if Wnts can be released adjacent to the receiving cells via cytonemes, comparable to synapses, the closely apposed membranes might facilitate the *trans*-transfer model.

As an alternative, the structure of WLS bound to Wnts might be consistent with a *cis*-transfer model (Nygaard et al., 2021). In this model, Wnt is transferred directly from WLS to FZD within the same membrane, without an absolute requirement for other Wnt carrier molecules such as glypicans (Fig. 4). This cis-transfer mechanism may function during autocrine signaling. In paracrine signaling it could be facilitated by fusion of Wnt-containing extracellular vesicles with the plasma membrane of target cells, resulting in both Wnt (carried by WLS or another carrier) and FZD being immediately adjacent to each other in the same membrane. There is evidence in Drosophila for the cis-transfer model, i.e., Wnt and WLS were transferred together across synapses on vesicles such as exosomes (Gross et al., 2012; Koles et al., 2012; Korkut et al., 2009). It is well documented that secreted exosomes can be absorbed by fusion with the plasma membrane of recipient cells to deliver their cargo molecules (Korkut et al., 2009; Thery et al., 2009). Several lines of evidence indicate that components of the extracellular matrix, specifically HSPGs, are involved in the transmission of exosomes to target cells and support the fusion process (Christianson & Belting, 2014; Christianson et al., 2013; Olsson et al., 2001; Stanford et al., 2010; Wilsie & Orlando, 2003)(. Most importantly, this vesicle fusion model can explain how the transfer of Wnts from WLS to FZD takes place. In the *cis*-transfer model, WLS and FZD sit on the same membrane enabling very close interaction of both molecules to facilitate the extraction of Wnt from WLS and the delivery to the receptor. Both WLS and FZD are GPCRs, and heterodimerization of GPCRs is well established (Prinster et al., 2005). Heterodimerization may enable the release of the palmitoleate residue out of the membrane cavity of WLS. Wnt hairpin 3 might function as a conformational switch facilitating a unidirectional transfer of Wnt from WLS to FZD. In the *cis*-transfer model, transfer of acylated Wnts from PORCN to WLS and then from WLS to FZD might take place in a nearly identical fashion, occurring laterally within the plane of the membrane. Moreover, since the WLS-Wnt complex may sit at the plasma membrane after fusion of the intracellular transport vesicles, the model provides a plausible explanation for how autocrine Wnt signaling takes place. Wnt could be transferred directly to an adjacent FZD receptor in a lateral manner both in autocrine and paracrine signaling.

In summary, how Wnts are actually delivered is unknown so far. Both the *cis*- and the *trans*-model might apply, depending on the cellular setup. For short range signaling, e.g., between adjacent cells or from close cytoneme-target cell interaction sites, the *trans*-model might transmit the Wnts. The extracellular vesicle model, which might occur with either cis or *trans*-transfer, could be suitable both for long- and short-range transport.

## 6. Unanswered questions

Despite extensive studies on how Wnts are processed and transported in Wnt producing cells, there are still many open questions. For example, little is known about how Wnts and Wnt-containing extracellular vesicles are guided within the cell to the site of release, e.g., to the tips of cytonemes (Dickins & Salinas, 2013). Since cytoneme formation involves PCP pathway-regulated cytoskeleton organization by actin polymerization and since the cytoskeletal filament systems are commonly involved in the intracellular transport of exosomes (Titus, 2018), it's possible that the release of Wnt-carrying vesicles are controlled in a similar fashion. However, the exact guiding mechanism has not yet been identified. Another unanswered question is, how do Wnt-releasing cells know where and when the Wnts should be delivered to target cells. Wnt signaling in neural development may give an answer to that. Wnts promote synaptogenesis during development and are necessary for synapse maintenance in adult nervous systems (Alvarez & Sabatini, 2007; Holtmaat & Svoboda, 2009). It's well established that Wnt pathway proteins are central to pre- and postsynaptic cell communication critical for the assembly of synaptic connections, i.e., the presynaptic release and the postsynaptic receptive machinery (Koles et al., 2012). It was speculated that postsynaptic release of Wnt signals guide incoming axons and trigger axon growth by modulating the cytoskeleton of presynaptic cells eventually resulting in synapse formation provoking presynaptic Wnt release (Dickins & Salinas, 2013). If similar mechnisms exist in other systems is so far unknown. This mechanism is similar to that described by Mattes and colleagues (Mattes et al., 2018), where the autocrine release of Wnt8a triggered the biogenesis of cytonemes transporting Wnt8a itself via the outgrowing cytonemes to the signal receiving cells.

## 7. Concluding remarks

Wnt signaling is essential in a wide variety of cellular processes from proliferation to cell fate specification occurring in embryonic development and adulthood. The precise delivery of hydrophobic Wnts from sending to receiving cells is important, and disturbances of this system frequently result in aberrant development and homeostasis. Many mechanisms including free diffusion, carrier proteins, extracellular vesicles and cytonemes have been proposed to explain how the palmitoleated Wnts can travel from the sending cell through the lipophobic extracellular space to receiving cells to activate downstream signaling events. Moreover, Wnt delivery must be achieved in a temporally and spatially defined manner. How this complex task is managed is yet unknown. Most likely, the 500 million years of evolution in Wnt signaling has allowed multiple mechanisms to deliver Wnt proteins to their desired site of action that can vary in a tissue-specific and context manner depending on the particular organism. While the potential transport mechanisms regulating the Wnt transport in the extracellular space from one cell to another has been discussed in a mutually exclusive manner it is likely that all these mechanisms including carrier proteins, exosomes and cytonemes act in concert to assure the proper delivery of Wnt to target cells.

## Acknowledgments

The Wnt literature is huge and shows no signs of slowing down. We apologize in advance to our colleagues if we lacked space or insight to cite all appropriate studies. LM was supported by Deutsche Forschungsgemeinschaft (MI 2595/1-1) and DMV is supported by Singapore Translational Research (STaR) Award MOH-000155.

## References

Alfaro, M. P., Vincent, A., Saraswati, S., Thorne, C. A., Hong, C. C., Lee, E., et al. (2010). sFRP2 suppression of bone morphogenic protein (BMP) and Wnt signaling mediates mesenchymal stem cell (MSC) self-renewal promoting engraftment and myocardial repair. *The Journal of Biological Chemistry*, *285*(46), 35645–35653. https://doi.org/10.1074/jbc.M110.135335.

Alok, A., Lei, Z., Jagannathan, N. S., Kaur, S., Harmston, N., Rozen, S. G., et al. (2017). Wnt proteins synergize to activate beta-catenin signaling. *Journal of Cell Science*, *130*(9), 1532–1544. https://doi.org/10.1242/jcs.198093.

Alvarez, V. A., & Sabatini, B. L. (2007). Anatomical and physiological plasticity of dendritic spines. *Annual Review of Neuroscience*, *30*, 79–91. https://doi.org/10.1146/annurev.neuro.30.051606.094222.

Ataman, B., Ashley, J., Gorczyca, M., Ramachandran, P., Fouquet, W., Sigrist, S. J., et al. (2008). Rapid activity-dependent modifications in synaptic structure and function require bidirectional Wnt signaling. *Neuron*, *57*(5), 705–718. https://doi.org/10.1016/j.neuron.2008.01.026.

Audet, M., & Stevens, R. C. (2019). Emerging structural biology of lipid G protein-coupled receptors. *Protein Science*, *28*(2), 292–304. https://doi.org/10.1002/pro.3509.

Baeg, G. H., Lin, X., Khare, N., Baumgartner, S., & Perrimon, N. (2001). Heparan sulfate proteoglycans are critical for the organization of the extracellular distribution of Wingless. *Development*, *128*(1), 87–94.

Baeg, G. H., Selva, E. M., Goodman, R. M., Dasgupta, R., & Perrimon, N. (2004). The Wingless morphogen gradient is established by the cooperative action of Frizzled and Heparan Sulfate Proteoglycan receptors. *Developmental Biology*, *276*(1), 89–100. https://doi.org/10.1016/j.ydbio.2004.08.023.

Baker, M. E. (1998). Albumin's role in steroid hormone action and the origins of vertebrates: Is albumin an essential protein? *FEBS Letters*, *439*(1-2), 9–12. https://doi.org/10.1016/s0014-5793(98)01346-5.

Banziger, C., Soldini, D., Schutt, C., Zipperlen, P., Hausmann, G., & Basler, K. (2006). Wntless, a conserved membrane protein dedicated to the secretion of Wnt proteins from signaling cells. *Cell*, *125*(3), 509–522. https://doi.org/10.1016/j.cell.2006.02.049.

Barrott, J. J., Cash, G. M., Smith, A. P., Barrow, J. R., & Murtaugh, L. C. (2011). Deletion of mouse Porcn blocks Wnt ligand secretion and reveals an ectodermal etiology of human focal dermal hypoplasia/Goltz syndrome. *Proceedings of the National Academy of Sciences of the United States of America*, *108*(31), 12752–12757. https://doi.org/10.1073/pnas.1006437108.

Bartscherer, K., Pelte, N., Ingelfinger, D., & Boutros, M. (2006). Secretion of Wnt ligands requires Evi, a conserved transmembrane protein. *Cell, 125*(3), 523–533. https://doi.org/10.1016/j.cell.2006.04.009.

Bazan, J. F., Janda, C. Y., & Garcia, K. C. (2012). Structural architecture and functional evolution of Wnts. *Developmental Cell, 23*(2), 227–232. https://doi.org/10.1016/j.devcel.2012.07.011.

Belenkaya, T. Y., Wu, Y., Tang, X., Zhou, B., Cheng, L., Sharma, Y. V., et al. (2008). The retromer complex influences Wnt secretion by recycling wntless from endosomes to the trans-Golgi network. *Developmental Cell, 14*(1), 120–131. https://doi.org/10.1016/j.devcel.2007.12.003.

Biechele, S., Cox, B. J., & Rossant, J. (2011). Porcupine homolog is required for canonical Wnt signaling and gastrulation in mouse embryos. *Developmental Biology, 355*(2), 275–285. https://doi.org/10.1016/j.ydbio.2011.04.029.

Binnerts, M. E., Kim, K. A., Bright, J. M., Patel, S. M., Tran, K., Zhou, M., et al. (2007). R-Spondin1 regulates Wnt signaling by inhibiting internalization of LRP6. *Proceedings of the National Academy of Sciences of the United States of America, 104*(37), 14700–14705. https://doi.org/10.1073/pnas.0702305104.

Bischoff, M., Gradilla, A. C., Seijo, I., Andres, G., Rodriguez-Navas, C., Gonzalez-Mendez, L., et al. (2013). Cytonemes are required for the establishment of a normal Hedgehog morphogen gradient in Drosophila epithelia. *Nature Cell Biology, 15*(11), 1269–1281. https://doi.org/10.1038/ncb2856.

Bodeen, W. J., Marada, S., Truong, A., & Ogden, S. K. (2017). A fixation method to preserve cultured cell cytonemes facilitates mechanistic interrogation of morphogen transport. *Development, 144*(19), 3612–3624. https://doi.org/10.1242/dev.152736.

Briscoe, J., & Therond, P. P. (2013). The mechanisms of Hedgehog signalling and its roles in development and disease. *Nature Reviews. Molecular Cell Biology, 14*(7), 416–429. https://doi.org/10.1038/nrm3598.

Brunt, L., Greicius, G., Rogers, S., Evans, B. D., Virshup, D. M., Wedgwood, K. C. A., et al. (2021). Vangl2 promotes the formation of long cytonemes to enable distant Wnt/beta-catenin signaling. *Nature Communications, 12*(1), 2058. https://doi.org/10.1038/s41467-021-22393-9.

Callejo, A., Bilioni, A., Mollica, E., Gorfinkiel, N., Andres, G., Ibanez, C., et al. (2011). Dispatched mediates Hedgehog basolateral release to form the long-range morphogenetic gradient in the Drosophila wing disk epithelium. *Proceedings of the National Academy of Sciences of the United States of America, 108*(31), 12591–12598. https://doi.org/10.1073/pnas.1106881108.

Cantarero Carmona, I., Luesma Bartolome, M. J., & Junquera Escribano, C. (2011). Identification of telocytes in the lamina propria of rat duodenum: Transmission electron microscopy. *Journal of Cellular and Molecular Medicine, 15*(1), 26–30. https://doi.org/10.1111/j.1582-4934.2010.01207.x.

Carpenter, A. C., Rao, S., Wells, J. M., Campbell, K., & Lang, R. A. (2010). Generation of mice with a conditional null allele for Wntless. *Genesis, 48*(9), 554–558. https://doi.org/10.1002/dvg.20651.

Ceafalan, L., Gherghiceanu, M., Popescu, L. M., & Simionescu, O. (2012). Telocytes in human skin—Are they involved in skin regeneration? *Journal of Cellular and Molecular Medicine, 16*(7), 1405–1420. https://doi.org/10.1111/j.1582-4934.2012.01580.x.

Chai, G., Szenker-Ravi, E., Chung, C., Li, Z., Wang, L., Khatoo, M., et al. (2021). A Human Pleiotropic Multiorgan Condition Caused by Deficient Wnt Secretion. *The New England Journal of Medicine, 385*(14), 1292–1301. https://doi.org/10.1056/NEJMoa2033911.

Chen, W., Huang, H., Hatori, R., & Kornberg, T. B. (2017). Essential basal cytonemes take up Hedgehog in the Drosophila wing imaginal disc. *Development, 144*(17), 3134–3144. https://doi.org/10.1242/dev.149856.

Christianson, H. C., & Belting, M. (2014). Heparan sulfate proteoglycan as a cell-surface endocytosis receptor. *Matrix Biology, 35*, 51–55. https://doi.org/10.1016/j.matbio.2013.10.004.
Christianson, H. C., Svensson, K. J., van Kuppevelt, T. H., Li, J. P., & Belting, M. (2013). Cancer cell exosomes depend on cell-surface heparan sulfate proteoglycans for their internalization and functional activity. *Proceedings of the National Academy of Sciences of the United States of America, 110*(43), 17380–17385. https://doi.org/10.1073/pnas.1304266110.
Chung, M. I., Bujnis, M., Barkauskas, C. E., Kobayashi, Y., & Hogan, B. L. M. (2018). Niche-mediated BMP/SMAD signaling regulates lung alveolar stem cell proliferation and differentiation. *Development, 145*(9). https://doi.org/10.1242/dev.163014.
Clague, M. J., Urbe, S., Aniento, F., & Gruenberg, J. (1994). Vacuolar ATPase activity is required for endosomal carrier vesicle formation. *The Journal of Biological Chemistry, 269*(1), 21–24.
Clevers, H. (2013). The intestinal crypt, a prototype stem cell compartment. *Cell, 154*(2), 274–284. https://doi.org/10.1016/j.cell.2013.07.004.
Clevers, H., & Nusse, R. (2012). Wnt/beta-catenin signaling and disease. *Cell, 149*(6), 1192–1205. https://doi.org/10.1016/j.cell.2012.05.012.
Coombs, G. S., Yu, J., Canning, C. A., Veltri, C. A., Covey, T. M., Cheong, J. K., et al. (2010). WLS-dependent secretion of WNT3A requires Ser209 acylation and vacuolar acidification. *Journal of Cell Science, 123*(Pt 19), 3357–3367. https://doi.org/10.1242/jcs.072132.
Corradi, L. S., Jesus, M. M., Fochi, R. A., Vilamaior, P. S., Justulin, L. A., Jr., Goes, R. M., et al. (2013). Structural and ultrastructural evidence for telocytes in prostate stroma. *Journal of Cellular and Molecular Medicine, 17*(3), 398–406. https://doi.org/10.1111/jcmm.12021.
Cretoiu, D., Radu, B. M., Banciu, A., Banciu, D. D., & Cretoiu, S. M. (2017). Telocytes heterogeneity: From cellular morphology to functional evidence. *Seminars in Cell & Developmental Biology, 64*, 26–39. https://doi.org/10.1016/j.semcdb.2016.08.023.
Cruciat, C. M., & Niehrs, C. (2013). Secreted and transmembrane wnt inhibitors and activators. *Cold Spring Harbor Perspectives in Biology, 5*(3). https://doi.org/10.1101/cshperspect.a015081.
Dai, J., Su, Y., Zhong, S., Cong, L., Liu, B., Yang, J., et al. (2020). Exosomes: Key players in cancer and potential therapeutic strategy. *Signal Transduction and Targeted Therapy, 5*(1), 145. https://doi.org/10.1038/s41392-020-00261-0.
de Lau, W., Peng, W. C., Gros, P., & Clevers, H. (2014). The R-spondin/Lgr5/Rnf43 module: Regulator of Wnt signal strength. *Genes & Development, 28*(4), 305–316. https://doi.org/10.1101/gad.235473.113.
Deane, H. W. (1964). Some electron microscopic observations on the Lamina propria of the gut, with comments on the close association of macrophages, plasma cells, and eosinophils. *The Anatomical Record, 149*, 453–473. https://doi.org/10.1002/ar.1091490315.
DeBruine, Z. J., Ke, J., Harikumar, K. G., Gu, X., Borowsky, P., Williams, B. O., et al. (2017). Wnt5a promotes Frizzled-4 signalosome assembly by stabilizing cysteine-rich domain dimerization. *Genes & Development, 31*(9), 916–926. https://doi.org/10.1101/gad.298331.117.
Dickins, E. M., & Salinas, P. C. (2013). Wnts in action: From synapse formation to synaptic maintenance. *Frontiers in Cellular Neuroscience, 7*, 162. https://doi.org/10.3389/fncel.2013.00162.
Dickinson, M. E., Flenniken, A. M., Ji, X., Teboul, L., Wong, M. D., White, J. K., et al. (2016). High-throughput discovery of novel developmental phenotypes. *Nature, 537*(7621), 508–514. https://doi.org/10.1038/nature19356.

Dieplinger, H., & Dieplinger, B. (2015). Afamin—A pleiotropic glycoprotein involved in various disease states. *Clinica Chimica Acta, 446*, 105–110. https://doi.org/10.1016/j.cca.2015.04.010.

Doumpas, N., Lampart, F., Robinson, M. D., Lentini, A., Nestor, C. E., Cantu, C., et al. (2019). TCF/LEF dependent and independent transcriptional regulation of Wnt/beta-catenin target genes. *The EMBO Journal, 38*(2). https://doi.org/10.15252/embj.201798873.

Durand, A., Donahue, B., Peignon, G., Letourneur, F., Cagnard, N., Slomianny, C., et al. (2012). Functional intestinal stem cells after Paneth cell ablation induced by the loss of transcription factor Math1 (Atoh1). *Proceedings of the National Academy of Sciences of the United States of America, 109*(23), 8965–8970. https://doi.org/10.1073/pnas.1201652109.

El Maadawi, Z. M. (2016). A tale of two cells: Telocyte and stem cell unique relationship. *Advances in Experimental Medicine and Biology, 913*, 359–376. https://doi.org/10.1007/978-981-10-1061-3_23.

Farin, H. F., Jordens, I., Mosa, M. H., Basak, O., Korving, J., Tauriello, D. V., et al. (2016). Visualization of a short-range Wnt gradient in the intestinal stem-cell niche. *Nature, 530*(7590), 340–343. https://doi.org/10.1038/nature16937.

Farin, H. F., Van Es, J. H., & Clevers, H. (2012). Redundant sources of Wnt regulate intestinal stem cells and promote formation of Paneth cells. *Gastroenterology, 143*(6), 1518–1529 e1517. https://doi.org/10.1053/j.gastro.2012.08.031.

Franch-Marro, X., Marchand, O., Piddini, E., Ricardo, S., Alexandre, C., & Vincent, J. P. (2005). Glypicans shunt the Wingless signal between local signalling and further transport. *Development, 132*(4), 659–666. https://doi.org/10.1242/dev.01639.

Fu, J., Jiang, M., Mirando, A. J., Yu, H. M., & Hsu, W. (2009). Reciprocal regulation of Wnt and Gpr177/mouse Wntless is required for embryonic axis formation. *Proceedings of the National Academy of Sciences of the United States of America, 106*(44), 18598–18603. https://doi.org/10.1073/pnas.0904894106.

Fuerer, C., Habib, S. J., & Nusse, R. (2010). A study on the interactions between heparan sulfate proteoglycans and Wnt proteins. *Developmental Dynamics, 239*(1), 184–190. https://doi.org/10.1002/dvdy.22067.

Galli, L. M., Barnes, T., Cheng, T., Acosta, L., Anglade, A., Willert, K., et al. (2006). Differential inhibition of Wnt-3a by Sfrp-1, Sfrp-2, and Sfrp-3. *Developmental Dynamics, 235*(3), 681–690. https://doi.org/10.1002/dvdy.20681.

Gherghiceanu, M., Manole, C. G., & Popescu, L. M. (2010). Telocytes in endocardium: Electron microscope evidence. *Journal of Cellular and Molecular Medicine, 14*(9), 2330–2334. https://doi.org/10.1111/j.1582-4934.2010.01133.x.

Goodman, R. M., Thombre, S., Firtina, Z., Gray, D., Betts, D., Roebuck, J., et al. (2006). Sprinter: A novel transmembrane protein required for Wg secretion and signaling. *Development, 133*(24), 4901–4911. https://doi.org/10.1242/dev.02674.

Gradilla, A. C., Gonzalez, E., Seijo, I., Andres, G., Bischoff, M., Gonzalez-Mendez, L., et al. (2014). Exosomes as Hedgehog carriers in cytoneme-mediated transport and secretion. *Nature Communications, 5*, 5649. https://doi.org/10.1038/ncomms6649.

Greco, V., Hannus, M., & Eaton, S. (2001). Argosomes: A potential vehicle for the spread of morphogens through epithelia. *Cell, 106*(5), 633–645. https://doi.org/10.1016/s0092-8674(01)00484-6.

Gregorieff, A., Pinto, D., Begthel, H., Destree, O., Kielman, M., & Clevers, H. (2005). Expression pattern of Wnt signaling components in the adult intestine. *Gastroenterology, 129*(2), 626–638. https://doi.org/10.1016/j.gastro.2005.06.007.

Greicius, G., Kabiri, Z., Sigmundsson, K., Liang, C., Bunte, R., Singh, M. K., et al. (2018). PDGFRalpha(+) pericryptal stromal cells are the critical source of Wnts and RSPO3 for murine intestinal stem cells in vivo. *Proceedings of the National Academy of*

*Sciences of the United States of America, 115*(14), E3173–E3181. https://doi.org/10.1073/pnas.1713510115.
Gross, J. C., Chaudhary, V., Bartscherer, K., & Boutros, M. (2012). Active Wnt proteins are secreted on exosomes. *Nature Cell Biology, 14*(10), 1036–1045. https://doi.org/10.1038/ncb2574.
Grumolato, L., Liu, G., Mong, P., Mudbhary, R., Biswas, R., Arroyave, R., et al. (2010). Canonical and noncanonical Wnts use a common mechanism to activate completely unrelated coreceptors. *Genes & Development, 24*(22), 2517–2530. https://doi.org/10.1101/gad.1957710.
Grzeschik, K. H., Bornholdt, D., Oeffner, F., Konig, A., del Carmen Boente, M., Enders, H., et al. (2007). Deficiency of PORCN, a regulator of Wnt signaling, is associated with focal dermal hypoplasia. *Nature Genetics, 39*(7), 833–835. https://doi.org/10.1038/ng2052.
Guder, C., Philipp, I., Lengfeld, T., Watanabe, H., Hobmayer, B., & Holstein, T. W. (2006). The Wnt code: Cnidarians signal the way. *Oncogene, 25*(57), 7450–7460. https://doi.org/10.1038/sj.onc.1210052.
Hall, E. T., Cleverdon, E. R., & Ogden, S. K. (2019). Dispatching sonic hedgehog: Molecular mechanisms controlling deployment. *Trends in Cell Biology, 29*(5), 385–395. https://doi.org/10.1016/j.tcb.2019.02.005.
Hall, E. T., Dillard, M. E., Stewart, D. P., Zhang, Y., Wagner, B., Levine, R. M., et al. (2021). Cytoneme delivery of Sonic Hedgehog from ligand-producing cells requires Myosin 10 and a Dispatched-BOC/CDON co-receptor complex. *eLife, 10*. https://doi.org/10.7554/eLife.61432.
Han, C., Yan, D., Belenkaya, T. Y., & Lin, X. (2005). Drosophila glypicans Dally and Dally-like shape the extracellular Wingless morphogen gradient in the wing disc. *Development, 132*(4), 667–679. https://doi.org/10.1242/dev.01636.
Hannoush, R. N. (2015). Synthetic protein lipidation. *Current Opinion in Chemical Biology, 28*, 39–46. https://doi.org/10.1016/j.cbpa.2015.05.025.
Hao, H. X., Jiang, X., & Cong, F. (2016). Control of Wnt receptor turnover by R-spondin-ZNRF3/RNF43 signaling module and its dysregulation in cancer. *Cancers (Basel), 8*(6). https://doi.org/10.3390/cancers8060054.
Hao, H. X., Xie, Y., Zhang, Y., Charlat, O., Oster, E., Avello, M., et al. (2012). ZNRF3 promotes Wnt receptor turnover in an R-spondin-sensitive manner. *Nature, 485*(7397), 195–200. https://doi.org/10.1038/nature11019.
Heinosalo, T., Gabriel, M., Kallio, L., Adhikari, P., Huhtinen, K., Laajala, T. D., et al. (2018). Secreted frizzled-related protein 2 (SFRP2) expression promotes lesion proliferation via canonical WNT signaling and indicates lesion borders in extraovarian endometriosis. *Human Reproduction, 33*(5), 817–831. https://doi.org/10.1093/humrep/dey026.
Herr, P., & Basler, K. (2012). Porcupine-mediated lipidation is required for Wnt recognition by Wls. *Developmental Biology, 361*(2), 392–402. https://doi.org/10.1016/j.ydbio.2011.11.003.
Hirai, H., Matoba, K., Mihara, E., Arimori, T., & Takagi, J. (2019). Crystal structure of a mammalian Wnt-frizzled complex. *Nature Structural & Molecular Biology, 26*(5), 372–379. https://doi.org/10.1038/s41594-019-0216-z.
Ho, H. Y., Susman, M. W., Bikoff, J. B., Ryu, Y. K., Jonas, A. M., Hu, L., et al. (2012). Wnt5a-Ror-Dishevelled signaling constitutes a core developmental pathway that controls tissue morphogenesis. *Proceedings of the National Academy of Sciences of the United States of America, 109*(11), 4044–4051. https://doi.org/10.1073/pnas.1200421109.
Hoang, B., Moos, M., Jr., Vukicevic, S., & Luyten, F. P. (1996). Primary structure and tissue distribution of FRZB, a novel protein related to Drosophila frizzled, suggest a role in skeletal morphogenesis. *The Journal of Biological Chemistry, 271*(42), 26131–26137. https://doi.org/10.1074/jbc.271.42.26131.

Holtmaat, A., & Svoboda, K. (2009). Experience-dependent structural synaptic plasticity in the mammalian brain. *Nature Reviews. Neuroscience*, *10*(9), 647–658. https://doi.org/10.1038/nrn2699.

Holzer, T., Liffers, K., Rahm, K., Trageser, B., Ozbek, S., & Gradl, D. (2012). Live imaging of active fluorophore labelled Wnt proteins. *FEBS Letters*, *586*(11), 1638–1644. https://doi.org/10.1016/j.febslet.2012.04.035.

Houart, C., Caneparo, L., Heisenberg, C., Barth, K., Take-Uchi, M., & Wilson, S. (2002). Establishment of the telencephalon during gastrulation by local antagonism of Wnt signaling. *Neuron*, *35*(2), 255–265. https://doi.org/10.1016/s0896-6273(02)00751-1.

Hu, Y. B., Yan, C., Mu, L., Mi, Y. L., Zhao, H., Hu, H., et al. (2019). Exosomal Wnt-induced dedifferentiation of colorectal cancer cells contributes to chemotherapy resistance. *Oncogene*, *38*(11), 1951–1965. https://doi.org/10.1038/s41388-018-0557-9.

Hua, Y., Yang, Y., Li, Q., He, X., Zhu, W., Wang, J., et al. (2018). Oligomerization of Frizzled and LRP5/6 protein initiates intracellular signaling for the canonical WNT/beta-catenin pathway. *The Journal of Biological Chemistry*, *293*(51), 19710–19724. https://doi.org/10.1074/jbc.RA118.004434.

Huang, H., & Kornberg, T. B. (2015). Myoblast cytonemes mediate Wg signaling from the wing imaginal disc and Delta-Notch signaling to the air sac primordium. *eLife*, *4*, e06114. https://doi.org/10.7554/eLife.06114.

Janda, C. Y., Waghray, D., Levin, A. M., Thomas, C., & Garcia, K. C. (2012). Structural basis of Wnt recognition by Frizzled. *Science*, *337*(6090), 59–64. https://doi.org/10.1126/science.1222879.

Jin, Y. R., & Yoon, J. K. (2012). The R-spondin family of proteins: Emerging regulators of WNT signaling. *The International Journal of Biochemistry & Cell Biology*, *44*(12), 2278–2287. https://doi.org/10.1016/j.biocel.2012.09.006.

Junyent, S., Garcin, C. L., Szczerkowski, J. L. A., Trieu, T. J., Reeves, J., & Habib, S. J. (2020). Specialized cytonemes induce self-organization of stem cells. *Proceedings of the National Academy of Sciences of the United States of America*, *117*(13), 7236–7244. https://doi.org/10.1073/pnas.1920837117.

Kabiri, Z., Greicius, G., Madan, B., Biechele, S., Zhong, Z., Zaribafzadeh, H., et al. (2014). Stroma provides an intestinal stem cell niche in the absence of epithelial Wnts. *Development*, *141*(11), 2206–2215. https://doi.org/10.1242/dev.104976.

Kadowaki, T., Wilder, E., Klingensmith, J., Zachary, K., & Perrimon, N. (1996). The segment polarity gene porcupine encodes a putative multitransmembrane protein involved in Wingless processing. *Genes & Development*, *10*(24), 3116–3128. https://doi.org/10.1101/gad.10.24.3116.

Kaiser, K., Gyllborg, D., Prochazka, J., Salasova, A., Kompanikova, P., Molina, F. L., et al. (2019). WNT5A is transported via lipoprotein particles in the cerebrospinal fluid to regulate hindbrain morphogenesis. *Nature Communications*, *10*(1), 1498. https://doi.org/10.1038/s41467-019-09298-4.

Karpus, O. N., Westendorp, B. F., Vermeulen, J. L. M., Meisner, S., Koster, J., Muncan, V., et al. (2019). Colonic CD90+ Crypt Fibroblasts Secrete Semaphorins to Support Epithelial Growth. *Cell Reports*, *26*(13), 3698–3708 e3695. https://doi.org/10.1016/j.celrep.2019.02.101.

Kazanskaya, O., Glinka, A., del Barco Barrantes, I., Stannek, P., Niehrs, C., & Wu, W. (2004). R-Spondin2 is a secreted activator of Wnt/beta-catenin signaling and is required for Xenopus myogenesis. *Developmental Cell*, *7*(4), 525–534. https://doi.org/10.1016/j.devcel.2004.07.019.

Kirkpatrick, C. A., & Selleck, S. B. (2007). Heparan sulfate proteoglycans at a glance. *Journal of Cell Science*, *120*(Pt 11), 1829–1832. https://doi.org/10.1242/jcs.03432.

Koizumi, K., Takano, K., Kaneyasu, A., Watanabe-Takano, H., Tokuda, E., Abe, T., et al. (2012). RhoD activated by fibroblast growth factor induces cytoneme-like cellular protrusions through mDia3C. *Molecular Biology of the Cell, 23*(23), 4647–4661. https://doi.org/10.1091/mbc.E12-04-0315.

Koles, K., Nunnari, J., Korkut, C., Barria, R., Brewer, C., Li, Y., et al. (2012). Mechanism of evenness interrupted (Evi)-exosome release at synaptic boutons. *The Journal of Biological Chemistry, 287*(20), 16820–16834. https://doi.org/10.1074/jbc.M112.342667.

Komekado, H., Yamamoto, H., Chiba, T., & Kikuchi, A. (2007). Glycosylation and palmitoylation of Wnt-3a are coupled to produce an active form of Wnt-3a. *Genes to Cells, 12*(4), 521–534. https://doi.org/10.1111/j.1365-2443.2007.01068.x.

Koo, B. K., Spit, M., Jordens, I., Low, T. Y., Stange, D. E., van de Wetering, M., et al. (2012). Tumour suppressor RNF43 is a stem-cell E3 ligase that induces endocytosis of Wnt receptors. *Nature, 488*(7413), 665–669. https://doi.org/10.1038/nature11308.

Korkut, C., Ataman, B., Ramachandran, P., Ashley, J., Barria, R., Gherbesi, N., et al. (2009). *Trans*-synaptic transmission of vesicular Wnt signals through Evi/Wntless. *Cell, 139*(2), 393–404. https://doi.org/10.1016/j.cell.2009.07.051.

Kornberg, T. B., & Roy, S. (2014). Cytonemes as specialized signaling filopodia. *Development, 141*(4), 729–736. https://doi.org/10.1242/dev.086223.

Kress, E., Rezza, A., Nadjar, J., Samarut, J., & Plateroti, M. (2009). The frizzled-related sFRP2 gene is a target of thyroid hormone receptor alpha1 and activates beta-catenin signaling in mouse intestine. *The Journal of Biological Chemistry, 284*(2), 1234–1241. https://doi.org/10.1074/jbc.M806548200.

Langton, P. F., Kakugawa, S., & Vincent, J. P. (2016). Making, Exporting, and Modulating Wnts. *Trends in Cell Biology, 26*(10), 756–765. https://doi.org/10.1016/j.tcb.2016.05.011.

Leyns, L., Bouwmeester, T., Kim, S. H., Piccolo, S., & De Robertis, E. M. (1997). Frzb-1 is a secreted antagonist of Wnt signaling expressed in the Spemann organizer. *Cell, 88*(6), 747–756. https://doi.org/10.1016/s0092-8674(00)81921-2.

Li, B., & Niswander, L. A. (2020). TMEM132A, a novel Wnt signaling pathway regulator through wntless (WLS) interaction. *Frontiers in Cell and Development Biology, 8*. https://doi.org/10.3389/fcell.2020.599890.

Li, L., Lin, M., Li, L., Wang, R., Zhang, C., Qi, G., et al. (2014). Renal telocytes contribute to the repair of ischemically injured renal tubules. *Journal of Cellular and Molecular Medicine, 18*(6), 1144–1156. https://doi.org/10.1111/jcmm.12274.

Lidke, D. S., Lidke, K. A., Rieger, B., Jovin, T. M., & Arndt-Jovin, D. J. (2005). Reaching out for signals: Filopodia sense EGF and respond by directed retrograde transport of activated receptors. *The Journal of Cell Biology, 170*(4), 619–626. https://doi.org/10.1083/jcb.200503140.

Ling, I. T., Rochard, L., & Liao, E. C. (2017). Distinct requirements of wls, wnt9a, wnt5b and gpc4 in regulating chondrocyte maturation and timing of endochondral ossification. *Developmental Biology, 421*(2), 219–232. https://doi.org/10.1016/j.ydbio.2016.11.016.

Lipkin, M. (1965). Cell replication in the gastrointestinal tract of man. *Gastroenterology, 48*, 616–624.

Luga, V., Zhang, L., Viloria-Petit, A. M., Ogunjimi, A. A., Inanlou, M. R., Chiu, E., et al. (2012). Exosomes mediate stromal mobilization of autocrine Wnt-PCP signaling in breast cancer cell migration. *Cell, 151*(7), 1542–1556. https://doi.org/10.1016/j.cell.2012.11.024.

Lum, L., & Clevers, H. (2012). Cell biology. The unusual case of Porcupine. *Science, 337*(6097), 922–923. https://doi.org/10.1126/science.1228179.

Luz, M., Spannl-Muller, S., Ozhan, G., Kagermeier-Schenk, B., Rhinn, M., Weidinger, G., et al. (2014). Dynamic association with donor cell filopodia and lipid-modification are essential features of Wnt8a during patterning of the zebrafish neuroectoderm. *PLoS One, 9*(1), e84922. https://doi.org/10.1371/journal.pone.0084922.

Maas, S. L. N., Breakefield, X. O., & Weaver, A. M. (2017). Extracellular vesicles: Unique intercellular delivery vehicles. *Trends in Cell Biology, 27*(3), 172–188. https://doi.org/10.1016/j.tcb.2016.11.003.

MacDonald, B. T., & He, X. (2012). Frizzled and LRP5/6 receptors for Wnt/beta-catenin signaling. *Cold Spring Harbor Perspectives in Biology*, *4*(12). https://doi.org/10.1101/cshperspect.a007880.

MacDonald, B. T., Hien, A., Zhang, X., Iranloye, O., Virshup, D. M., Waterman, M. L., et al. (2014). Disulfide bond requirements for active Wnt ligands. *The Journal of Biological Chemistry*, *289*(26), 18122–18136. https://doi.org/10.1074/jbc.M114.575027.

Madan, B., Walker, M. P., Young, R., Quick, L., Orgel, K. A., Ryan, M., et al. (2016). USP6 oncogene promotes Wnt signaling by deubiquitylating Frizzleds. *Proceedings of the National Academy of Sciences of the United States of America*, *113*(21), E2945–E2954. https://doi.org/10.1073/pnas.1605691113.

Mastri, M., Shah, Z., Hsieh, K., Wang, X., Wooldridge, B., Martin, S., et al. (2014). Secreted Frizzled-related protein 2 as a target in antifibrotic therapeutic intervention. *American Journal of Physiology. Cell Physiology*, *306*(6), C531–C539. https://doi.org/10.1152/ajpcell.00238.2013.

Mattes, B., Dang, Y., Greicius, G., Kaufmann, L. T., Prunsche, B., Rosenbauer, J., et al. (2018). Wnt/PCP controls spreading of Wnt/beta-catenin signals by cytonemes in vertebrates. *eLife*, *7*. https://doi.org/10.7554/eLife.36953.

McCarthy, N., Manieri, E., Storm, E. E., Saadatpour, A., Luoma, A. M., Kapoor, V. N., et al. (2020). Distinct mesenchymal cell populations generate the essential intestinal BMP signaling gradient. *Cell Stem Cell*, *26*(3), 391–402 e395. https://doi.org/10.1016/j.stem.2020.01.008.

McGough, I. J., Vecchia, L., Bishop, B., Malinauskas, T., Beckett, K., Joshi, D., et al. (2020). Glypicans shield the Wnt lipid moiety to enable signalling at a distance. *Nature*, *585*(7823), 85–90. https://doi.org/10.1038/s41586-020-2498-z.

McNew, J. A., Sogaard, M., Lampen, N. M., Machida, S., Ye, R. R., Lacomis, L., et al. (1997). Ykt6p, a prenylated SNARE essential for endoplasmic reticulum-Golgi transport. *The Journal of Biological Chemistry*, *272*(28), 17776–17783. https://doi.org/10.1074/jbc.272.28.17776.

Meiringer, C. T., Auffarth, K., Hou, H., & Ungermann, C. (2008). Depalmitoylation of Ykt6 prevents its entry into the multivesicular body pathway. *Traffic*, *9*(9), 1510–1521. https://doi.org/10.1111/j.1600-0854.2008.00778.x.

Mihara, E., Hirai, H., Yamamoto, H., Tamura-Kawakami, K., Matano, M., Kikuchi, A., et al. (2016). Active and water-soluble form of lipidated Wnt protein is maintained by a serum glycoprotein afamin/alpha-albumin. *eLife*, *5*. https://doi.org/10.7554/eLife.11621.

Mii, Y., & Taira, M. (2009). Secreted Frizzled-related proteins enhance the diffusion of Wnt ligands and expand their signalling range. *Development*, *136*(24), 4083–4088. https://doi.org/10.1242/dev.032524.

Misra, K., & Matise, M. P. (2010). A critical role for sFRP proteins in maintaining caudal neural tube closure in mice via inhibition of BMP signaling. *Developmental Biology*, *337*(1), 74–83. https://doi.org/10.1016/j.ydbio.2009.10.015.

Morello, R., Bertin, T. K., Schlaubitz, S., Shaw, C. A., Kakuru, S., Munivez, E., et al. (2008). Brachy-syndactyly caused by loss of Sfrp2 function. *Journal of Cellular Physiology*, *217*(1), 127–137. https://doi.org/10.1002/jcp.21483.

Moti, N., Yu, J., Boncompain, G., Perez, F., & Virshup, D. M. (2019). Wnt traffic from endoplasmic reticulum to filopodia. *PLoS One*, *14*(2), e0212711. https://doi.org/10.1371/journal.pone.0212711.

Mulligan, K. A., Fuerer, C., Ching, W., Fish, M., Willert, K., & Nusse, R. (2012). Secreted Wingless-interacting molecule (Swim) promotes long-range signaling by maintaining Wingless solubility. *Proceedings of the National Academy of Sciences of the United States of America*, *109*(2), 370–377. https://doi.org/10.1073/pnas.1119197109.

Nabhan, A. N., Brownfield, D. G., Harbury, P. B., Krasnow, M. A., & Desai, T. J. (2018). Single-cell Wnt signaling niches maintain stemness of alveolar type 2 cells. *Science, 359*(6380), 1118–1123. https://doi.org/10.1126/science.aam6603.

Najdi, R., Proffitt, K., Sprowl, S., Kaur, S., Yu, J., Covey, T. M., et al. (2012). A uniform human Wnt expression library reveals a shared secretory pathway and unique signaling activities. *Differentiation, 84*(2), 203–213. https://doi.org/10.1016/j.diff.2012.06.004.

Naschberger, A., Orry, A., Lechner, S., Bowler, M. W., Nurizzo, D., Novokmet, M., et al. (2017). Structural evidence for a role of the multi-functional human glycoprotein afamin in Wnt transport. *Structure, 25*(12), 1907–1915 e1905. https://doi.org/10.1016/j.str.2017.10.006.

Neumann, S., Coudreuse, D. Y., van der Westhuyzen, D. R., Eckhardt, E. R., Korswagen, H. C., Schmitz, G., et al. (2009). Mammalian Wnt3a is released on lipoprotein particles. *Traffic, 10*(3), 334–343. https://doi.org/10.1111/j.1600-0854.2008.00872.x.

Niehrs, C. (2012). The complex world of WNT receptor signalling. *Nature Reviews. Molecular Cell Biology, 13*(12), 767–779. https://doi.org/10.1038/nrm3470.

Nile, A. H., Mukund, S., Stanger, K., Wang, W., & Hannoush, R. N. (2017). Unsaturated fatty acyl recognition by Frizzled receptors mediates dimerization upon Wnt ligand binding. *Proceedings of the National Academy of Sciences of the United States of America, 114*(16), 4147–4152. https://doi.org/10.1073/pnas.1618293114.

Nusse, R., & Clevers, H. (2017). Wnt/beta-catenin signaling, disease, and emerging therapeutic modalities. *Cell, 169*(6), 985–999. https://doi.org/10.1016/j.cell.2017.05.016.

Nygaard, R., Yu, J., Kim, J., Ross, D. R., Parisi, G., Clarke, O. B., et al. (2021). Structural basis of WLS/Evi-mediated Wnt transport and secretion. *Cell, 184*(1), 194–206 e114. https://doi.org/10.1016/j.cell.2020.11.038.

Ohkawara, B., Yamamoto, T. S., Tada, M., & Ueno, N. (2003). Role of glypican 4 in the regulation of convergent extension movements during gastrulation in *Xenopus laevis*. *Development, 130*(10), 2129–2138. https://doi.org/10.1242/dev.00435.

Oishi, I., Suzuki, H., Onishi, N., Takada, R., Kani, S., Ohkawara, B., et al. (2003). The receptor tyrosine kinase Ror2 is involved in non-canonical Wnt5a/JNK signalling pathway. *Genes to Cells, 8*(7), 645–654. https://doi.org/10.1046/j.1365-2443.2003.00662.x.

Olsson, U., Egnell, A. C., Lee, M. R., Lunden, G. O., Lorentzon, M., Salmivirta, M., et al. (2001). Changes in matrix proteoglycans induced by insulin and fatty acids in hepatic cells may contribute to dyslipidemia of insulin resistance. *Diabetes, 50*(9), 2126–2132. https://doi.org/10.2337/diabetes.50.9.2126.

Panakova, D., Sprong, H., Marois, E., Thiele, C., & Eaton, S. (2005). Lipoprotein particles are required for Hedgehog and Wingless signalling. *Nature, 435*(7038), 58–65. https://doi.org/10.1038/nature03504.

Pani, A. M., & Goldstein, B. (2018). Direct visualization of a native Wnt in vivo reveals that a long-range Wnt gradient forms by extracellular dispersal. *eLife*, 7. https://doi.org/10.7554/eLife.38325.

Pepinsky, R. B., Zeng, C., Wen, D., Rayhorn, P., Baker, D. P., Williams, K. P., et al. (1998). Identification of a palmitic acid-modified form of human Sonic hedgehog. *The Journal of Biological Chemistry, 273*(22), 14037–14045. https://doi.org/10.1074/jbc.273.22.14037.

Peradziryi, H., Kaplan, N. A., Podleschny, M., Liu, X., Wehner, P., Borchers, A., et al. (2011). PTK7/Otk interacts with Wnts and inhibits canonical Wnt signalling. *The EMBO Journal, 30*(18), 3729–3740. https://doi.org/10.1038/emboj.2011.236.

Pfeiffer, S., Ricardo, S., Manneville, J. B., Alexandre, C., & Vincent, J. P. (2002). Producing cells retain and recycle Wingless in Drosophila embryos. *Current Biology, 12*(11), 957–962. https://doi.org/10.1016/s0960-9822(02)00867-9.

Popescu, L. M., Gherghiceanu, M., Suciu, L. C., Manole, C. G., & Hinescu, M. E. (2011). Telocytes and putative stem cells in the lungs: Electron microscopy, electron tomography and laser scanning microscopy. *Cell and Tissue Research*, *345*(3), 391–403. https://doi.org/10.1007/s00441-011-1229-z.

Popescu, L. M., Manole, C. G., Gherghiceanu, M., Ardelean, A., Nicolescu, M. I., Hinescu, M. E., et al. (2010). Telocytes in human epicardium. *Journal of Cellular and Molecular Medicine*, *14*(8), 2085–2093. https://doi.org/10.1111/j.1582-4934.2010.01129.x.

Port, F., & Basler, K. (2010). Wnt trafficking: New insights into Wnt maturation, secretion and spreading. *Traffic*, *11*(10), 1265–1271. https://doi.org/10.1111/j.1600-0854.2010.01076.x.

Port, F., Kuster, M., Herr, P., Furger, E., Banziger, C., Hausmann, G., et al. (2008). Wingless secretion promotes and requires retromer-dependent cycling of Wntless. *Nature Cell Biology*, *10*(2), 178–185. https://doi.org/10.1038/ncb1687.

Porter, J. A., Young, K. E., & Beachy, P. A. (1996). Cholesterol modification of hedgehog signaling proteins in animal development. *Science*, *274*(5285), 255–259. https://doi.org/10.1126/science.274.5285.255.

Potten, C. S., Kovacs, L., & Hamilton, E. (1974). Continuous labelling studies on mouse skin and intestine. *Cell and Tissue Kinetics*, 7(3), 271–283.

Prinster, S. C., Hague, C., & Hall, R. A. (2005). Heterodimerization of g protein-coupled receptors: Specificity and functional significance. *Pharmacological Reviews*, *57*(3), 289–298. https://doi.org/10.1124/pr.57.3.1.

Qi, G., Lin, M., Xu, M., Manole, C. G., Wang, X., & Zhu, T. (2012). Telocytes in the human kidney cortex. *Journal of Cellular and Molecular Medicine*, *16*(12), 3116–3122. https://doi.org/10.1111/j.1582-4934.2012.01582.x.

Ramirez-Weber, F. A., & Kornberg, T. B. (1999). Cytonemes: Cellular processes that project to the principal signaling center in Drosophila imaginal discs. *Cell*, *97*(5), 599–607. https://doi.org/10.1016/s0092-8674(00)80771-0.

Rios-Esteves, J., & Resh, M. D. (2013). Stearoyl CoA desaturase is required to produce active, lipid-modified Wnt proteins. *Cell Reports*, *4*(6), 1072–1081. https://doi.org/10.1016/j.celrep.2013.08.027.

Rojas-Rios, P., Guerrero, I., & Gonzalez-Reyes, A. (2012). Cytoneme-mediated delivery of hedgehog regulates the expression of bone morphogenetic proteins to maintain germline stem cells in Drosophila. *PLoS Biology*, *10*(4), e1001298. https://doi.org/10.1371/journal.pbio.1001298.

Routledge, D., & Scholpp, S. (2019). Mechanisms of intercellular Wnt transport. *Development*, *146*(10). https://doi.org/10.1242/dev.176073.

Roy, S., Hsiung, F., & Kornberg, T. B. (2011). Specificity of Drosophila cytonemes for distinct signaling pathways. *Science*, *332*(6027), 354–358. https://doi.org/10.1126/science.1198949.

Ruivo, C. F., Adem, B., Silva, M., & Melo, S. A. (2017). The biology of cancer exosomes: Insights and new perspectives. *Cancer Research*, 77(23), 6480–6488. https://doi.org/10.1158/0008-5472.CAN-17-0994.

Sagar, P. F., Wiegreffe, C., & Scaal, M. (2015). Communication between distant epithelial cells by filopodia-like protrusions during embryonic development. *Development*, *142*(4), 665–671. https://doi.org/10.1242/dev.115964.

Sakane, H., Yamamoto, H., Matsumoto, S., Sato, A., & Kikuchi, A. (2012). Localization of glypican-4 in different membrane microdomains is involved in the regulation of Wnt signaling. *Journal of Cell Science*, *125*(Pt 2), 449–460. https://doi.org/10.1242/jcs.091876.

Sanders, T. A., Llagostera, E., & Barna, M. (2013). Specialized filopodia direct long-range transport of SHH during vertebrate tissue patterning. *Nature*, *497*(7451), 628–632. https://doi.org/10.1038/nature12157.

Sasaki, N., Sachs, N., Wiebrands, K., Ellenbroek, S. I., Fumagalli, A., Lyubimova, A., et al. (2016). Reg4 + deep crypt secretory cells function as epithelial niche for Lgr5 + stem cells in colon. *Proceedings of the National Academy of Sciences of the United States of America*, *113*(37), E5399–E5407. https://doi.org/10.1073/pnas.1607327113.

Sato, T., van Es, J. H., Snippert, H. J., Stange, D. E., Vries, R. G., van den Born, M., et al. (2011). Paneth cells constitute the niche for Lgr5 stem cells in intestinal crypts. *Nature*, *469*(7330), 415–418. https://doi.org/10.1038/nature09637.

Schlessinger, K., Hall, A., & Tolwinski, N. (2009). Wnt signaling pathways meet Rho GTPases. *Genes & Development*, *23*(3), 265–277. https://doi.org/10.1101/gad.1760809.

Schubert, A., & Boutros, M. (2021). Extracellular vesicles and oncogenic signaling. *Molecular Oncology*, *15*(1), 3–26. https://doi.org/10.1002/1878-0261.12855.

Servage, K. A., Stefanius, K., Gray, H. F., & Orth, K. (2020). Proteomic profiling of small extracellular vesicles secreted by human pancreatic cancer cells implicated in cellular transformation. *Scientific Reports*, *10*(1), 7713. https://doi.org/10.1038/s41598-020-64718-6.

Sharma, R. P., & Chopra, V. L. (1976). Effect of the Wingless (wg1) mutation on wing and haltere development in Drosophila melanogaster. *Developmental Biology*, *48*(2), 461–465. https://doi.org/10.1016/0012-1606(76)90108-1.

Shoshkes-Carmel, M., Wang, Y. J., Wangensteen, K. J., Toth, B., Kondo, A., Massasa, E. E., et al. (2018). Subepithelial telocytes are an important source of Wnts that supports intestinal crypts. *Nature*, *557*(7704), 242–246. https://doi.org/10.1038/s41586-018-0084-4.

Sirianni, F. E., Chu, F. S., & Walker, D. C. (2003). Human alveolar wall fibroblasts directly link epithelial type 2 cells to capillary endothelium. *American Journal of Respiratory and Critical Care Medicine*, *168*(12), 1532–1537. https://doi.org/10.1164/rccm.200303-371OC.

Snyder, J. C., Rochelle, L. K., Marion, S., Lyerly, H. K., Barak, L. S., & Caron, M. G. (2015). Lgr4 and Lgr5 drive the formation of long actin-rich cytoneme-like membrane protrusions. *Journal of Cell Science*, *128*(6), 1230–1240. https://doi.org/10.1242/jcs.166322.

Song, H. H., Shi, W., Xiang, Y. Y., & Filmus, J. (2005). The loss of glypican-3 induces alterations in Wnt signaling. *The Journal of Biological Chemistry*, *280*(3), 2116–2125. https://doi.org/10.1074/jbc.M410090200.

Spiering, D., & Hodgson, L. (2011). Dynamics of the Rho-family small GTPases in actin regulation and motility. *Cell Adhesion & Migration*, *5*(2), 170–180. https://doi.org/10.4161/cam.5.2.14403.

Stanford, K. I., Wang, L., Castagnola, J., Song, D., Bishop, J. R., Brown, J. R., et al. (2010). Heparan sulfate 2-O-sulfotransferase is required for triglyceride-rich lipoprotein clearance. *The Journal of Biological Chemistry*, *285*(1), 286–294. https://doi.org/10.1074/jbc.M109.063701.

Stanganello, E., Hagemann, A. I., Mattes, B., Sinner, C., Meyen, D., Weber, S., et al. (2015). Filopodia-based Wnt transport during vertebrate tissue patterning. *Nature Communications*, *6*, 5846. https://doi.org/10.1038/ncomms6846.

Stanganello, E., & Scholpp, S. (2016). Role of cytonemes in Wnt transport. *Journal of Cell Science*, *129*(4), 665–672. https://doi.org/10.1242/jcs.182469.

Sun, J., Yu, S., Zhang, X., Capac, C., Aligbe, O., Daudelin, T., et al. (2017). A Wntless-SEC12 complex on the ER membrane regulates early Wnt secretory vesicle assembly and mature ligand export. *Journal of Cell Science*, *130*(13), 2159–2171. https://doi.org/10.1242/jcs.200634.

Szymanski, K. M., Binns, D., Bartz, R., Grishin, N. V., Li, W. P., Agarwal, A. K., et al. (2007). The lipodystrophy protein seipin is found at endoplasmic reticulum lipid droplet junctions and is important for droplet morphology. *Proceedings of the National Academy of Sciences of the United States of America*, *104*(52), 20890–20895. https://doi.org/10.1073/pnas.0704154104.

Takada, R., Satomi, Y., Kurata, T., Ueno, N., Norioka, S., Kondoh, H., et al. (2006). Monounsaturated fatty acid modification of Wnt protein: Its role in Wnt secretion. *Developmental Cell, 11*(6), 791–801. https://doi.org/10.1016/j.devcel.2006.10.003.

Tassew, N. G., Charish, J., Shabanzadeh, A. P., Luga, V., Harada, H., Farhani, N., et al. (2017). Exosomes mediate mobilization of autocrine Wnt10b to promote axonal regeneration in the injured CNS. *Cell Reports, 20*(1), 99–111. https://doi.org/10.1016/j.celrep.2017.06.009.

Thery, C., Ostrowski, M., & Segura, E. (2009). Membrane vesicles as conveyors of immune responses. *Nature Reviews. Immunology, 9*(8), 581–593. https://doi.org/10.1038/nri2567.

Titus, M. A. (2018). Myosin-driven intracellular transport. *Cold Spring Harbor Perspectives in Biology, 10*(3). https://doi.org/10.1101/cshperspect.a021972.

Topczewski, J., Sepich, D. S., Myers, D. C., Walker, C., Amores, A., Lele, Z., et al. (2001). The zebrafish glypican knypek controls cell polarity during gastrulation movements of convergent extension. *Developmental Cell, 1*(2), 251–264. https://doi.org/10.1016/s1534-5807(01)00005-3.

Torpe, N., Gopal, S., Baltaci, O., Rella, L., Handley, A., Korswagen, H. C., et al. (2019). A protein disulfide isomerase controls neuronal migration through regulation of Wnt secretion. *Cell Reports, 26*(12), 3183–3190 e3185. https://doi.org/10.1016/j.celrep.2019.02.072.

Uren, A., Reichsman, F., Anest, V., Taylor, W. G., Muraiso, K., Bottaro, D. P., et al. (2000). Secreted frizzled-related protein-1 binds directly to Wingless and is a biphasic modulator of Wnt signaling. *The Journal of Biological Chemistry, 275*(6), 4374–4382. https://doi.org/10.1074/jbc.275.6.4374.

van den Heuvel, M., Harryman-Samos, C., Klingensmith, J., Perrimon, N., & Nusse, R. (1993). Mutations in the segment polarity genes wingless and porcupine impair secretion of the wingless protein. *The EMBO Journal, 12*(13), 5293–5302.

van Loon, K., Huijbers, E. J. M., & Griffioen, A. W. (2021). Secreted frizzled-related protein 2: A key player in noncanonical Wnt signaling and tumor angiogenesis. *Cancer Metastasis Reviews, 40*(1), 191–203. https://doi.org/10.1007/s10555-020-09941-3.

Vannucchi, M. G., Traini, C., Manetti, M., Ibba-Manneschi, L., & Faussone-Pellegrini, M. S. (2013). Telocytes express PDGFRalpha in the human gastrointestinal tract. *Journal of Cellular and Molecular Medicine, 17*(9), 1099–1108. https://doi.org/10.1111/jcmm.12134.

Wang, M., Marco, P., Capra, V., & Kibar, Z. (2019). Update on the role of the non-canonical Wnt/planar cell polarity pathway in neural tube defects. *Cell, 8*(10). https://doi.org/10.3390/cells8101198.

Wang, X., Reid Sutton, V., Omar Peraza-Llanes, J., Yu, Z., Rosetta, R., Kou, Y. C., et al. (2007). Mutations in X-linked PORCN, a putative regulator of Wnt signaling, cause focal dermal hypoplasia. *Nature Genetics, 39*(7), 836–838. https://doi.org/10.1038/ng2057.

Wawrzak, D., Metioui, M., Willems, E., Hendrickx, M., de Genst, E., & Leyns, L. (2007). Wnt3a binds to several sFRPs in the nanomolar range. *Biochemical and Biophysical Research Communications, 357*(4), 1119–1123. https://doi.org/10.1016/j.bbrc.2007.04.069.

Willert, K., Brown, J. D., Danenberg, E., Duncan, A. W., Weissman, I. L., Reya, T., et al. (2003). Wnt proteins are lipid-modified and can act as stem cell growth factors. *Nature, 423*(6938), 448–452. https://doi.org/10.1038/nature01611.

Willert, K. H. (2008). Isolation and application of bioactive Wnt proteins. *Methods in Molecular Biology, 468*, 17–29. https://doi.org/10.1007/978-1-59745-249-6_2.

Wilsie, L. C., & Orlando, R. A. (2003). The low density lipoprotein receptor-related protein complexes with cell surface heparan sulfate proteoglycans to regulate proteoglycan-mediated lipoprotein catabolism. *The Journal of Biological Chemistry, 278*(18), 15758–15764. https://doi.org/10.1074/jbc.M208786200.

Xavier, C. P., Melikova, M., Chuman, Y., Uren, A., Baljinnyam, B., & Rubin, J. S. (2014). Secreted Frizzled-related protein potentiation versus inhibition of Wnt3a/beta-catenin signaling. *Cellular Signalling*, *26*(1), 94–101. https://doi.org/10.1016/j.cellsig.2013.09.016.

Xie, Y., Zamponi, R., Charlat, O., Ramones, M., Swalley, S., Jiang, X., et al. (2013). Interaction with both ZNRF3 and LGR4 is required for the signalling activity of R-spondin. *EMBO Reports*, *14*(12), 1120–1126. https://doi.org/10.1038/embor.2013.167.

Yamamoto, H., Awada, C., Hanaki, H., Sakane, H., Tsujimoto, I., Takahashi, Y., et al. (2013). The apical and basolateral secretion of Wnt11 and Wnt3a in polarized epithelial cells is regulated by different mechanisms. *Journal of Cell Science*, *126*(Pt 13), 2931–2943. https://doi.org/10.1242/jcs.126052.

Yamazaki, Y., Palmer, L., Alexandre, C., Kakugawa, S., Beckett, K., Gaugue, I., et al. (2016). Godzilla-dependent transcytosis promotes Wingless signalling in Drosophila wing imaginal discs. *Nature Cell Biology*, *18*(4), 451–457. https://doi.org/10.1038/ncb3325.

Yan, D., Wu, Y., Feng, Y., Lin, S. C., & Lin, X. (2009). The core protein of glypican Dally-like determines its biphasic activity in wingless morphogen signaling. *Developmental Cell*, *17*(4), 470–481. https://doi.org/10.1016/j.devcel.2009.09.001.

Yang, Y., & Mlodzik, M. (2015). Wnt-Frizzled/planar cell polarity signaling: Cellular orientation by facing the wind (Wnt). *Annual Review of Cell and Developmental Biology*, *31*, 623–646. https://doi.org/10.1146/annurev-cellbio-100814-125315.

Yu, J., Chia, J., Canning, C. A., Jones, C. M., Bard, F. A., & Virshup, D. M. (2014). WLS retrograde transport to the endoplasmic reticulum during Wnt secretion. *Developmental Cell*, *29*(3), 277–291. https://doi.org/10.1016/j.devcel.2014.03.016.

Yu, J., Liao, P.-J., Xu, W., Jones, J. R., Everman, D. B., Flanagan-Steet, H., et al. (2021). Structural model of PORCN illuminates disease-associated variants and drug binding sites. *bioRxiv*. https://doi.org/10.1101/2021.07.19.452875.

Zheng, Y., Li, H., Manole, C. G., Sun, A., Ge, J., & Wang, X. (2011). Telocytes in trachea and lungs. *Journal of Cellular and Molecular Medicine*, *15*(10), 2262–2268. https://doi.org/10.1111/j.1582-4934.2011.01404.x.

Zhong, Q., Zhao, Y., Ye, F., Xiao, Z., Huang, G., Xu, M., et al. (2021). Cryo-EM structure of human Wntless in complex with Wnt3a. *Nature Communications*, *12*(1), 4541. https://doi.org/10.1038/s41467-021-24731-3.

Zhong, Z., Zylstra-Diegel, C. R., Schumacher, C. A., Baker, J. J., Carpenter, A. C., Rao, S., et al. (2012). Wntless functions in mature osteoblasts to regulate bone mass. *Proceedings of the National Academy of Sciences of the United States of America*, *109*(33), E2197–E2204. https://doi.org/10.1073/pnas.1120407109.

CHAPTER FOUR

# Dynamic regulation of human epidermal differentiation by adhesive and mechanical forces

**Sebastiaan Zijl[a], Vasiliki Salameti[a], Blaise Louis[a,‡], Victor A. Negri[a,†,‡], and Fiona M. Watt[a,b,*]**
[a]Centre for Stem Cells and Regenerative Medicine, King's College London, London, United Kingdom
[b]Directors' Research Unit, European Molecular Biology Laboratory, Heidelberg, Germany
*Corresponding author: e-mail address: fiona.watt@kcl.ac.uk

## Contents



## Abstract

The interfollicular epidermis is the multilayered epithelium that forms the outer layer of the skin. It is maintained by stem cells that are attached to a basement membrane, which lies on top of the underlying connective tissue, the dermis. Cells undergo terminal differentiation as they detach from the basement membrane and move toward the outer epidermal surface. Over time, many of the molecular regulators of this process have been identified. It is now is clear that these pathways also receive critical input from the physical properties of the tissue. In this review, we describe how changes in these factors regulate differentiation and how new insights from single cell RNA sequencing could provide validation or challenge to the existing experimental models.

† Present address: Bioscience Asthma, Research and Early Development, Respiratory & Immunology, BioPharmaceuticals R&D, AstraZeneca, Cambridge, United Kingdom.
‡ Equal contribution.

*Current Topics in Developmental Biology*, Volume 150
ISSN 0070-2153
https://doi.org/10.1016/bs.ctdb.2022.03.004

## 1. Introduction

The epidermis is the outer layer of mammalian skin and is comprised of a multilayered epithelium and associated adnexal structures: the hair follicles, sweat glands and sebaceous glands. The relative abundance of the adnexal structures varies according to species and body site. This review will focus on human interfollicular epidermis and will not consider adnexal structures. Human epidermis is one of the tissues in the human body with a high cell turnover rate. Stem cells that are attached to the basement membrane self-renew throughout adulthood and give rise to the differentiating cells of the spinous and granular layers, before losing their nucleus and cytoplasmic organelles and forming cornified cells (Fig. 1). The layers of cornified cells are an important physical barrier against environmental stresses, keeping out invading pathogens and preventing dehydration of the underlying tissues. Because of these crucial functions and its capability to self-renew, the epidermis has been long been used as a model to study stem cell behavior (Watt, 2014).

Research on human epidermis accelerated in 1975, when a successful *in vitro* culture system for primary epidermal cells (keratinocytes) was established (Rheinwald & Green, 1975) based on clonal expansion on a feeder layer of J2 3T3 cells. From then the epidermal stem cell field rapidly developed, which in the 1980s led to the demonstration that sheets of cultured epidermis could be used to transplant burn victims (Gallico, O'connor, Compton, Kehinde, & Green, 1984; O'Connor, Mulliken, Banks-Schlegel, Kehinde, & Green, 1981). The succeeding decades brought many studies on

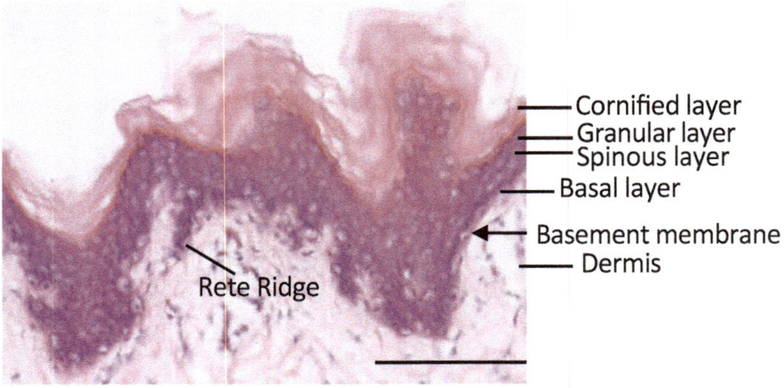


**Fig. 1** The organization of human interfollicular epidermis. H&E stained section of adult breast skin. Scale bar: 200 μm.

both human and mouse epidermal cells, providing many insights into epidermal stem cell behavior and differentiation. Clonal growth of keratinocytes has frequently been used as a quantitative assay of stem cells (Barrandon & Green, 1987; Hirsch et al., 2017; Jones & Watt, 1993). Initially, a candidate approach was taken to enrich for clonogenic cells by flow cytometry with antibodies to cell surface markers (Jones & Watt, 1993). Subsequently, single-cell transcriptomics was performed on cultured human keratinocytes (Jensen & Watt, 2006; Tan et al., 2013) and stem cell markers identified in that way were tested in colony formation experiments. Most recently, single-cell RNA sequencing of keratinocytes isolated directly from human skin has confirmed the existence of different subpopulations of basal cells that were predicted from the in vitro studies (Cheng et al., 2018; Reynolds et al., 2021; Wang et al., 2020).

## 2. Epidermal adhesion mechanisms

The multilayered structure of the epidermis is dependent on a variety of mechanisms that mediate adhesion to the basement membrane and adhesion between cells (Fig. 2). Integrins mediate adhesion of basal keratinocytes to the basement membrane (Watt, 2002) and integrin-ECM adhesion occurs at the basal plasma membrane. However, integrins are also present on the lateral membranes of basal keratinocytes, where they are in an inactive state. Beta1 integrins are located in focal adhesions, while the $\alpha6\beta4$ integrin is a component of hemidesmosomes. Changes in integrin expression are associated with placing cells in culture, wound healing and cancer (Watt, 2002). Although integrin expression is primarily confined to the basal layer (Fig. 2), suprabasal integrin expression is a feature of epidermal hyperproliferation, for example, in psoriasis and during wound healing (Hertle, Kubler, Leigh, & Watt, 1992). Single-cell RNA sequencing suggests that suprabasal integrin-positive cells in psoriasis are not expressing integrin mRNA (Reynolds et al., 2021); rather, suprabasal protein expression is likely to reflect the long half-life of integrin proteins in epithelial cells (Ly et al., 2018) and the rapid transit of keratinocytes through the suprabasal layers in this hyperproliferative epidermal disorder.

Epidermal cell-cell adhesion is mediated primarily by desmosomes, adherens junctions, gap junctions and tight junctions (Fig. 2). Keratinocytes express two cadherins that are the adhesive receptors of epidermal adherens junctions: P-cadherin, which is confined to the basal layer, and E-cadherin, which is expressed in all the living cell layers (Kobielak & Boddupally, 2014; Lai-Cheong, Arita, & McGrath, 2007). Desmosome-mediated adhesion is

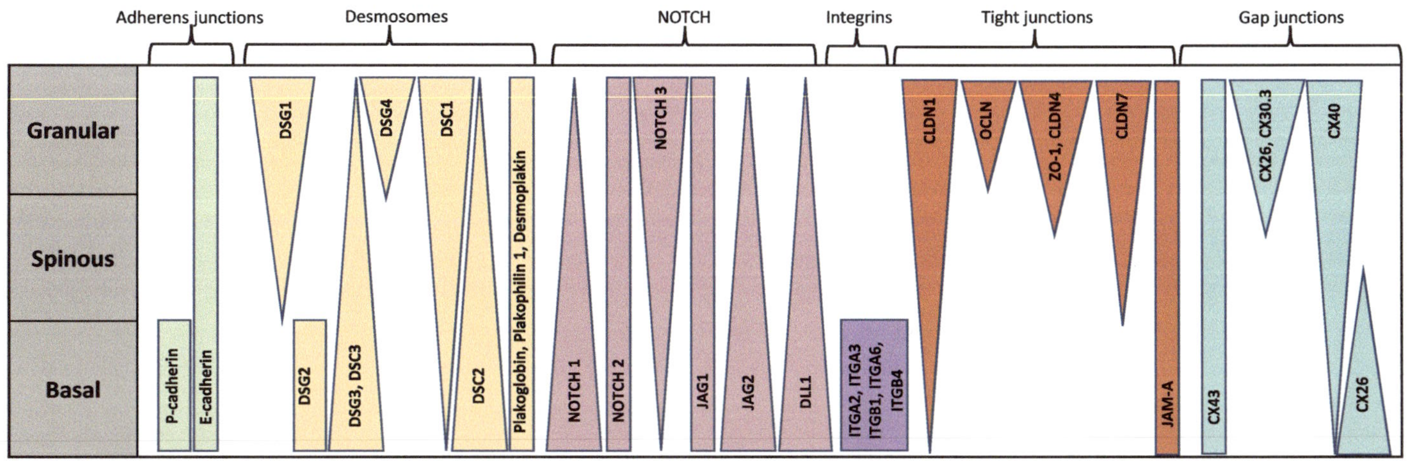


**Fig. 2** Schematic showing the major protein components of adherens junctions, desmosomes, gap and tight junctions, integrins, Notch ligands and Notch receptors in keratinocytes.

dependent on desmocollins and desmogleins (Hegazy, Perl, Svoboda, & Green, 2021; Kobielak & Boddupally, 2014; Lai-Cheong et al., 2007) while connexins are the key component of gap junctions (Chanson, Watanabe, O'Shaughnessy, Zoso, & Martin, 2018; Richard, 2000). The main structural proteins of tight junctions are claudins (Lai-Cheong et al., 2007). Notch ligands and receptors are not primarily considered adhesive; however, the Notch ligand Delta-like 1 (Dll1) plays a role in stem cell clustering within the epidermal basal layer and in modulating ECM adhesion and motility (Watt, Estrach, & Ambler, 2008).

Genetic disorders of human skin are linked to all of the adhesive junctions described above (Lai-Cheong et al., 2007; Uitto, Has, Vahidnezhad, Youssefian, & Bruckner-Tuderman, 2017), the phenotypes demonstrating the importance of the different adhesion structures in regulating differentiation and stratification.

## 3. Keratinocyte–substrate interactions at the single cell level

A key observation regarding the keratinocyte differentiation stimulus is that when single cell suspensions of human epidermal keratinocytes are held in semi-solid medium for several days the cells exit the cell cycle and undergo terminal differentiation (Green, 1977; Fig. 3A). Suspension-induced differentiation can be inhibited by the addition of extracellular matrix proteins or functional antibodies to β1 integrins, leading to the concept that integrins not only anchor cells to the basement membrane but also regulate the onset of differentiation (Adams & Watt, 1989). Over the years, a number of studies have documented the changes in gene expression, protein and lipid composition associated with suspension-induced differentiation (Mishra et al., 2017; Vietri Rudan, Mishra, Klose, Eggert, & Watt, 2020). A network of interacting protein phosphatases that act as a molecular switch between the stem cell state and the differentiated cell state has also been found (Mishra et al., 2017).

The concept of restricted cell–substrate interactions as a differentiation stimulus received further support from early studies in which single epidermal cells were captured on circular micropatterned islands (Watt, Jordan, & O'Neill, 1988; Fig. 3B). Human epidermal stem cells initiate terminal differentiation when spreading is restricted on ECM-coated micropatterned islands, soft hydrogels or hydrogel-nanoparticle composites with high nanoparticle spacing (Connelly et al., 2010; Trappmann et al., 2012).

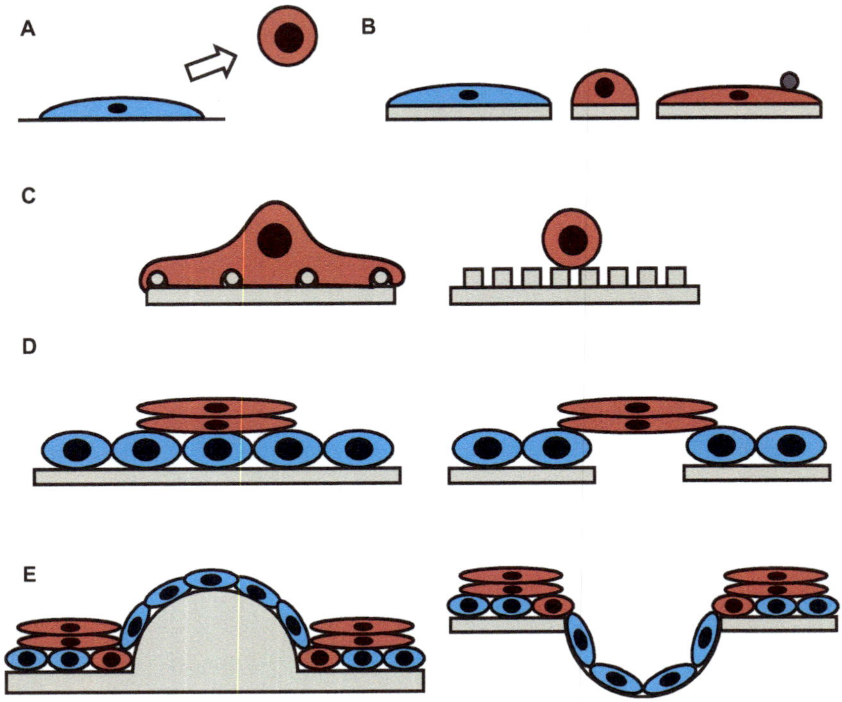


**Fig. 3** Schematic illustrating different experimental set-ups to analyze the effect of substrate adhesion on terminal differentiation of human epidermal keratinocytes in culture. (A) Suspension-induced differentiation. (B) Micropatterned islands accommodating single cells. The small grey circle represents bead with immobilized Jagged1 protein. (C) Substrates with topographical features in the μm range. (D) 100 μm diameter islands with a complete adhesive surface (left) or a 40 μm non-adhesive center (right). (E) Substrates that mimic the topography of the epidermal-dermal interface, created with PDMS (left) or via the application of a vacuum to a PLGA membrane (right). Blue: stem cell (undifferentiated); red: involucrin-expressing cell (differentiated).

Depending on the extrinsic differentiation signal, different downstream signaling pathways are involved. Maintenance of the stem cell compartment downstream of integrin-mediated signaling requires ERKMAPK signalling and activation of this pathway overcomes the dominant negative effects of integrin downregulation (Hiratsuka, Bordeu, Pruessner, & Watt, 2020; Trappmann et al., 2012; Zhu, Haase, & Watt, 1999). The signaling mechanisms that are activated when cell spreading is restricted on micropatterned substrates include actin polymerization and the activation of transcription factor Serum Response Factor (SRF) (Connelly et al., 2010). The localization of the transcriptional coactivator megakaryocytic acute leukemia (MAL; also known as MRTF-A and MKL1) controls the activity of SRF. MAL translocates to the nucleus in keratinocytes confined to small

(20 μm) micropatterned islands, thereby activating SRF and inducing the transcription of JUN and FOS, members of the AP-1 family. AP-1 transcription factors regulate the expression of many genes that are expressed in differentiating keratinocytes, including the cornified envelope precursor, involucrin (Connelly et al., 2010).

To explore the effect of substrate topography, human keratinocytes enriched for stem cells were seeded on a library with over 2000 different topographies in the micrometer range (Zijl et al., 2019). Twenty-four hours later the proportion of differentiated cells was determined. In this way it was possible to identify topographies that support the differentiation of highly spread cells (Fig. 3C). Inhibition of Rho kinase inhibition or treatment with blebbistatin blocked the differentiation of spread cells, whereas SRF inhibition did not, suggesting a role for the actin cytoskeleton in differentiation.

## 4. Role of cell–cell adhesion in regulating differentiation: A reductionist approach

To determine whether cell–cell adhesion modulates the differentiation stimulus of loss of substrate adhesion, Zhu and Watt (1996) overexpressed a dominant negative E-cadherin mutant, consisting of the extracellular domain of H-2Kd and the transmembrane and cytoplasmic domains of E-cadherin, into human keratinocytes. As predicted, cell-cell adhesion and stratification were inhibited. However, unexpectedly the dominant negative mutant stimulated terminal differentiation even in single cell suspension. Expression of stabilized, N-terminally truncated β-catenin relieved the differentiation stimulatory effect of overexpressing the E-cadherin cytoplasmic domain (Zhu & Watt, 1999), suggesting that high levels of cytoplasmic β-catenin could be a feature of epidermal stem cells. This was one of the first *in vitro* studies to indicate that β-catenin might regulate the size of the stem cell compartment, a finding that has now been confirmed and extended in many cell types and tissues (Nusse & Clevers, 2017; Silva-Vargas et al., 2005).

Flow sorting of keratinocytes on the basis of both β1 integrins and Dsg3 expression enriches for clonogenic cells (Wan et al., 2003). There is also evidence to suggest that desmosomal components can directly regulate keratinocyte differentiation. The guanine nucleotide exchange factor (GEF) breakpoint cluster region (Bcr) is a major upstream regulator of RhoA activity, stress fibers, and focal adhesion formation in keratinocytes, and modulates differentiation downstream of the desmosome protein Dsg1 (Dubash et al., 2013). Loss of Bcr reduces expression of differentiation markers and abrogates signaling via SRF and its coactivator MAL. Loss of Bcr or MAL reduces Dsg1 expression while ectopic expression of Dsg1 rescues the defects in differentiation found upon loss of Bcr or MAL signaling (Dubash et al., 2013).

In human interfollicular epidermis, stem cells lie in clusters surrounded by less clonogenic daughters and express high levels of the Notch ligand Dll-1 (Jones, Harper, & Watt, 1995; Lowell, Jones, Le Roux, Dunne, & Watt, 2000). Experiments in culture indicate that high Dll1 expression by epidermal stem cells has three effects: a protective effect on stem cells by blocking Notch signaling; enhanced cohesiveness of stem-cell clusters, which may discourage intermingling with neighboring cells and signaling to cells at the edges of the clusters to leave the stem cell compartment and differentiate (Lowell et al., 2000). The adhesive effects of Dll1 are independent of SuH-dependent Notch signaling. Dll1 promotes keratinocyte cohesiveness by restricting motility (Lowell & Watt, 2001). Mutation of the Delta1 PDZ-binding domain abolishes Dll1-mediated keratinocyte cohesiveness, stimulates Notch transcriptional activity and promotes epidermal differentiation (Estrach, Legg, & Watt, 2007). Dll1 binds to the adaptor protein syntenin. Syntenin, like Dll1, is upregulated in stem cell clusters. Knockdown of syntenin in cells overexpressing full-length Dll1 has the same effects on Notch signaling, epidermal differentiation and adhesion as overexpressing Dll1 with a mutated PDZ-binding domain. Syntenin has previously been reported to regulate membrane traffic, and mutation of the Dll1 PDZ-binding domain or knockdown of syntenin leads to rapid internalization of Dll1 (Estrach et al., 2007).

In contrast to Dll1, the Notch ligand Jagged1 triggers differentiation when presented on an adhesive substrate or on polystyrene beads and overrides the differentiation inhibitory effect of cell spreading (Negri et al., 2019; Fig. 3B). Dll1 overexpression abrogates the pro-differentiation effect of Jagged1 in a cell autonomous fashion. It thus appears that Dll1 expression by stem cells not only stimulates differentiation of neighboring cells in trans, but also acts cell autonomously to inhibit differentiation.

## 5. Cross-talk between cell–cell and cell–ECM adhesion mechanisms: Building a multilayered epidermis

The onset of terminal differentiation is not only associated with downregulation of integrins but also with changes in expression of multiple cell–cell adhesion molecules, including downregulation of P-cadherin, and increased expression of E-cadherin (Fig. 2). In the epidermis the onset of terminal differentiation normally coincides with inhibition of integrin function and expression, thereby ensuring that differentiating cells are selectively expelled from the basal layer. However, when stratification of cultured human epidermal keratinocytes is prevented by reducing the calcium concentration of the medium to inhibit assembly of adherens junctions and desmosomes, keratinocytes initiate terminal differentiation while still attached to the culture substrate (Watt & Green, 1982). In low calcium medium keratinocytes coexpress integrins and terminal differentiation markers; however, within 6 h of raising the extracellular calcium concentration, terminally differentiating cells no longer express integrins and they undergo selective migration out of the basal layer. Antibodies to P- and E-cadherin that block calcium-induced stratification prevent integrin downregulation, suggesting that cadherins play a role in the downregulation of integrin expression that is associated with terminal differentiation (Hodivala & Watt, 1994). In keratinocytes expressing a dominant negative E-cadherin mutant there is a fivefold decrease in the level of endogenous cadherins and a threefold increase in the level of β-catenin (Zhu & Watt, 1996).

Complementing experiments in which intercellular adhesion is modulated by extracellular calcium, it is possible to generate a stratified microepidermis with fewer than 10 human keratinocytes by applying the same polymer brush microengineering that has been used to capture single keratinocytes (Fig. 3D) (Gautrot et al., 2012). Epidermal stem cells are captured on 100 μm diameter circular collagen-coated disks. Within 24 h they assemble a stratified microtissue, in which differentiated cells have a central suprabasal location. The epidermis forms correctly even when the substrates have a non-adhesive center of up to 40 μm diameter (Fig. 3D). Cell–cell and cell–matrix adhesive interactions together result in correct microepidermis assembly. This requires actin polymerization, adherens junctions and desmosomes, but not myosin II-mediated contractility or coordinated cell movement.

In human skin, the interface between the epidermis and dermis is not flat but undulates, with projections of the epidermis into the dermis being called rete ridges (Fig. 1). The dimensions of the undulations change as a function of age and disease. Epidermal stem cell clusters lie in specific locations relative to the undulations (Jones et al., 1995; Tan et al., 2013). The clusters are known to form on flat substrates (Jones et al., 1995). However, whether the physical topography of the epidermal-dermal junction directly affects clustering was unknown until recently. To explore this, we created collagen-coated polydimethylsiloxane (PDMS) elastomer substrates that mimic the topographical features of the human epidermal-dermal junction in healthy, aged and hyperproliferative skin (Viswanathan et al., 2016; Fig. 3E). We found that cells seeded on the substrates patterned according to topography and separate cues determined the locations of stem cells, differentiated cells and proliferating cells.

To discover more about how the topography of the epidermal-dermal junction influences stem cell patterning, we developed a rig in which epidermal cells are cultured on a collagen-coated poly(d,L-lactide-*co*-glycolide) (PLGA) membrane (Helling et al., 2019). This dynamic platform is a new tool for investigating changes in the skin with age and disease. When a vacuum is applied the membrane is induced to undulate. Stem cells cluster in response to the vacuum, but whereas on the PDMS topographies the stem cells cluster on the tips of the features, on vacuum-induced topographies the stem cells cluster at the base (Fig. 3E). Rho kinase inhibition prevents clustering, indicating a role for Rho family members in the process. This experiment suggests that it is the sides of the features that are important for stem cell patterning, not the tips or troughs (Mobasseri et al., 2019).

## 6. Physical forces and the control of differentiation of individual keratinocytes

Cells sense and respond to mechanical forces through a process called mechanotransduction, whereby mechanical forces are translated into biochemical signals. Mechanotransduction generally depends on the cytoskeleton and adhesive junctions and involves integration with soluble factors (Discher, Mooney, & Zandstra, 2009; Romani, Valcarcel-Jimenez, Frezza, & Dupont, 2021). The differentiation response of single keratinocytes to confinement on micropatterned substrates (Watt et al., 1988) was early evidence that epidermal cells sense and respond to their physical environment (Fig. 4). The signal transduction pathway by which keratinocytes differentiate when spreading is restricted on micropatterned islands is mediated by the actin cytoskeleton and SRF transcriptional activity (Connelly et al., 2010). There is also evidence that responses of keratinocytes to substrate geometry depend on differences in the diffusion dynamics of the cytoskeletal adapter protein vinculin (Gautrot et al., 2014) and that differential expression of different integrins is involved (Di Cio, Bøggild, Connelly, Sutherland, & Gautrot, 2017). Furthermore there is evidence that YAP signaling is regulated by restricted cell spreading

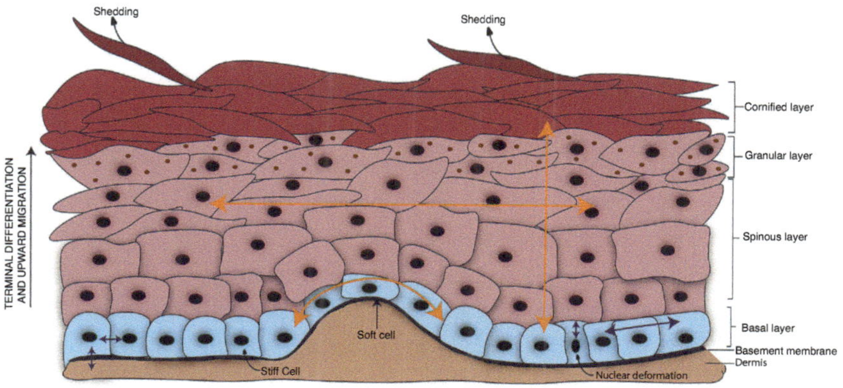


**Fig. 4** Schematic representing different physical forces acting on the epidermis. Tissue scale forces (Biggs et al., 2020) are represented by orange arrows, while forces acting at the cellular or intracellular level are represented by blue arrows.

(Totaro et al., 2017), although in our studies confinement of keratinocytes on small islands was not sufficient to cause efficient cytoplasmic translocation of YAP (Walko et al., 2017).

While it is well established that cells can sense the bulk stiffness of the surface to which they are attached (Discher et al., 2009), we found that when single keratinocytes differentiated on soft hydrogels they sensed the distance of collagen anchoring points, and thus that stem cells exert a mechanical force on collagen fibers and gauge the feedback to make cell-fate decisions (Trappmann et al., 2012) (Fig. 4). Although differentiation occurs as a result of restricted cell spreading on micropatterned islands and soft hydrogels, the signal transduction pathways are different, with differentiation on hydrogels being linked to a failure of integrins to cluster in focal adhesions and decreased ERK MAPK activity (Trappmann et al., 2012). The keratinocyte response to soft hydrogels may be cell-type specific, because ECM tethers are not essential for differentiation of osteogenic and adipogenic stem cells, with bulk matrix stiffness being the major mechanical regulator of differentiation (Wen et al., 2014). The keratin cytoskeleton could potentially account for differences in the response of keratinocytes and other cell types to the physical properties of their environment, since keratin filaments provide mechanical resilience (Ramms et al., 2013), yet are dynamic and regulated by the mechanical properties of the niche (Pora et al., 2020).

There is good evidence that physical forces affect nuclear architecture, chromatin and gene expression in keratinocytes (Tajik et al., 2016). Nuclear membrane proteins translate forces into the nucleus, affecting the distribution of chromatin, resulting in changes in gene expression, with different genes being differentially affected (Le et al., 2016). Applying a stretching force to keratinocytes in culture leads to accumulation of actin around the nucleus and recruitment of phospho-Myosin light chain 2 (pMLC2) (Le et al., 2016). This, in turn, reduces the levels of nuclear actin and decreases occupancy of RNA Polymerase-2 (RNA Pol) on differentiation genes. The effects are dependent on the nuclear membrane protein Emerin and suggest a physical connection between the cytoplasm and the nucleus in regulating differentiation. The keratin cyto-linker protein plectin protects against nuclear deformation (Almeida et al., 2015) (Fig. 4). Mechanical stretch deforms the nucleus (Almeida et al., 2015; Nava et al., 2020), which cells initially counteract via a calcium-dependent nuclear softening driven by loss of H3K9me3-marked heterochromatin (Nava et al., 2020). The resulting changes in chromatin rheology and architecture insulate genetic material from mechanical force and are required to prevent DNA damage. It is interesting that microtopography induced differentiation, like the effect of stretching, depends on the polymerization of actin and the activity of pMLC2 (Zijl et al., 2019).

## 7. Physical forces acting at the level of epidermal assembly

The epidermis is subject to multiple physical forces both within the tissue and at the whole tissue level (Biggs, Kim, Miroshnikova, & Wickström, 2020) (Fig. 4). Within the epidermis the selective movement of committed cells out of the basal layer and into the first spinous layer depends on integrated assembly and reorganization of cell–matrix and cell–cell junctions (Hodivala & Watt, 1994; Mertz et al., 2013; Noethel et al., 2018). Intercellular adhesions modulate forces transmitted to the ECM and cadherins are essential for this mechanical cooperativity, modulating the physical cohesion between contractile cells (Mertz et al., 2013; Fig. 4). Focal adhesions and adherens junctions act as mechanosensitive elements in response to cyclic strain. Paxillin and talin affect mechanosensitivity in focal adhesions, while in adherens junctions α-catenin is the main mechanosensor via binding to vinculin (Noethel et al., 2018). Persistent, high-amplitude stretch induces supracellular alignment of tissue to redistribute mechanical energy before it reaches the nucleus. This tissue-scale mechanoadaptation is mediated by cell–cell contacts and allows cells and tissues to switch off nuclear mechanotransduction to restore the initial chromatin state (Miroshnikova, Cohen, Ezhkova, & Wickström, 2019; Nava et al., 2020).

The actin-based cytoskeleton and its associated adherens junctions are well-established contributors to mechanosensing and -transduction machinery. However, the desmosome-keratin filament network is also involved (Broussard et al., 2017). Strengthening the interaction between desmosomes and keratin filaments increases cell–substrate and cell–cell forces and cell stiffness in pairs and sheets of cells, effects that are nevertheless dependent on the actin cytoskeleton. In keratinocytes subjected to stretch forces, keratin filaments act as a mechanical buffer and play a role in force transmission from cellular junctions to the cell body (Ahrens et al., 2019; Fig. 4).

Within the epidermal basal layer cell divisions lead to tissue crowding and local changes in force distribution, which in turn suppress the rate of cell divisions (Jiang et al., 2021). B-plexin transmembrane receptors are required for crowding-induced mechanical forces during mouse embryonic skin development. B-plexins mediate mechanoresponses to crowding through stabilization of adhesive cell junctions and lowering of cortical stiffness (Jiang et al., 2021). Conversely, stretching the epidermis stimulates stem cell renewal and modulates gene-regulatory networks (Aragona et al., 2020). Surface tension governed by differential adhesion can drive fluid particle mixtures to sort into separate regions, a process known as demixing (Sahu et al., 2020). Experiments with mixtures of wild-type and E-cadherin-deficient keratinocytes on a substrate are consistent with the predicted phenomenon of microdemixing, which biology may exploit to create subtle patterning (Sahu et al., 2020).

An early suggestion that post-mitotic keratinocytes are forced out of the basal layer by pressure from neighboring cells in mitosis (Leblond, Greulich, & Pereira, 1964) is not consistent with the observation that when proliferation in guinea-pig epidermis is inhibited by UV irradiation, upward migration still occurs (Etoh, Taguchi, & Tabachnick, 1975). We have speculated that with time a post-mitotic basal keratinocyte will increase in size; with no additional space to accommodate it on the basement membrane the proportion of the cell's total surface area that is attached to the ECM will be reduced, thereby triggering terminal differentiation (Watt, 1988; Fig. 3B). Since the adhesiveness of keratinocytes to the basement membrane is reduced during terminal differentiation (Watt, 2002), while their adhesiveness to differentiating keratinocytes is increased (Watt, 1984), there will be a positive-feedback system in which cells are at the same time induced to differentiate and migrate out of the basal layer (Watt, 1988).

In determining the patterning of stem cells in the basal layer there is clearly a role for Dll1 signaling (Watt et al., 2008), yet physical forces are also important (Fig. 4). When we examined the patterning of keratinocytes on collagen-coated polydimethylsiloxane (PDMS) elastomer substrates that mimic the undulations of the epidermal-dermal junction (Fig. 3E) (Viswanathan et al., 2016), we found that the stiffness of cells on the tips is lower than cells on the base (Mobasseri et al., 2019). We also found that cell–cell junctions are stiffer than the cell body. The differences in stiffness of keratinocytes in different locations depend on Rho kinase activity and intercellular adhesion (Fig. 4). These findings have led us to propose that epidermal stem cell patterning is controlled by mechanical forces exerted at intercellular junctions in response to the slope of the undulations (Mobasseri et al., 2019).

The role of communication between the basal and suprabasal epidermal layers in maintaining a homeostatic balance between proliferation and differentiation has been appreciated for many years (Carroll, Romero, & Watt, 1995; Owens & Watt, 2003; Read & Watt, 1988). However, the involvement of physical forces has only recently been elucidated (Fig. 4; Ning, Muroyama, Li, & Lechler, 2021). Increasing contractility of differentiated mouse keratinocytes results in non-cell-autonomous hyperproliferation of stem cells and prevents commitment to a hair follicle lineage (Ning et al., 2021). Lineage tracing in mouse epidermis indicates that differentiated cells are the progeny of the underlying basal cells (Donati et al., 2017); nevertheless, this relationship is not fixed, and differentiated cells can move relative to the underlying basal cells during wound healing in the mouse (Donati et al., 2017) and in cultured human epidermis (Mobasseri et al., 2019).

## 8. Conclusions and future directions

As this review has shown, there has been considerable progress in understanding how epidermal stem cell renewal and differentiation are controlled by extrinsic factors. We believe that the next phase of research requires a focus on signal integration. Over the years many individual signaling pathways have been identified as playing a role in the control of epidermal proliferation, stem cell renewal and differentiation, including beta-catenin, ERK MAPK, p38 mitogen-activated protein kinase (MAPK), phosphoinositide-3—(PI3K) and PKC, and it is now apparent that different classes of lipids also play a role in the regulation and maintenance of epidermal homeostasis (Mishra et al., 2017; Vietri Rudan & Watt, 2021).

Some transcriptional regulators, including YAP/TAZ, SRF and MAL, integrate mechanical cues with the response to soluble signals to control multiple aspects of cell behavior (Connelly et al., 2010; Totaro, Panciera, & Piccolo, 2018). We now appreciate that the same outcome—initiation of terminal differentiation—can be achieved via different signaling pathways depending on the extrinsic differentiation stimulus (Trappmann et al., 2012). Conversely, transcriptional regulators such as YAP/TAZ are at the nexus of multiple pathways, including Notch and β-catenin (Totaro et al., 2017; Walko et al., 2017).

Just as biochemical signaling pathways can no longer be thought of as linear, the classic structural elements of keratinocytes exhibit far more crosstalk than appreciated previously. This is illustrated by recent research on desmosomal proteins, which not only act as anchorage for keratin filaments but also interact with other cytoskeletal proteins, adherens junctions and gap junctions (Cohen-Barak et al., 2020). Dsg1 suppresses EGF receptor signaling (Harmon et al., 2013; Valenzuela-Iglesias et al., 2019) and contributes to the control of actin polymerization at cell-cell borders, thereby promoting detachment from the substrate (Nekrasova et al., 2018). Dsg1 compartmentalization in lipid domains affects tension at adherens junctions and, conversely, the Mammalian target of rapamycin (mTOR) pathway, which is important in nutrient sensing and cell growth (Liu & Sabatini, 2020), regulates epidermal differentiation (Asrani et al., 2017; Ding et al., 2016; Elkenani et al., 2016) in part via its effects on cytoskeletal tension (Asrani et al., 2017).

Two key technologies that will unlock further insights into epidermal differentiation and homeostasis are the ability to visualize signaling pathway dynamics in real time (Hiratsuka et al., 2020) and integration of experimental data with transcriptional profiling of single cells isolated directly from the skin (Reynolds et al., 2021). In addition, new methods for identifying genetic variants that regulate genome-wide chromatin states will allow further examination of the molecular switches that underlie cell state transitions (Rubin et al., 2019).

## Acknowledgments

The authors gratefully acknowledge the financial support from the UK Regenerative Medicine Platform (UKRMP; MR/R015635/1), the Centre for Cellular Signal Patterns (CellPAT) initiative (Danish National Research Foundation), Medical Research Council (MR/PO18823/1), Biotechnology and Biological Sciences Research Council (BB/S506898/1) and Wellcome Trust (206439/Z/17/Z).

## References

Adams, J. C., & Watt, F. M. (1989). Fibronectin inhibits the terminal differentiation of human keratinocytes. *Nature, 340*, 307–309.

Ahrens, D., Rubner, W., Springer, R., Hampe, N., Gehlen, J., Magin, T. M., et al. (2019). A combined AFM and lateral stretch device enables microindentation analyses of living cells at high strains. *Methods Protocols, 2*, 43.

Almeida, F. V., Walko, G., McMillan, J. R., McGrath, J. A., Wiche, G., Barber, A. H., et al. (2015). The cytolinker plectin regulates nuclear mechanotransduction in keratinocytes. *Journal of Cell Science, 128*, 4475–4486.

Aragona, M., Sifrim, A., Malfait, M., Song, Y., Van Herck, J., Dekoninck, S., et al. (2020). Mechanisms of stretch-mediated skin expansion at single-cell resolution. *Nature, 584*, 268–273.

Asrani, K., Sood, A., Torres, A., Georgess, D., Phatak, P., Kaur, H., et al. (2017). MTORC1 loss impairs epidermal adhesion via TGF-β/rho kinase activation. *The Journal of Clinical Investigation, 127*, 4001–4017.

Barrandon, Y., & Green, H. (1987). Three clonal types of keratinocyte with different capacities for multiplication. *Proceedings of the National Academy of Sciences of the United States of America, 84*, 2302–2306.

Biggs, L. C., Kim, C. S., Miroshnikova, Y. A., & Wickström, S. A. (2020). Mechanical forces in the skin: Roles in tissue architecture, stability, and function. *The Journal of Investigative Dermatology, 140*, 284–290.

Broussard, J. A., Yang, R., Huang, C., Nathamgari, S. S. P., Beese, A. M., Godsel, L. M., et al. (2017). The desmoplakin-intermediate filament linkage regulates cell mechanics. *Molecular Biology of the Cell, 28*, 3156–3164.

Carroll, J. M., Romero, M. R., & Watt, F. M. (1995). Suprabasal integrin expression in the epidermis of transgenic mice results in developmental defects and a phenotype resembling psoriasis. *Cell, 83*, 957–968.

Chanson, M., Watanabe, M., O'Shaughnessy, E. M., Zoso, A., & Martin, P. E. (2018). Connexin communication compartments and wound repair in epithelial tissue. *International Journal of Molecular Sciences, 19*, 1354.

Cheng, J. B., Sedgewick, A. J., Finnegan, A. I., Harirchian, P., Lee, J., Kwon, S., et al. (2018). Transcriptional programming of normal and inflamed human epidermis at single-cell resolution. *Cell Reports, 25*, 871–883.

Cohen-Barak, E., Godsel, L. M., Koetsier, J. L., Hegazy, M., Kushnir-Grinbaum, D., Hammad, H., et al. (2020). The role of Desmoglein 1 in gap junction turnover revealed through the study of SAM syndrome. *The Journal of Investigative Dermatology, 140*, 556–567.

Connelly, J. T., Gautrot, J. E., Trappmann, B., Tan, D. W., Donati, G., Huck, W. T., et al. (2010). Actin and serum response factor transduce physical cues from the microenvironment to regulate epidermal stem cell fate decisions. *Nature Cell Biology, 12*, 711–718.

Di Cio, S., Bøggild, T. M. L., Connelly, J., Sutherland, D. S., & Gautrot, J. E. (2017). Differential integrin expression regulates cell sensing of the matrix nanoscale geometry. *Acta Biomaterialia, 50*, 280–292.

Ding, X., Bloch, W., Iden, S., Rüegg, M. A., Hall, M. N., Leptin, M., et al. (2016). mTORC1 and mTORC2 regulate skin morphogenesis and epidermal barrier formation. *Nature Communications*, 7, 13226.

Discher, D. E., Mooney, D. J., & Zandstra, P. W. (2009). Growth factors, matrices, and forces combine and control stem cells. *Science, 324*, 1673–1677.

Donati, G., Rognoni, E., Hiratsuka, T., Liakath-Ali, K., Hoste, E., Kar, G., et al. (2017). Wounding induces dedifferentiation of epidermal $Gata6^+$ cells and acquisition of stem cell properties. *Nature Cell Biology, 19*, 603–613.

Dubash, A. D., Koetsier, J. L., Amargo, E. V., Najor, N. A., Harmon, R. M., & Green, K. J. (2013). The GEF Bcr activates RhoA/MAL signaling to promote keratinocyte differentiation via desmoglein-1. *The Journal of Cell Biology, 202*, 653–666.

Elkenani, M., Nyamsuren, G., Raju, P., Liakath-Ali, K., Hamdaoui, A., Kata, A., et al. (2016). Pelota regulates epidermal differentiation by modulating BMP and PI3K/AKT signaling pathways. *The Journal of Investigative Dermatology, 136*, 1664–1671.

Estrach, S., Legg, J., & Watt, F. M. (2007). Syntenin mediates Delta1-induced cohesiveness of epidermal stem cells in culture. *Journal of Cell Science, 120*, 2944–2952.

Etoh, H., Taguchi, Y., & Tabachnick, J. (1975). Movement of beta-irradiated epidermal basal cells to the spinous-granular layers in the absence of cell division. *The Journal of Investigative Dermatology, 64*, 431–435.

Gallico, G. G., O'connor, N. E., Compton, C. C., Kehinde, O., & Green, H. (1984). Permanent coverage of large burn wounds with autologous cultured human epithelium. *The New England Journal of Medicine, 311*, 448–451.

Gautrot, J. E., MalmStröm, J., Sundh, M., Margadant, C., Sonnenberg, A., & Sutherland, D. S. (2014). The nanoscale geometrical maturation of focal adhesions controls stem cell differentiation and mechanotransduction. *Nano Letters, 14*, 3945–3952.

Gautrot, J. E., Wang, C., Liu, X., Goldie, S. J., Trappmann, B., Huck, W. T., et al. (2012). Mimicking normal tissue architecture and perturbation in cancer with engineered micro-epidermis. *Biomaterials, 33*, 5221–5229.

Green, H. (1977). Terminal differentiation of cultured human epidermal cells. *Cell, 11*, 405–416.

Harmon, R. M., Simpson, C. L., Johnson, J. L., Koetsier, J. L., Dubash, A. D., Najor, N. A., et al. (2013). Desmoglein-1/erbin interaction suppresses erk activation to support epidermal differentiation. *The Journal of Clinical Investigation, 123*, 1556–15570.

Hegazy, M., Perl, A. L., Svoboda, S. A., & Green, K. J. (2021). Desmosomal cadherins in health and disease. *Annual Review of Pathology*. https://doi.org/10.1146/annurev-pathol-042320-092912.

Helling, A. L., Viswanathan, P., Cheliotis, K. S., Mobasseri, S. A., Yang, Y., El Haj, A. J., et al. (2019). Dynamic culture substrates that mimic the topography of the epidermal-dermal junction. *Tissue Engineering. Part A, 25*, 214–223.

Hertle, M. D., Kubler, M. D., Leigh, I. M., & Watt, F. M. (1992). Aberrant integrin expression during epidermal wound healing in in psoriatic epidermis. *The Journal of Clinical Investigation, 89*, 1892–1901.

Hiratsuka, T., Bordeu, I., Pruessner, G., & Watt, F. M. (2020). Regulation of ERK basal and pulsatile activity control proliferation and exit from the stem cell compartment in mammalian epidermis. *Proceedings of the National Academy of Sciences of the United States of America, 117*, 17796–17807.

Hirsch, T., Rothoeft, T., Teig, N., Bauer, J. W., Pellegrini, G., De Rosa, L., et al. (2017). Regeneration of the entire human epidermis using transgenic stem cells. *Nature, 2551*, 327–332.

Hodivala, K. J., & Watt, F. M. (1994). Evidence that cadherins play a role in the downregulation of integrin expression that occurs during keratinocyte terminal differentiation. *The Journal of Cell Biology, 124*, 589–600.

Jensen, K. B., & Watt, F. M. (2006). Single-cell expression profiling of human epidermal stem and transit-amplifying cells: Lrig1 is a regulator of stem cell quiescence. *Proceedings of the National Academy of Sciences of the United States of America, 103*, 11958–11963.

Jiang, C., Javed, A., Kaiser, L., Nava, M. M., Xu, R., Brandt, D. T., et al. (2021). Mechanochemical control of epidermal stem cell divisions by B-plexins. *Nature Communications, 12*, 1308.

Jones, P. H., Harper, S., & Watt, F. M. (1995). Stem cell patterning and fate in human epidermis. *Cell, 80*, 83–93.

Jones, P. H., & Watt, F. M. (1993). Separation of human epidermal stem cells from transit amplifying cells on the basis of differences in integrin function and expression. *Cell*, *73*, 713–724.

Kobielak, A., & Boddupally, K. (2014). Junctions and inflammation in the skin. *Cell Communication & Adhesion*, *21*, 141–147.

Lai-Cheong, J. E., Arita, K., & McGrath, J. A. (2007). Genetic diseases of junctions. *The Journal of Investigative Dermatology*, *127*, 2713–2725.

Le, H. Q., Ghatak, S., Yeung, C. Y., Tellkamp, F., Günschmann, C., Dieterich, C., et al. (2016). Mechanical regulation of transcription controls Polycomb-mediated gene silencing during lineage commitment. *Nature Cell Biology*, *18*, 864–875.

Leblond, C. P., Greulich, R. C., & Pereira, J. P. M. (1964). Relationship of cell formation and cell migration in the renewal of stratified squamous epithelia. In W. Montagna (Ed.), *Advances in biology of skin. Volume 5: Wound healing* (pp. 39–67). Pergamon Press.

Liu, G. Y., & Sabatini, D. M. (2020). mTOR at the nexus of nutrition, growth, ageing and disease. *Nature Reviews. Molecular Cell Biology*, *21*, 183–203.

Lowell, S., Jones, P., Le Roux, I., Dunne, J., & Watt, F. M. (2000). Stimulation of human epidermal differentiation by delta-notch signalling at the boundaries of stem-cell clusters. *Current Biology*, *10*, 491–500.

Lowell, S., & Watt, F. M. (2001). Delta regulates keratinocyte spreading and motility independently of differentiation. *Mechanisms of Development*, *107*, 133–140.

Ly, T., Endo, A., Brenes, A., Gierlinski, M., Afzal, V., Pawellek, A., et al. (2018). Proteome-wide analysis of protein abundance and turnover remodelling during oncogenic transformation of human breast epithelial cells. *Wellcome Open Research*, *3*, 51.

Mertz, A. F., Che, Y., Banerjee, S., Goldstein, J. M., Rosowski, K. A., Revilla, S. F., et al. (2013). Cadherin-based intercellular adhesions organize epithelial cell-matrix traction forces. *Proceedings of the National Academy of Sciences of the United States of America*, *110*, 842–847.

Miroshnikova, Y. A., Cohen, I., Ezhkova, E., & Wickström, S. A. (2019). Epigenetic gene regulation, chromatin structure, and force-induced chromatin remodelling in epidermal development and homeostasis. *Current Opinion in Genetics & Development*, *55*, 46–51.

Mishra, A., Oulès, B., Pisco, A. O., Ly, T., Liakath-Ali, K., Walko, G., et al. (2017). A protein phosphatase network controls the temporal and spatial dynamics of differentiation commitment in human epidermis. *eLife*, *6*, e2735.

Mobasseri, S. A., Zijl, S., Salameti, V., Walko, G., Stannard, A., Garcia-Manyes, S., et al. (2019). Patterning of human epidermal stem cells on undulating elastomer substrates reflects differences in cell stiffness. *Acta Biomaterialia*, *87*, 256–264.

Nava, M. M., Miroshnikova, Y. A., Biggs, L. C., Whitefield, D. B., Metge, F., Boucas, J., et al. (2020). Heterochromatin-driven nuclear softening protects the genome against mechanical stress-induced damage. *Cell*, *181*, 800–817.

Negri, V. A., Logtenberg, M. E. W., Renz, L. M., Oulès, B., Walko, G., & Watt, F. M. (2019). Delta-like 1-mediated cis-inhibition of Jagged1/2 signalling inhibits differentiation of human epidermal cells in culture. *Scientific Reports*, *9*, 10825.

Nekrasova, O., Harmon, R. M., Broussard, J. A., Koetsier, J. L., Godsel, L. M., Fitz, G. N., et al. (2018). Desmosomal cadherin association with Tctex-1 and cortactin-Arp2/3 drives perijunctional actin polymerization to promote keratinocyte delamination. *Nature Communications*, *9*, 1053.

Ning, W., Muroyama, A., Li, H., & Lechler, T. (2021). Differentiated daughter cells regulate stem cell proliferation and fate through intra-tissue tension. *Cell Stem Cell*, *28*, 436–452.

Noethel, B., Ramms, L., Dreissen, G., Hoffmann, M., Springer, R., Rübsam, M., et al. (2018). Transition of responsive mechanosensitive elements from focal adhesions to adherens junctions on epithelial differentiation. *Molecular Biology of the Cell*, *29*, 2317–2325.

Nusse, R., & Clevers, H. (2017). Wnt/beta-catenin signaling, disease, and emerging therapeutic modalities. *Cell, 169*, 985–999.

O'Connor, N. E., Mulliken, J. B., Banks-Schlegel, S., Kehinde, O., & Green, H. (1981). Grafting of burns with cultured epithelium prepared from autologous epidermal cells. *Lancet, 1*(8211), 75–78.

Owens, D. M., & Watt, F. M. (2003). Contribution of stem cells and differentiated cells to epidermal tumours. *Nature Reviews. Cancer, 3*, 444–451.

Pora, A., Yoon, S., Dreissen, G., Hoffmann, B., Merkel, R., Windoffer, R., et al. (2020). Regulation of keratin network dynamics by the mechanical properties of the environment in migrating cells. *Scientific Reports, 10*, 4574.

Ramms, L., Fabris, G., Windoffer, R., Schwarz, N., Springer, R., Zhou, C., et al. (2013). Keratins as the main component for the mechanical integrity of keratinocytes. *Proceedings of the National Academy of Sciences of the United States of America, 110*, 18513–18518.

Read, J., & Watt, F. M. (1988). A model for in vitro studies of epidermal homeostasis: Proliferation and involucrin synthesis by cultured human keratinocytes during recovery after stripping off the suprabasal layers. *The Journal of Investigative Dermatology, 90*, 739–743.

Reynolds, G., Vegh, P., Fletcher, J., Poyner, E. F. M., Stephenson, E., Goh, I., et al. (2021). Developmental cell programs are co-opted in inflammatory skin disease. *Science, 371* (eaba6500).

Rheinwald, J. G., & Green, H. (1975). Serial cultivation of strains of human epidemal keratinocytes: The formation keratinizing colonies from single cells. *Cell, 6*, 331–343.

Richard, G. (2000). Connexins: A connection with the skin. *Experimental Dermatology, 9*, 77–96.

Romani, P., Valcarcel-Jimenez, L., Frezza, C., & Dupont, S. (2021). Crosstalk between mechanotransduction and metabolism. *Nature Reviews. Molecular Cell Biology, 22*, 22–38.

Rubin, A. J., Parker, K. R., Satpathy, A. T., Qi, Y., Wu, B., Ong, A. J., et al. (2019). Coupled single-cell CRISPR screening and epigenomic profiling reveals causal gene regulatory networks. *Cell, 176*, 361–376.

Sahu, P., Sussman, D. M., Rübsam, M., Mertz, A. F., Horsley, V., Dufresne, E. R., et al. (2020). Small-scale demixing in confluent biological tissues. *Soft Matter, 16*, 3325–3337.

Silva-Vargas, V., Lo, C. C., Giangreco, A., Ofstad, T., Prowse, D. M., Braun, K. M., et al. (2005). Β-catenin and hedgehog signal strength can specify number and location of hair follicles in adult epidermis without recruitment of bulge stem cells. *Developmental Cell, 9*, 121–131.

Tajik, A., Zhang, Y., Wei, F., Sun, J., Jia, Q., Zhou, W., et al. (2016). Transcription upregulation via force-induced direct stretching of chromatin. *Nature Materials, 15*, 1287–1296.

Tan, D. W., Jensen, K. B., Trotter, M. W., Connelly, J. T., Broad, S., & Watt, F. M. (2013). Single-cell gene expression profiling reveals functional heterogeneity of undifferentiated human epidermal cells. *Development, 140*, 1433–1444.

Totaro, A., Castellan, M., Battilana, G., Zanconato, F., Azzolin, L., Giulitti, S., et al. (2017). YAP/TAZ link cell mechanics to notch signalling to control epidermal stem cell fate. *Nature Communications, 8*, 15206.

Totaro, A., Panciera, T., & Piccolo, S. (2018). YAP/TAZ upstream signals and downstream responses. *Nature Cell Biology, 20*, 888–899.

Trappmann, B., Gautrot, J. E., Connelly, J. T., Strange, D. G., Li, Y., Oyen, M. L., et al. (2012). Extracellular-matrix tethering regulates stem-cell fate. *Nature Materials, 11*, 642–649.

Uitto, J., Has, C., Vahidnezhad, H., Youssefian, L., & Bruckner-Tuderman, L. (2017). Molecular pathology of the basement membrane zone in heritable blistering diseases: The paradigm of epidermolysis bullosa. *Matrix Biology, 57-58*, 76–85.

Valenzuela-Iglesias, A., Burks, H. E., Arnette, C. R., Yalamanchili, A., Nekrasova, O., Godsel, L. M., et al. (2019). Desmoglein 1 regulates invadopodia by suppressing EGFR/Erk signaling in an Erbin-dependent manner. *Molecular Cancer Research*, *17*, 1195–1206.

Vietri Rudan, M., Mishra, A., Klose, C., Eggert, U. S., & Watt, F. M. (2020). Human epidermal stem cell differentiation is modulated by specific lipid subspecies. *Proceedings of the National Academy of Sciences of the United States of America*, *117*, 22173–22182.

Vietri Rudan, M., & Watt, F. M. (2021). Mammalian epidermis: A compendium of lipid functionality. *Frontiers in Physiology*, *12*, 804824.

Viswanathan, P., Guvendiren, M., Chua, W., Telerman, S. B., Liakath-Ali, K., Burdick, J. A., et al. (2016). Mimicking the topography of the epidermal-dermal interface with elastomer substrates. *Integrative Biology*, *8*, 21–29.

Walko, G., Woodhouse, S., Pisco, A. O., Rognoni, E., Liakath-Ali, K., Lichtenberger, B. M., et al. (2017). A genome-wide screen identifies YAP/WBP2 interplay conferring growth advantage on human epidermal stem cells. *Nature Communications*, *8*, 14744.

Wan, H., Stone, M. G., Simpson, C., Reynolds, L. E., Marshall, J. F., Hart, I. R., et al. (2003). Desmosomal proteins, including desmoglein 3, serve as novel negative markers for epidermal stem cell-containing population of keratinocytes. *Journal of Cell Science*, *116*, 4239–4248.

Wang, S., Drummond, M. L., Guerrero-Juarez, C. F., Tarapore, E., MacLean, A. L., Stabell, A. R., et al. (2020). Single cell transcriptomics of human epidermis identifies basal stem cell transition states. *Nature Communications*, *11*, 4239.

Watt, F. M. (1984). Selective migration of terminally differentiating cells from the basal layer of cultured human epidermis. *The Journal of Cell Biology*, *98*, 16–21.

Watt, F. M. (1988). The epidermal keratinocyte. *BioEssays*, *8*, 163–167.

Watt, F. M. (2002). Role of integrins in regulating epidermal adhesion, growth and differentiation. *The EMBO Journal*, *21*, 3919–3926.

Watt, F. M. (2014). Mammalian skin cell biology: At the interface between laboratory and clinic. *Science*, *346*, 937–940.

Watt, F. M., Estrach, S., & Ambler, C. A. (2008). Epidermal notch signalling: Differentiation, cancer and adhesion. *Current Opinion in Cell Biology*, *20*, 171–179.

Watt, F. M., & Green, H. (1982). Stratification and terminal differentiation of cultured epidermal cells. *Nature*, *295*, 434–436.

Watt, F. M., Jordan, P. W., & O'Neill, C. H. (1988). Cell shape controls terminal differentiation of human epidermal keratinocytes. *Proceedings of the National Academy of Sciences of the United States of America*, *85*, 5576–5580.

Wen, J. H., Vincent, L. G., Fuhrmann, A., Choi, Y. S., Hribar, K. C., Taylor-Weiner, H., et al. (2014). Interplay of matrix stiffness and protein tethering in stem cell differentiation. *Nature Materials*, *13*, 979–987.

Zhu, A. J., Haase, I., & Watt, F. M. (1999). Signaling via β1 integrins and mitogen-activated protein kinase determines human epidermal stem cell fate in vitro. *Proceedings of the National Academy of Sciences of the United States of America*, *96*, 6728–6733.

Zhu, A. J., & Watt, F. M. (1996). Expression of a dominant negative cadherin mutant inhibits proliferation and stimulates terminal differentiation of human epidermal keratinocytes. *Journal of Cell Science*, *109*, 3013–3023.

Zhu, A. J., & Watt, F. M. (1999). Beta-catenin signalling modulates proliferative potential of human epidermal keratinocytes independently of intercellular adhesion. *Development*, *126*, 2285–2298.

Zijl, S., Vasilevich, A. S., Viswanathan, P., Helling, A. L., Beijer, N. R. M., Walko, G., et al. (2019). Micro-scaled topographies direct differentiation of human epidermal stem cells. *Acta Biomaterialia*, *84*, 133–145.

CHAPTER FIVE

# Cell signaling pathways controlling an axis organizing center in the zebrafish

**William D. Jones and Mary C. Mullins***
Department of Cell and Developmental Biology, University of Pennsylvania Perelman School of Medicine, Philadelphia, PA, United States
*Corresponding author: e-mail address: mullins@pennmedicine.upenn.edu

## Contents



## Abstract

Body axis formation in vertebrate development entails the remarkable feat of patterning a myriad of specialized cell types and organ progenitors from a field of unpatterned, multipotent cells. This feat is achieved largely by secreted cell-cell signaling molecules, enabling cells at different positions within the embryo to adopt distinct fates. During patterning of the vertebrate embryonic axes, a multitude of cell fates is induced by a

*Current Topics in Developmental Biology*, Volume 150
ISSN 0070-2153
https://doi.org/10.1016/bs.ctdb.2022.03.005

surprisingly small set of signaling pathways: Wnt, Nodal, Bone Morphogenetic Protein (BMP), and Fibroblast Growth Factor (FGF) signaling. These signals function as morphogens, specifying multiple cell fates in a concentration-dependent mechanism, and must therefore be distributed non-uniformly throughout the embryo. A primary signaling center that sets up spatial asymmetries in these signaling pathways to break the symmetry of the vertebrate embryo is known as the dorsal organizer. Discovered nearly a century ago by Hilde Mangold and Hans Spemann in the newt, the organizer has the remarkable ability to induce a secondary body axis when grafted ectopically into a host embryo. Here, we review the cell-cell signaling pathways that control the establishment of the dorsal organizer and its inductive functions in the zebrafish *Danio rerio*, a vertebrate model highly amenable to genetic manipulation. The organizer's remarkable inductive abilities continue to provide a fascinating source of scientific inquiry in the field of developmental biology.

## 1. Introduction

The body plan of early vertebrate embryos is set up by a small handful of signaling pathways, including Wnt, Nodal, Bone Morphogenetic Protein (BMP), and Fibroblast Growth Factor (FGF) signaling. Ultimately, these signaling pathways modulate gene expression in cells of the early embryo, activating and repressing target genes to promote a particular cell fate while repressing alternative fates (reviewed in Barolo & Posakony, 2002). These few signaling pathways generate myriad cell types through their ability to act as morphogens. A morphogen directly induces two or more distinct cell fates within a field of unpatterned cells in a concentration-dependent manner: that is, cells exposed to different concentrations of the morphogen adopt distinct fates. This mechanism requires the morphogen concentration to be non-uniform throughout a field of cells (Green & Sharpe, 2015; Wolpert, 1969, 2011). This can be accomplished by a small prior asymmetry in the field, such as a source of a diffusible ligand at the edge of a field of cells, or a source of an inhibitor targeting a broadly expressed ligand. This small asymmetry sets up a concentration gradient, with the highest concentration of ligand at one edge of the field and the lowest concentration of ligand at the other.

In all vertebrate embryos studied so far, the dorsal side of the body axis is established by a group of cells called the dorsal organizer, which sets up asymmetries in multiple morphogen gradients in the embryo. The organizer was first discovered by Hilde Mangold and her mentor Hans Spemann in *Triturus* newts (Spemann & Mangold, 2001). Spemann had previously shown that dividing the gastrula-stage newt in half could give rise to two identical twins with complete body axes. However, the divided embryo only formed identical twins if both halves contained the domain that gives rise to the dorsal lip of the blastopore, the location in the embryo where cells first move internally via involution to form the germ layers during gastrulation (Moriyama & De Robertis, 2018; Spemann, 1962). Mangold then showed that the dorsal lip of the blastopore—later dubbed the dorsal organizer—can induce a complete secondary body axis when it is grafted to a host embryo in an ectopic location (Fig. 1). This property of the organizer has also been demonstrated in the mouse (Beddington, 1994), chick (Waddington, 1932), and zebrafish (Koshida, Shinya, Mizuno, Kuroiwa, & Takeda, 1998; Saúde, Woolley, Martin, Driever, & Stemple, 2000; Shih & Fraser, 1996), indicating that this function is highly conserved throughout vertebrates. In this chapter, we discuss how the zebrafish organizer is established and how it promotes dorsal fates in the embryo. While the dorsal organizer was discovered in amphibians and has been studied extensively in

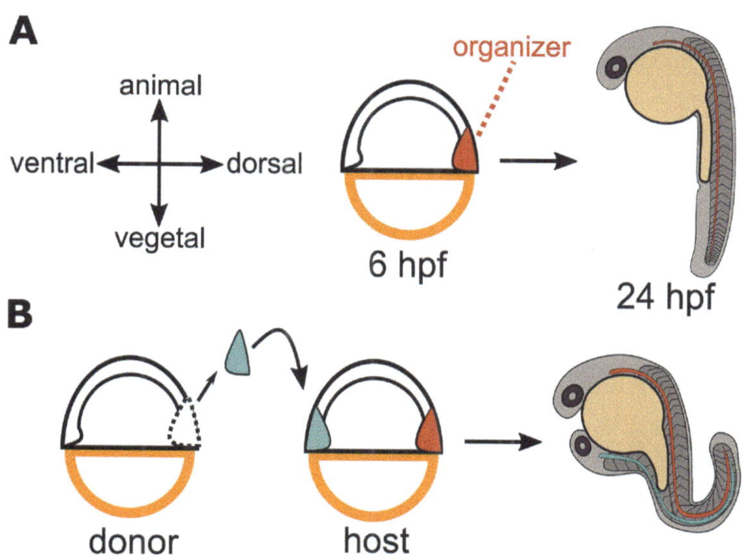


**Fig. 1** The dorsal organizer is sufficient to duplicate the body axis when grafted into a host embryo. (A) The zebrafish dorsal organizer (labeled in red) is roughly equivalent, but not identical, to the shield, a dorsal marginal thickening that appears by approximately 6 hours post-fertilization (hpf) as convergence movements begin. (B) Hilde Mangold showed in the newt that grafting an organizer from a donor embryo into a recipient embryo induces a secondary body axis. This axis-inducing ability of the organizer is conserved in zebrafish, as shown in this schematic. The shield gives rise to the axial tissues of the zebrafish, principally the notochord (labeled), prechordal plate, floor plate of the neural tube, and hatching gland (not shown).

*Xenopus laevis*, here we focus on zebrafish, where large-scale mutant screens and reverse genetics have enabled loss-of-function genetic analyses that so far have been difficult to perform in *Xenopus*. The zebrafish egg is also externally fertilized and the embryo translucent, allowing development to be more easily analyzed and visualized than in mammalian models.

# 2. Defining the zebrafish dorsal organizer

## 2.1 Functions of the organizer

Based on the experiments of Mangold and Spemann, a rigorous functional definition of the dorsal organizer can be stated: the organizer is the collection of cells that are both necessary to establish a complete body axis and sufficient to duplicate the body axis when grafted to a host embryo. To form a body axis, the organizer exerts dorsalizing inductive effects upon the three primary germ layers of the embryo: the ectoderm, which gives rise to neural tissue and the epidermis; the mesoderm, which gives rise to blood, muscles, and notochord; and the endoderm, which gives rise to the gut. The organizer has three primary activities: first, in the mesoderm the organizer promotes dorsal fates such as notochord and inhibits ventral fates such as blood; second, in the ectoderm the organizer promotes neural fates and inhibits non-neural fates, like the epidermis; and third, the organizer activates convergence and extension movements that draw cells into the growing body axis and cause the axis to thicken and elongate.

There is also evidence that the organizer anteriorizes the endoderm, promoting dorsally-derived anterior endoderm at the expense of ventrally-derived posterior endoderm (reviewed in Harland & Gerhart, 1997). For example, in zebrafish the dorsal-to-ventral arrangement of endodermal precursors in the late blastula tends to correspond to their anterior-to-posterior position after the completion of organogenesis (Kimmel, Warga, & Schilling, 1990; Warga & Nüsslein-Volhard, 1999). During gastrulation, the organizer may promote dorsally-located anterior endoderm at the expense of more ventrally-derived posterior endoderm in zebrafish by inhibiting BMP signaling (Tiso, Filippi, Pauls, Bortolussi, & Argenton, 2002). However, the organizer's specific effect on the endoderm is not well studied and is not further discussed here.

In this chapter, we focus on the cell signaling pathways that establish the organizer and underlie its dorsalizing functions in the mesoderm and ectoderm. While dorsalizing mesoderm and neuralizing ectoderm are sometimes discussed separately, here we label them both as "dorsalizing" activities, as in both cases the organizer promotes dorsally-derived tissues at the expense of ventrally-derived ones. The signaling pathways controlling the organizer's third activity—convergence and extension movements—have been recently reviewed elsewhere (Roszko, Sawada, & Solnica-Krezel, 2009; Williams & Solnica-Krezel, 2020).

## 2.2 The zebrafish embryonic shield is not identical to the organizer

In zebrafish, the organizer is roughly equivalent, but not identical, to a structure called the embryonic shield. The shield is a dorsal thickening of cells that becomes morphologically apparent at the early gastrula stage, approximately 6 hours post-fertilization (hpf), as cells begin to converge dorsally (Kimmel, Ballard, Kimmel, Ullmann, & Schilling, 1995; Koshida et al., 1998; Saúde et al., 2000; Shih & Fraser, 1996). The shield is located at the dorsal margin of the embryo, where the blastoderm meets the yolk cell. Fate-mapping experiments have demonstrated that the cells of the shield give rise to dorsal midline tissues in the zebrafish, principally the notochord, prechordal plate, floor plate of the neural tube, and hatching gland (Kimmel et al., 1990; Melby, Warga, & Kimmel, 1996; Saúde et al., 2000; Shih & Fraser, 1996).

The organizer is often equated to the shield because grafting the shield to a host embryo induces a secondary axis (Fig. 1). However, equating the organizer to the shield has two caveats. First, the organizer begins to function well before the shield becomes morphologically apparent at 6 hpf. In particular, when the shield is removed from a zebrafish embryo, dorsal midline derivatives are lost, but an otherwise normal body axis forms with correct anteroposterior and dorsoventral patterning (Saúde et al., 2000; Shih & Fraser, 1996). Similarly, some mutant and morphant embryos lacking a morphological shield can exhibit delayed organizer induction and can ultimately form a nearly complete body axis, showing that a morphological shield is dispensable for most organizer activity (Dougan, Warga, Kane, Schier, & Talbot, 2003; Fan et al., 2007; Fekany et al., 1999; Kelly, Chin, Leatherman, Kozlowski, & Weinberg, 2000; Koos & Ho, 1999). In contrast, genetic manipulations that block organizer formation (discussed below) cause a severely ventralized phenotype including complete loss of the head, the notochord, and other dorsal derivatives. Thus, the shield is sufficient but may not be necessary to establish a body axis in zebrafish, as the organizer exerts its inductive effects before the shield becomes morphologically apparent. A second caveat of equating the organizer with the shield is that some cells of the organizer actually lie adjacent to the shield

(Saúde et al., 2000). Thus, even at the early gastrula stage, the organizer is not completely identical to the shield. In spite of these caveats, the shield is still a useful, albeit imperfect, morphological definition of the organizer.

## 2.3 Defining the organizer with molecular markers

While the functional definition of the organizer is perhaps the most rigorous, identifying the organizer using molecular markers is much more convenient experimentally. Commonly utilized markers of the organizer in zebrafish include the homeobox transcription factor *goosecoid* (*gsc*), which marks prechordal plate mesoderm and was among the first organizer marker genes discovered in amphibians, and the prospective notochord marker *notochord homeobox* (*noto*, also known as *floating head*) (Blumberg, Wright, De Robertis, & Cho, 1991; Halpern et al., 1995; Schulte-Merker et al., 1994; Stachel, Grunwald, & Myers, 1993; Talbot et al., 1995). While these molecular markers provide a convenient and clear method to identify the organizer, neither ubiquitously and specifically marks the entire organizer at all developmental stages. For example, *gsc* and *noto* mark mostly overlapping domains at late blastula stages but largely distinct domains by early gastrula stages, with *noto* confined to more superficial cells and *gsc* to deep cells (Gritsman, Talbot, & Schier, 2000; Melby et al., 1996; Saúde et al., 2000). These distinct sub-regions of the organizer have different inductive properties in grafting experiments in the zebrafish: grafts of deep, primarily *gsc*-expressing cells induce anterior structures, while grafts of superficial, primarily *noto*-expressing cells induce posterior structures (Saúde et al., 2000). Both *gsc*- and *noto*-expressing regions of the organizer are needed to induce a complete secondary axis, showing that neither *gsc* nor *noto* marks the entire organizer (Saúde et al., 2000). Similarly, grafting of anterior and posterior sub-regions of the amphibian organizer—sometimes called "head" and "trunk" organizer—gives rise to secondary axes containing primarily anterior and posterior structures, respectively (Spemann, 1962; Zoltewicz & Gerhart, 1997) (reviewed in Saxén & Toivonen, 1962).

The organizer also expresses secreted factors that pattern cells along the dorsoventral axis. During the first searches for secreted factors produced by the organizer, it was assumed that the organizer would be a source of many signaling ligands that control dorsoventral patterning. Surprisingly, it was discovered that many factors secreted by the organizer are not signaling ligands but rather signaling inhibitors that antagonize broadly expressed signals in the embryo (reviewed in De Robertis, 2009). Thus, the organizer dorsalizes the embryo in part by inhibiting broadly expressed ventralizing ligands, which has led to the hypothesis that dorsal mesoderm and neuroectoderm could be "default" fates (reviewed in Ozair, Kintner, & Brivanlou, 2013). Secreted inhibitors expressed by the organizer include Chordin, Noggin 1, and Follistatin-like 1b, which inhibit BMP signaling (Dal-Pra, Fürthauer, Van-Celst, Thisse, & Thisse, 2006; Hammerschmidt et al., 1996; Hemmati-Brivanlou, Kelly, & Melton, 1994; Sasai et al., 1994; Smith & Harland, 1992), and Dickkopf 1b and Frizzled-related proteins, which antagonize Wnt/β-catenin signaling (Glinka et al., 1998; Hashimoto et al., 2000; Lu, Thisse, & Thisse, 2011; Shinya, Eschbach, Clark, Lehrach, & Furutani-Seiki, 2000; Tendeng & Houart, 2006; Wang, Krinks, Lin, Luyten, & Moos, 1997).

## 3. Establishment of the zebrafish dorsal organizer

The single-celled zebrafish embryo is initially radially symmetric about the animal-vegetal axis, requiring a symmetry-breaking event to establish the organizer and produce an embryo with distinct dorsal and ventral sides. Animal-vegetal polarity is established in the oocyte, where a structure called the Balbiani body deposits mRNAs and proteins vegetally (reviewed in Fuentes et al., 2020). After a zebrafish egg is laid, cytoplasmic streaming segregates the yolk from the cytoplasm, establishing a non-yolky animal pole, called the blastodisc, and a yolky vegetal pole. Dorsoventral polarity is established after fertilization by molecules known as dorsal determinants, which accumulate on the prospective dorsal side of the embryo. In this section, we discuss how dorsal determinants establish the zebrafish organizer. These dorsal determinants, initially localized vegetally in the egg, are transported by microtubules to the prospective dorsal side of the embryo following fertilization. There, they activate a Wnt/β-catenin signaling pathway to establish the dorsal organizer.

### 3.1 Microtubule arrays in the yolk cortex transport dorsal determinants to the future dorsal blastomeres

#### *3.1.1 The yolk cell contains the dorsal determinants*

Dorsal determinants are initially localized to the vegetal pole of the zebrafish embryo. In support of this notion, removing the vegetal half of the yolk cell during the first 20 minutes post-fertilization (mpf) produces embryos with a severely ventralized phenotype and blocks expression of the dorsal organizer marker *gsc* (Koshida et al., 1998; Mizuno, Yamaha, Kuroiwa, & Takeda,

1999; Ober & Schulte-Merker, 1999). However, removal of the vegetal yolk between 80 and 90 mpf does not affect *gsc* expression, indicating that a sufficient amount of dorsal determinant has translocated toward the dorsal blastomeres between 20 and 80 mpf (that is, before the 16-cell stage) (Mizuno et al., 1999). Furthermore, grafting a second yolk cell onto a blastula-stage host embryo can induce a second source of *gsc* in the host (Mizuno et al., 1999; Ober & Schulte-Merker, 1999), but this dorsal-inducing ability is lost if the donor yolk cell has had its vegetal half removed before 20 mpf (Mizuno et al., 1999). Together, these observations show that dorsal determinants are initially present in the vegetal half of the yolk and suggest that they are transported out of the vegetal pole within the first few cell divisions (Fig. 2).

Early experiments established that dorsal determinant transport depends on microtubules in the yolk cell. Destabilizing microtubules prior to the first cell division with ultraviolet radiation, cold treatment, or the drug nocodazole all prevent formation of the dorsal organizer (Jesuthasan & Strähle, 1997; Tran et al., 2012). Interestingly, transport of the dorsal determinants depends on two separate populations of microtubules: a short-lived parallel microtubule array established at the vegetal pole cortex (hereafter called "vegetal microtubules"), and a later, longer-lived population of longitudinal microtubules stretching animal-vegetally within the cytoplasm along the yolk lateral cortex (hereafter called "longitudinal microtubules") (Fig. 2).

### 3.1.2 Vegetal microtubules and Kinesin 1 motors drive the initial asymmetric transport of the dorsal determinants

The vegetal microtubules are established in the yolk cell prior to the first cell division and run parallel to each other along the vegetal pole cortex. These microtubules are short-lived, persisting only between ~15–30 mpf, and drive an approximately 20° translocation of specific mRNAs and proteins along the yolk cortex (Jesuthasan & Strähle, 1997; Tran et al., 2012). This mechanism is analogous to the cortical rotation observed in *Xenopus* embryos, in which the entire cortex of the single-celled embryo rotates approximately 30° relative to the denser cytoplasm underneath it (reviewed in Gerhart, 2010; Houston, 2012).

The vegetal microtubules are thought to transport the dorsal determinants from the center of the vegetal yolk cortex to an off-center position, moving the dorsal determinants toward the future dorsal side of the embryo (Fig. 2). Four molecules have been shown to be transported in this way

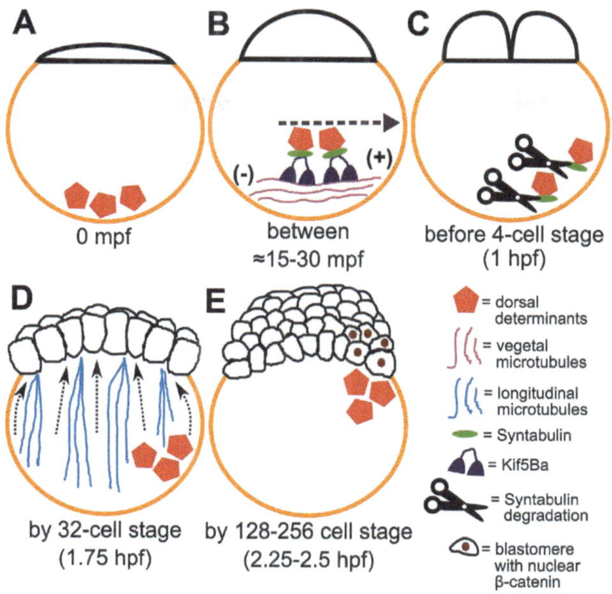


**Fig. 2** The zebrafish dorsal determinants are transported to dorsal blastomeres by two separate populations of microtubules. (A) The zebrafish egg contains maternally-deposited dorsal determinants located at the vegetal pole. (B) Between 15 and 30 min post-fertilization (mpf), vegetal microtubules are present at the vegetal pole cortex. Plus-end (+) directed Kinesin 1 motors (Kif5Ba) and the Syntabulin linker protein transport the dorsal determinants along the vegetal microtubules to one side of the embryo. (C) By the 4-cell stage, Syntabulin rapidly degrades, which is thought to release the dorsal determinants. (D) By the 32-cell stage, longitudinal microtubules mediate cytoplasmic streaming that moves the dorsal determinants animally along the yolk cortex toward the prospective dorsal blastomeres. (E) By the 128- to 256-cell stage, the dorsal determinants stabilize β-catenin 2 in prospective dorsal blastomeres, enabling β-catenin 2 to enter prospective dorsal nuclei and activate Wnt/β-catenin target genes.

during the first 30 mpf: Syntabulin protein, *grip2a* mRNA, *wnt8a* mRNA, and *huluwa* mRNA (Ge et al., 2014; Lu et al., 2011; Nojima et al., 2010; Yan et al., 2018). Transport of Syntabulin, *grip2a*, and *wnt8a* is blocked when embryos are incubated in the microtubule depolymerizing drug nocodazole, causing Syntabulin, *grip2a*, and *wnt8a* to remain at the vegetal pole (Ge et al., 2014; Lu et al., 2011; Nojima et al., 2010). The vegetal microtubules were specifically disrupted in elegant experiments by placing nocodazole-soaked beads next to the vegetal yolk, which destabilizes only the microtubules nearest the bead (Tran et al., 2012). Bead-treated embryos exhibited no dorsal nuclear translocation of β-catenin, demonstrating that the vegetal microtubules are essential to establish the dorsal side of the embryo.

While the precise mechanism of dorsal determinant transport along the vegetal microtubules remains unknown, some molecular components of this process have been identified via maternal-effect loss-of-function mutants. The *tokkaebi* ventralized maternal-effect mutant revealed that Syntabulin, a Kinesin 1 cargo linker protein, is maternally required for organizer formation (Nojima et al., 2004, 2010) (Fig. 3). Syntabulin protein moves from the center of the vegetal pole to an off-center position between 20 and 50 mpf in a microtubule-dependent manner (Nojima et al., 2010). This indicates that the dorsal determinants are probably linked via Syntabulin to Kinesin 1 microtubule motors. In support of this model, embryos maternally deficient in Kinesin 1 (Kif5Ba) are strongly ventralized (Campbell, Heim, Smith, & Marlow, 2015), like maternal *syntabulin* mutants. Furthermore, consistent with transport of the dorsal determinant by Kinesin 1, which is a plus-end directed microtubule motor (reviewed in Verhey, Kaul, & Soppina, 2011), the plus end of the vegetal microtubule arrays is oriented toward the future dorsal side of the embryo (Tran et al., 2012). Interestingly, Syntabulin protein appears to be actively degraded around the 2-cell stage while its mRNA persists (Nojima et al., 2010). Thus, degradation of this linker protein could release the dorsal determinants after being transported asymmetrically by the vegetal microtubules, allowing the dorsal determinants to complete their translocation toward the prospective dorsal blastomeres via the lateral longitudinal microtubules (next section).

Three maternal-effect mutants disrupt the vegetal microtubules, but their exact function in dorsal determinant transport remains unknown. First, in the maternal-effect mutant *brom bones*, which disrupts the gene *polypyrimidine tract binding protein 1a* (*ptbp1a*), embryos exhibit a disorganized vegetal microtubule array, leading to a ventralized phenotype. *ptbp1a* encodes an RNA-binding protein of the heterogeneous nuclear ribonucleo protein (hnRNP) family. In *ptbp1a* mutant embryos, cortical granules in the early embryo fail to be exocytosed after fertilization, preventing proper organization of the vegetal microtubules. The severity of the *ptbp1a* phenotype is enhanced by cold treatment, which further destabilizes the microtubules (Mei, Lee, Marlow, Miller, & Mullins, 2009). Second, maternal-zygotic mutants of *dachsous1b*, which encodes a cadherin, exhibit various defects including misoriented and thickened vegetal microtubule bundles. These disorganized microtubules lead to defects in microtubule transport and reduced or patchy dorsal gene expression (Li-Villarreal et al., 2015). Third, in the maternal-effect mutant *hecate*, which disrupts

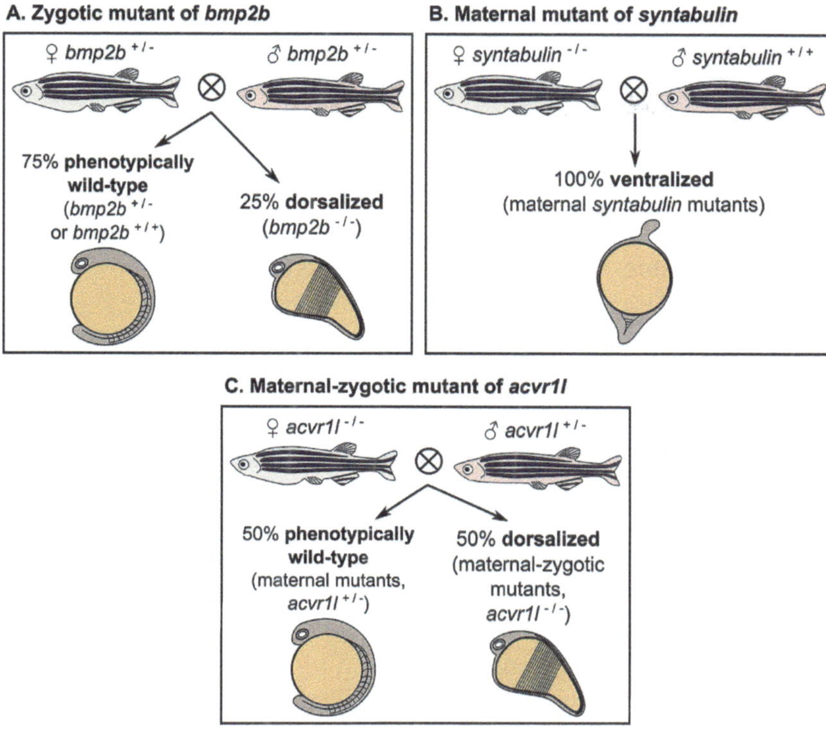


**Fig. 3** Comparing zygotic and maternal-effect mutants. In each diagram, + represents a wild-type allele and - represents a recessive null allele. (A) For recessive zygotic mutant alleles, intercrossing adult heterozygotes produces 25% homozygous mutant progeny. Zygotic *bmp2b* homozygous mutant embryos are strongly dorsalized, with dorsolaterally-derived somites expanded ventrolaterally, encircling the yolk (Mullins et al., 1996). (B) For a recessive maternal-effect mutation, all progeny of homozygous mutant females display the mutant phenotype, regardless of the genotype of the male parent. Maternal mutants of *syntabulin* are strongly ventralized because Syntabulin is required for dorsal determinant transport (Nojima et al., 2004). (C) In crosses between homozygous mutant females of the Type I BMP receptor *acvr1l* and heterozygous males, 50% of the progeny are maternal-zygotic mutants of *acvr1l*. Because *acvr1l* homozygous mutant fish display a mildly dorsalized phenotype and die by 3 dpf, homozygous embryos must be rescued by injection of *acvr1l* RNA to generate homozygous mutant females. Maternal-zygotic mutants of *acvr1l* display a strongly dorsalized phenotype (Mintzer et al., 2001). However, zygotic expression of *acvr1l* fully rescues maternal deficiency of Acvr1l (left, maternal mutant).

*glutamate receptor interacting protein 2a* (*grip2a*), embryos display disorganized vegetal microtubule arrays and fail to establish a dorsal organizer (Ge et al., 2014; Lyman Gingerich, Westfall, Slusarski, & Pelegri, 2005). In the *grip2a* mutant, *wnt8a* transcript and Syntabulin protein remain localized at the center of the vegetal pole and fail to be transported off-center by the vegetal microtubules (Ge et al., 2014). While the exact roles of Ptbp1a, Dachsous1b, and Grip2a in dorsal determinant transport remain unclear, these mutants underscore the importance of properly organized vegetal microtubules for this process.

### 3.1.3 Longitudinal microtubules complete the transport of dorsal determinants to the prospective dorsal blastomeres

While the vegetal microtubules move the dorsal determinants off-center at the vegetal pole, longitudinal microtubules are thought to move the dorsal determinants animally to the prospective dorsal blastomeres. In support of this model, disrupting microtubules by cold exposure as late as the 64-cell stage (2 hpf) reduces or blocks *gsc* expression (Jesuthasan & Strähle, 1997), while vegetal microtubules persist only from 15 to 30 mpf (Jesuthasan & Strähle, 1997; Tran et al., 2012). Thus, microtubules continue to transport the dorsal determinants after the vegetal microtubules dissociate. Longitudinal microtubules are thought to mediate this second phase of dorsal determinant transport. The longitudinal microtubules, which run along the yolk cell cortex between the region abutting the blastomere margin and the vegetal pole, are established by the 16-cell stage and persist through the 256-cell stage (Jesuthasan & Strähle, 1997; Mei et al., 2009). These microtubules generate vegetal-to-animal cytoplasmic streaming movements, as labeled polystyrene beads injected vegetally at the 2–4 cell stage can be transported along the cortex to the marginal blastomeres by the 128-cell stage (Jesuthasan & Strähle, 1997). Strikingly, two beads injected slightly off-center on opposite sides of the embryo are transported animally along opposite sides of the cortex (Ge et al., 2014). Thus, unlike the vegetal microtubules, which only mediate dorsally-directed transport, the longitudinal microtubules mediate vegetal-to-animal transport throughout the yolk cell cortex. Transcripts of *wnt8a* and *huluwa*, which are moved off-center at the vegetal pole by the vegetal microtubule array, continue to move animally along the cortex during the cleavage stages, suggesting that they are transported by the longitudinal microtubules (Lu et al., 2011; Yan et al., 2018). However, the longitudinal microtubules have not yet been specifically disrupted experimentally, so further experiments are needed to probe their role in dorsal determinant transport.

Further evidence for a second step in dorsal determinant transport mediated by longitudinal microtubules has been provided by the *hecate* (*grip2a*) maternal-effect mutant. As discussed above, *hecate* embryos display disorganized vegetal microtubules and fail to form an organizer. Strikingly, in *hecate* embryos, fluorescent beads injected near the vegetal pole are still transported animally despite the disorganization of vegetal microtubules (Ge et al., 2014). Hence, the *hecate* mutant exclusively disrupts dorsally-directed transport along the vegetal microtubules, while animally-directed transport by the longitudinal microtubules remains active. However, it remains unclear exactly how Grip2a functions molecularly in vegetal microtubule organization and dorsal determinant transport.

Additional experiments will continue to shed light on the dorsal determinant transport pathway in zebrafish. For example, in a newly described maternal-effect mutant called *dullahan*, which has yet to be identified molecularly, embryos display an enlarged cytoplasmic domain between the blastomeres and the yolk and also exhibit a reduced or absent dorsal organizer (Abrams et al., 2020). Maternal loss of *dullahan* could impede dorsal determinant transport as a secondary effect similar to *ptbp1a*, or *dullahan* could encode a novel component of the dorsal determinant transport pathway. Further experiments are also needed to reveal the precise timing and mechanism of dorsal determinant translocation from the vegetal to the longitudinal microtubules, and finally the mechanism of delivery to the prospective dorsal blastomeres. For example, live imaging of *wnt8a, grip2a,* and *huluwa* transcripts or Syntabulin protein would better characterize the path and timing of dorsal determinant transport.

## 3.2 Stabilization of β-catenin establishes the dorsal side of the embryo

After reaching the prospective dorsal side of the embryo, the dorsal determinants activate a canonical Wnt/β-catenin signaling pathway, ultimately stabilizing β-catenin in dorsal blastomere nuclei. In canonical Wnt/β-catenin signaling, Wnt ligands stabilize β-catenin by binding to Frizzled/LRP heterodimeric receptors, which sequester to the cell membrane components of the β-catenin destruction complex, consisting of Axin, Adenomatous Polyposis Coli (APC), Casein Kinase 1 (CK1), and Glycogen Synthase Kinase 3 (GSK-3) (reviewed in Nusse & Clevers, 2017) (Fig. 4). With the destruction complex sequestered to the membrane, β-catenin can then enter the nucleus, where it acts as a transcriptional co-activator with TCF/Lef transcription factors to initiate transcription of Wnt pathway target genes. In contrast, when Wnt ligands are absent, cytoplasmic β-catenin is phosphorylated by GSK-3 and CK1 and ubiquitinated by the E3 ubiquitin ligase βTrCP, promoting its degradation by the proteasome (Fig. 4). In zebrafish, nuclear

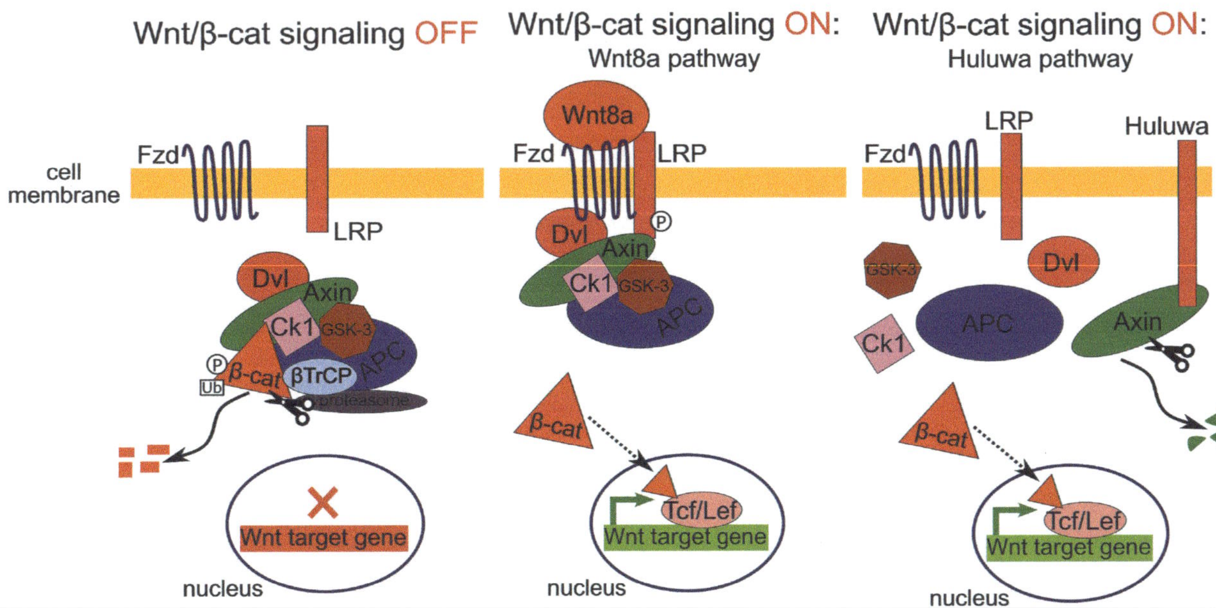


**Fig. 4** Maternal Wnt/β-catenin signaling and zygotic Wnt8a signaling may stabilize nuclear β-catenin by distinct mechanisms. Canonical Wnt/β-catenin signaling works by stabilizing an intracellular pool of β-catenin (β-cat). (Left) In the absence of a Wnt ligand, a destruction complex consisting of Axin, Casein Kinase 1 (CK1), Glycogen Synthase Kinase 3 (GSK-3), Adenomatous Polyposis Coli (APC), and the E3 ubiquitin ligase βTrCP phosphorylates (P) and ubiquitinates (Ub) cytoplasmic β-catenin, targeting it for degradation by the proteasome. (Center) When a Wnt ligand binds the Frizzled (Fzd) receptor and LRP co-receptor, LRP is phosphorylated and directly binds Axin, sequestering the destruction complex to the membrane. This stabilizes cytoplasmic β-catenin, enabling it to enter the nucleus and complex with Tcf/Lef transcription factors to activate Wnt/β-catenin target gene transcription. (Right) The candidate dorsal determinant Huluwa is thought to act independently of Wnt ligand and Frizzled/LRP co-receptors. Huluwa, a transmembrane protein, binds directly to Axin and promotes its degradation, which may prevent the destruction complex from degrading β-catenin without Frizzled/LRP receptor activation. Note that destruction complex-bound β-catenin and activated LRP are both phosphorylated at multiple sites, although only one is shown (see Nusse & Clevers, 2017).

localization of β-catenin is detected in the dorsal blastomeres by the early blastula stage (128- to 256-cell stage, ~2.25 hpf) and persists at least through the mid-blastula stage (dome, 4.33 hpf) (Dougan et al., 2003; Kelly et al., 2000; Nojima et al., 2010; Schneider, Steinbeisser, Warga, & Hausen, 1996) but is no longer evident by the onset of gastrulation (Schneider et al., 1996).

The involvement of a Wnt/β-catenin signaling pathway in establishing the dorsal side of the embryo has been demonstrated through both loss- and gain-of-function experiments. Classic experiments in *Xenopus* and later in zebrafish showed that exposing embryos to lithium cations causes strong dorsalization and can induce multiple dorsal axes (Shao et al., 2012; Stachel et al., 1993). In this context, lithium inhibits GSK-3, ectopically stabilizing β-catenin and dorsalizing the embryo (Klein & Melton, 1996). Similarly, morpholino depletion of Wnt intracellular inhibitor Axin1, which binds to cytoplasmic β-catenin and promotes its degradation, dorsalizes zebrafish embryos and can produce multiple dorsal axes (Heisenberg et al., 2001; Schneider, Slusarski, & Houston, 2012). Overexpression of Wnt components including β-catenin, the Wnt8a ligand, or a Frizzled receptor also dorsalizes zebrafish embryos (Kelly et al., 2000; Kelly, Erezyilmaz, & Moon, 1995). Furthermore, maternal loss-of-function of *β-catenin 2* (*ichabod*) in zebrafish abrogates organizer formation and causes strong ventralization (Kelly et al., 2000). Thus, canonical Wnt/β-catenin signaling establishes the dorsal side of the embryo. Genes activated by β-catenin on the dorsal side of the embryo include the homeodomain transcription factor *dharma* (*bozozok/nieuwkoid*) (Leung, Soll, Arnold, Kemler, & Driever, 2003; Ryu et al., 2001), the *nodal-related 1* (*ndr1/squint*) ligand (Kelly et al., 2000; Shimizu et al., 2000), and the Wnt antagonist *dickkopf1b* (*dkk1b*) (Shinya et al., 2000; Tanaka, Hosokawa, Weinberg, & Maegawa, 2017), discussed in further detail below. More recently, RNA-seq comparing ventralized *β-catenin 2* mutant embryos to *β-catenin 2* mutant embryos rescued with *β-catenin 2* RNA has revealed additional candidate target genes of β-catenin in dorsoventral patterning (Fodor et al., 2013).

While the dorsal determinant promotes nuclear localization of β-catenin, several factors in zebrafish have been reported to antagonize this process, restricting the specification of the dorsal organizer. For example, Forkhead box O3b (Foxo3b) (Xie, Liu, Hu, & Xiao, 2011), Caveolin 1 (Cav1) (Mo et al., 2010), Leucine zipper, putative tumor suppressor 2a (Lzts2a) (Li, Li, Long, & Cui, 2011), and Transducer of ERBB2, 1a (Tob1a) (Xiong et al., 2006) have all been reported to interact with β-catenin in zebrafish, and their depletion causes dorsalization. However, maternal mutants of these factors are needed to demonstrate their role in dorsoventral patterning. Chemokine (C—C motif) receptor 7 (Ccr7), a G-protein coupled receptor, has been suggested to promote β-catenin degradation by mobilizing intracellular calcium stores (Wu, Shin, Sepich, & Solnica-Krezel, 2012). Depletion of Ccr7 was reported to cause a dorsalized phenotype (Wu et al., 2012). However, recently generated maternal-zygotic Ccr7 mutants are not dorsalized, although they are sensitized to β-catenin 2 overexpression at a dose that has no effect on wild-type embryos (Malhotra, Shin, Solnica-Krezel, & Raz, 2018). Finally, it has recently been shown that Nanog, which has a well-known role in zygotic genome activation, is also required to restrict β-catenin activity (He et al., 2020). Nanog interacts with the Tcf7 transcriptional co-activator, preventing β-catenin from binding Tcf and blocking Wnt/β-catenin target gene activation (He et al., 2020). Maternal-zygotic *nanog* mutants exhibit a dorsalized phenotype due to ectopic activation of β-catenin target genes (He et al., 2020). Hence, restriction of β-catenin activity by various factors prevents ectopic dorsal fate specification in zebrafish.

## 3.3 Possible identities of the dorsal determinant

Since the dorsal determinant specifies the dorsal side of the embryo through canonical Wnt/β-catenin signaling, it has been hypothesized that the dorsal determinant is a Wnt ligand. Indeed, in *Xenopus* embryos, Wnt5a and Wnt11 complexes are thought to act as the dorsal determinants by activating both canonical and non-canonical Wnt signaling (Cha, Tadjuidje, Tao, Wylie, & Heasman, 2008). In zebrafish, the dorsal determinant should fulfill several criteria: it must be maternally expressed, it must be localized at the vegetal pole initially and localized asymmetrically at the margin by early blastula stages, and it must be both necessary and sufficient to promote dorsal fates. These criteria have prompted systematic searches for maternally-expressed, vegetally-localized Wnt ligands that could be the dorsal determinant(s) (Hino et al., 2018; Lu et al., 2011).

One Wnt that could be a dorsal determinant is Wnt8a, which is maternally expressed and can dorsalize embryos when overexpressed. Furthermore, *wnt8a* transcripts are initially localized at the vegetal pole and then transported asymmetrically to one side of the embryo by microtubules (Lu et al., 2011). Since loss of *wnt8a* is lethal zygotically (Lekven, Thorpe, Waxman, & Moon, 2001), it was challenging to test its maternal function as a candidate dorsal determinant. However, maternally mutant *wnt8a* embryos have recently been generated by germ-line replacement, a technique in which germ cells from zygotically mutant *wnt8a* embryos were transplanted into germ cell-deficient wild-type host embryos (Ciruna et al., 2002). Surprisingly, *wnt8a* maternal mutant embryos exhibit a wild-type phenotype, arguing that Wnt8a is not the dorsal determinant (Hino et al., 2018). An alternate Wnt candidate dorsal determinant is *wnt6a*, which is also maternally deposited at the vegetal pole and can also dorsalize embryos when overexpressed (Hino et al., 2018). However, maternal *wnt6a* mutants have not yet been generated. It is also possible that *wnt8a* and *wnt6a* function redundantly as dorsal determinants, requiring double maternal mutant generation to test this hypothesis.

If a Wnt ligand is the dorsal determinant, then blocking the activity of the Frizzled-LRP co-receptor complex should ablate the dorsal organizer and cause a ventralized phenotype. The role of Frizzled-LRP activity in the early embryo has been tested by both loss- and gain-of-function experiments. Interpreting these experiments is somewhat difficult, however, because a Wnt/β-catenin signaling pathway acts twice in dorso ventral patterning and in opposite ways. The early maternal Wnt/β-catenin pathway induces the dorsal organizer, whereas a slightly later zygotic Wnt/β-catenin pathway, mediated by ventrolaterally expressed Wnt8a, represses the dorsal organizer and keeps it confined to a dorsal region. Thus, disrupting Frizzled-LRP activity may affect both Wnt-mediated dorsal determination and the later ventralizing pathway. Consistent with these two roles, overexpression of the Wnt antagonist Dickkopf 1b, which binds directly to the Wnt co-receptor LRP and promotes LRP endocytosis, can reduce the expression of early dorsal marker *dharma* (Hino et al., 2018), but ultimately dorsalizes rather than ventralizes embryos (Hashimoto et al., 2000; Hino et al., 2018; Shinya et al., 2000; Yan et al., 2018). Similarly, overexpression of dominant-negative LRP5 dorsalizes embryos (Yan et al., 2018). On the other hand, depletion of Frizzled related protein, a Wnt antagonist that blocks interactions between Wnt ligands and Frizzled, causes dorsalization at blastula stages (Lu et al., 2011), consistent with a role in antagonizing Frizzled in organizer induction. Finally, the zebrafish Dishevelled (Dvl) proteins, Dvl2 and Dvl3a, are not required maternally or zygotically for dorsal fate specification (Xing et al., 2018). The exact role of Dishevelled proteins in Wnt signaling is still a subject of debate, but they contain Frizzled-binding domains and stabilize β-catenin downstream of Wnt/Frizzled receptor binding (Mlodzik, 2016). The observation that disrupting Frizzled-LRP activity fails to ventralize the embryo could indicate that the dorsal determinant is not a Wnt ligand. Alternatively, ventralized phenotypes might be rescued to a dorsalized phenotype because disrupting zygotic Wnt/Frizzled interactions downregulates the organizer-inhibiting Wnt8a pathway in later blastula stages (Hino et al., 2018). It therefore remains unclear whether a Wnt ligand is the dorsal determinant.

One candidate dorsal determinant that is not a Wnt ligand is the recently discovered transmembrane protein Huluwa, identified via a spontaneous maternal-effect mutant in zebrafish. Similar to the progeny of *β-catenin 2* mutant females, the progeny of homozygous *huluwa* mutant females exhibit severe ventralization. *huluwa* mRNA is localized to the Balbiani body of zebrafish oocytes and to the vegetal pole of late-stage oocytes and the early embryo, as predicted for the dorsal determinant. Huluwa overexpression stimulates nuclear translocation of β-catenin and induces a secondary body axis. While dorsal fate induction by Huluwa is β-catenin-dependent, the dorsalization of embryos by Huluwa misexpression cannot be blocked by overexpressing Dickkopf, dominant negative Wnt8a, or dominant-negative LRP5. Hence, dorsalization of the embryo by Huluwa does not require Frizzled-LRP activity, suggesting that Huluwa may function downstream of or in parallel to Frizzled-LRP. Mechanistically, Huluwa is thought to stabilize β-catenin by binding and degrading Axin1, a scaffold protein for the cytoplasmic β-catenin destruction complex (Yan et al., 2018) (Fig. 4).

While Huluwa is a strong candidate for the dorsal determinant, significant questions remain. When and where is the initially vegetally localized *huluwa* mRNA translated into protein? Does Huluwa protein act in the yolk cell membrane, or is the mRNA or protein transported into prospective dorsal blastomeres and the protein inserted into the cell membrane to bind Axin1? Furthermore, if *wnt8a* and *wnt6a* are also transported dorsally by microtubules, do they activate Wnt/β-catenin signaling in parallel to Huluwa? Finally, if a Wnt ligand is the dorsal determinant, how is it secreted from the yolk cell to bind to Frizzled receptors on prospective dorsal blastomeres? And if a Wnt ligand is not a dorsal determinant, what are the functions of the vegetally-localized *wnt* transcripts and why do they not activate Wnt signaling dorsally? Further experiments are needed to answer these critical questions surrounding dorsal determinant function.

## 3.4 β-Catenin activates the expression of *dharma* and Nodal-related ligands to establish the dorsal organizer

### *3.4.1 Dharma represses* bmp2b *expression downstream of β-catenin*

One transcriptional target of β-catenin downstream of the dorsal determinant is the homeodomain transcription factor *dharma* (also known as *bozozok* and *nieuwkoid*) (Fekany et al., 1999; Koos & Ho, 1998; Shimizu et al., 2000;

Sirotkin, Dougan, Schier, & Talbot, 2000; Yamanaka et al., 1998). *dharma* is first expressed in dorsal blastomeres shortly after the mid-blastula transition, when the major wave of zygotic genome activation begins (Koos & Ho, 1998; Yamanaka et al., 1998). Expression of *dharma* is activated in the dorsal blastomeres by β-catenin (Kelly et al., 2000; Shimizu et al., 2000), which is thought to engage TCF/Lef transcription factors at the promoter of *dharma* (Leung, Soll, et al., 2003; Ryu et al., 2001). In *dharma* mutant embryos, the dorsal organizer is strongly reduced, blocking notochord formation and reducing anterior neuroectoderm (Fekany et al., 1999; Fekany-Lee, Gonzalez, Miller-Bertoglio, & Solnica-Krezel, 2000; Koos & Ho, 1999). Furthermore, overexpression of Dharma causes expansion of the organizer (Koos & Ho, 1998; Yamanaka et al., 1998). Hence, *dharma* is both necessary and sufficient for organizer gene expression. Importantly, however, in *dharma* mutant embryos not all organizer gene expression is lost (Fekany et al., 1999; Fekany-Lee et al., 2000; Koos & Ho, 1998; Shimizu et al., 2000; Sirotkin et al., 2000; Yamanaka et al., 1998), indicating that another signal must act in parallel to *dharma* to establish the organizer (Nodal signaling, discussed in Section 3.4.2 below).

A critical role of Dharma is the direct transcriptional repression of *bmp2b* and repression of the transcription factors *vox*, *vent*, and *ved* in a small patch of dorsal cells, mediating an initial asymmetry that enables organizer induction (Imai et al., 2001; Kawahara, Wilm, Solnica-Krezel, & Dawid, 2000; Koos & Ho, 1999; Leung et al., 2003; Melby, Beach, Mullins, & Kimelman, 2000). Ventral domains of the embryo are patterned by a secreted heterodimer of Bmp2b (also known as Swirl) and Bmp7a (also known as Snailhouse) (Little & Mullins, 2009; Tajer, Dutko, Little, & Mullins, 2021), which are initially expressed broadly in early blastula-stage embryos (Nikaido, Tada, Saji, & Ueno, 1997; Schmid et al., 2000). The expression of both *bmp2b* and *bmp7a* subsequently becomes further ventrally restricted, first due to transcriptional repression by FGF signaling by late blastula stages and later due to the activity of BMP antagonists secreted by the organizer during gastrulation (Fürthauer, Thisse, & Thisse, 1997; Fürthauer, Van Celst, Thisse, & Thisse, 2004; Londin, Niemiec, & Sirotkin, 2005; Maegawa, Varga, & Weinberg, 2006). Dharma is responsible in part for setting up this early asymmetry by directly repressing *bmp* gene expression in the cells of the dorsal organizer beginning at sphere stage (4 hpf) (Koos & Ho, 1999; Leung, Bischof, et al., 2003). Furthermore, Dharma represses expression of the transcription factors *vox*, *vent*, and *ved*, which repress the organizer and at later mid-gastrula stages are also transcriptional targets of BMP signaling

(Gilardelli, Pozzoli, Sordino, Matassi, & Cotelli, 2004; Imai et al., 2001; Kawahara et al., 2000; Koos & Ho, 1999; Leung, Bischof, et al., 2003; Melby et al., 2000; Ramel, Buckles, Baker, & Lekven, 2005). Dharma acts as a transcriptional repressor through an N-terminal Engrailed Homology 1 (Eh1) motif, which enables Dharma to bind directly to transcriptional co-repressor Groucho (Shimizu et al., 2002). By repressing *bmp2b* and *vox*, *vent*, and *ved* in the dorsal midline, Dharma enables organizer induction.

Dharma protein has also been identified as a target for post-translational polyubiquitination by the E3 ubiquitin ligase Lnx2b (Ro & Dawid, 2009). Lnx2b directly binds to Dharma protein and mediates its degradation by the proteasome. Consistent with this activity, depletion of Lnx2b expands the organizer, dorsalizing the embryo, while overexpression of Lnx2b causes a loss of dorsal midline tissues. *lnx2b* is maternally expressed and is distributed ubiquitously in the early embryo through the late blastula stage (Ro & Dawid, 2009). Further experiments are needed to determine how Lnx2b activity is regulated in dorsal domains to repress the organizer.

### 3.4.2 Nodal ligands act as morphogens to activate dorsal gene expression downstream of β-catenin

In addition to Dharma, Nodal signaling acts downstream of β-catenin in organizer induction (Feldman et al., 1998; Shimizu et al., 2000; Sirotkin et al., 2000). Nodal signaling is broadly required for the induction of mesendoderm around the zebrafish margin, including the dorsal midline mesoderm that makes up the organizer (Feldman et al., 1998; Feldman, Dougan, Schier, & Talbot, 2000; Kimmel et al., 1990; Melby et al., 1996). These functions are mediated by two Nodal ligands expressed during gastrulation in zebrafish: Nodal-related 1 (Ndr1, also known as Squint) and Ndr2 (also known as Cyclops). A third zebrafish Nodal ligand, called Southpaw, is required during somitogenesis stages for left-right patterning (Long, Ahmad, & Rebagliati, 2003). Although *ndr1* is first expressed maternally (Gore et al., 2005; Goudarzi, Berg, Pieper, & Schier, 2019; Lim et al., 2012), maternal *ndr1* mutants are phenotypically wild-type (Bennett et al., 2007; Goudarzi et al., 2019). Furthermore, blocking Nodal receptor activity with small molecule inhibitors beginning at the mid-blastula transition prevents mesoderm specification, showing that any maternal Nodal ligands do not function until after the mid-blastula transition (Hagos & Dougan, 2007). Nodal ligands heterodimerize with Gdf3 (Vg1) in inducing mesendoderm (further discussed below).

*ndr1* is first zygotically expressed in dorsal blastomeres at the 1000-cell stage (Feldman et al., 1998; Rebagliati, Toyama, Fricke, Haffter, & Dawid, 1998). By the late blastula stage, both *ndr1* and *ndr2* are expressed around the margin, including in an extraembryonic structure called the yolk syncytial layer (YSL), which acts as a source of Nodal ligands (Feldman et al., 1998; Rebagliati et al., 1998; Sampath et al., 1998). The YSL arises during the tenth cell division in zebrafish, during which some of the most vegetal blastomeres in direct contact with the yolk—the marginal blastomeres—collapse their cell membranes and incorporate their contents into the adjacent yolk cell (Kimmel et al., 1995; Kimmel & Law, 1985). The initial mid-blastula expression and dorsal bias of *ndr1* suggest that *ndr1* is a direct target of β-catenin. Consistent with this, overexpressing β-catenin upregulates *ndr1* (Shimizu et al., 2000), while loss of maternal *β-catenin 2* blocks early dorsal expression of *ndr1* (Kelly et al., 2000). Interestingly, however, the later marginal expression of *ndr1* does not require maternal *β-catenin 2* (Kelly et al., 2000), indicating that another factor is sufficient to initiate this later phase of *ndr1* expression. Unlike *ndr1*, *ndr2* has no reported early dorsal expression prior to its expression around the margin but becomes restricted to the shield by the onset of gastrulation (Rebagliati et al., 1998; Sampath et al., 1998). While overexpression of β-catenin can induce *ndr2* expression (Dougan et al., 2003), it has not been reported whether dorsal *ndr2* expression requires β-catenin.

The two Nodal ligands function partially redundantly to induce the mesendoderm. *ndr1;ndr2* double mutants exhibit a severe phenotype with loss of dorsal midline mesendodermal derivatives, including trunk notochord and head mesoderm (Dougan et al., 2003; Feldman et al., 1998). However, *ndr1;ndr2* double mutants still display anterior neural derivatives and some *chordin* expression (Dougan et al., 2003; Feldman et al., 1998), showing that some organizer activity remains in the absence of Nodal signaling. In contrast, double loss of Nodal signaling and *dharma* completely blocks organizer induction and anterior neural fates (Shimizu et al., 2000; Sirotkin et al., 2000), showing that Nodal signaling and Dharma cooperate to induce the organizer. Consistent with a role in organizer induction, overexpression of Ndr1 induces *gsc* and *noto* expression (Chen & Schier, 2001; Feldman et al., 1998; Rebagliati et al., 1998) and can also induce a partial secondary body axis with a secondary notochord but not a second head (Rebagliati et al., 1998).

Nodal ligands, which are members of the TGFβ family along with BMPs, signal by binding heterotetrameric receptor serine-threonine kinase complexes (reviewed in Heldin & Moustakas, 2016; Zinski, Tajer, & Mullins, 2018). It has recently been demonstrated that Nodal ligands signal as obligate heterodimers with Growth differentiation factor 3 (Gdf3, also known as Vg1) (Bisgrove, Su, & Yost, 2017; Montague & Schier, 2017; Pelliccia, Jindal, & Burdine, 2017). *gdf3* is maternally expressed and ubiquitously distributed in the zebrafish embryo through mid-gastrula stages (Helde & Grunwald, 1993). Nodal signaling also requires the EGF-CFC co-receptor One-eyed pinhead (Oep, also known as Teratocarcinoma-derived growth factor 1) (Gritsman et al., 1999). After Nodal ligand binding, the receptor complex phosphorylates signal transducer Smad2, which translocates to the nucleus, associates with the transcriptional co-activator Smad4, and promotes Nodal target gene transcription by activating transcription factors such as FoxH1 (reviewed in Zinski et al., 2018). While Smad4 was thought to be an essential component of both Nodal and BMP signaling, recently generated maternal-zygotic Smad4 mutants exhibit wild-type levels of Nodal signaling, albeit with partially delayed and less stable signaling, calling into question the role of Smad4 in Nodal signaling (Guglielmi et al., 2021).

Nodal ligands are thought to act as morphogens during gastrulation, patterning cell fates in a dose-dependent manner (Chen & Schier, 2001; Dubrulle et al., 2015; Gritsman et al., 2000; Harvey & Smith, 2009; Thisse, Wright, & Thisse, 2000). In general, Nodal activity forms a gradient of high activity near the margin and low activity near the animal pole. This signaling gradient has been modeled quantitatively by immunological detection of endogenous phosphorylated Smad2 or by measuring the nuclear localization of tagged Smad2 (Dubrulle et al., 2015; Harvey & Smith, 2009; Lord, Carte, Abitua, & Schier, 2021). In general, cells at the highest levels of Nodal signaling adopt an endodermal fate, cells at lower Nodal levels adopt a mesodermal fate, and cells without Nodal signaling adopt an ectodermal fate (Dougan et al., 2003; Thisse et al., 2000). The Nodal signaling gradient is shaped in part by both positive and negative feedback: *ndr1* transcription in the YSL is positively regulated by a highly conserved Nodal Response Element in its first intron (Fan et al., 2007), while the Nodal feedback inhibitor Lefty limits Nodal signaling activity to domains near the margin (Chen & Schier, 2002; Müller et al., 2012; Rogers et al., 2017; van Boxtel et al., 2015). Recently, it has been appreciated that the distribution of Nodal co-receptor Oep also influences the Nodal signaling gradient. In particular, Oep-dependent binding of Nodal ligands at the cell membrane limits Nodal ligand diffusion, maintaining high Nodal signaling levels near the margin and lower Nodal levels more animally (Lord et al., 2021). Additionally, the timing of Oep expression affects Nodal signaling duration, as reduced duration of Oep expression in ectodermal domains attenuates Nodal responsiveness, preventing the mesendoderm from expanding animally (Vopalensky, Pralow, & Vastenhouw, 2018).

It remains unresolved how Nodal ligands, which are expressed around the margin, induce the organizer only dorsally. One possibility is that organizer induction could require higher Nodal signaling levels than general mesendoderm induction. For example, high levels of Ndr1 and Ndr2 overexpression induce prechordal plate and organizer marker *gsc* and pan-mesodermal marker *tbxta* (also known as *no tail*), while lower levels only induce *tbxta* (Chen & Schier, 2001; Gritsman et al., 2000). It has also been reported that levels of phosphorylated Smad2 are higher in dorsal than ventrolateral cells of the margin (Harvey & Smith, 2009). Expression of *oep* at the onset of gastrulation exhibits a strong dorsal bias (Zhang, Talbot, & Schier, 1998), so higher dorsal Oep levels could also contribute to higher Nodal signaling levels dorsally to promote organizer induction. However, by comparing *tbxta* expression in mutants in which one or both copies of *ndr1* and *ndr2* are deficient, the mesendoderm is more sensitive to loss of Nodal signaling dorsally than ventrolaterally (Dougan et al., 2003). Furthermore, ventrolateral mesoderm does not expand dorsally in *ndr1;ndr2* double mutants (Dougan et al., 2003). These observations are not consistent with the notion that a dorsoventral gradient of Nodal signaling distinguishes dorsal from dorsolateral mesendoderm. Similarly, a transgenic zebrafish Nodal reporter shows similar Nodal signaling levels in dorsal and ventrolateral mesoderm, further suggesting that different Nodal signaling levels alone do not distinguish between the dorsal organizer fate and general mesendoderm fate (Sako et al., 2016).

Other studies have shown that the duration of Nodal signaling can also control target gene expression. For example, experiments varying the timing of Oep activity in maternal-zygotic *oep* mutants suggested that induction of *gsc* requires longer Nodal signaling duration than induction of *flh* (Gritsman et al., 2000). Similarly, ablating Nodal signaling using small molecule inhibitors at different developmental time-points showed that duration of Nodal signaling distinguishes different cell fates (Hagos & Dougan, 2007). Finally, recent experiments using photoactivatable Nodal receptors in the zebrafish have shown that a long duration of Nodal signaling can promote organizer induction and repress endoderm (Sako et al., 2016). Thus, the longer Nodal duration at the dorsal margin than at the dorsolateral margin may play a role in distinguishing dorsal midline mesoderm from dorsolateral mesoderm.

Finally, Nodal may require the input of other signaling pathways, in addition to Dharma, to distinguish dorsal midline from dorsolateral mesendoderm. For example, different embryonic structures can be induced by different ratios of Nodal to BMP activity (Fauny, Thisse, & Thisse, 2009; Soh, Pomreinke, & Müller, 2020). In *Xenopus*, the genes *gsc* and *chd*, which are expressed in the organizer, are under the control of both Nodal and Wnt signaling (Reid, Zhang, Sheets, & Kessler, 2012). Finally, a recent study also demonstrates that FGF signaling may act downstream of Nodal to distinguish mesoderm and endoderm (van Boxtel, Economou, Heliot, & Hill, 2018). In summary, Nodal signaling levels, signaling duration, and interactions with BMP, Wnt, or other signaling pathways may all contribute to how Nodal distinguishes between dorsal midline and dorsolateral mesoderm.

### 3.4.3 The yolk syncytial layer is a source of mesendoderm-inducing activity that shares some properties with the amphibian Nieuwkoop center

The extraembryonic YSL (reviewed in Carvalho & Heisenberg, 2010) is the source of Nodal ligands Ndr1 and Ndr2 that induce the mesendoderm (Erter, Solnica-Krezel, & Wright, 1998; Feldman et al., 1998; Hong, Jang, Brown, McBride, & Feldman, 2011; Rebagliati et al., 1998; Sampath et al., 1998) and shares some properties with the Nieuwkoop center of amphibians. Classic transplantation experiments by Pieter Nieuwkoop in the axolotl revealed the existence of a group of vegetal blastomeres—now called the Nieuwkoop center—that have the capacity to induce dorsal mesendoderm including the dorsal organizer (Nieuwkoop, 1969). Nieuwkoop observed that, in mid- to late-blastula axolotl embryos, blastomeres isolated from the animal half of the embryo—so-called animal caps—gave rise only to ectoderm, while blastomeres isolated from the vegetal half of the embryo gave rise only to yolky endoderm. Only when the animal and vegetal explants were brought into direct contact did the resulting graft give rise to mesoderm and complete endodermal derivatives, which arose at the interface between them. Furthermore, Nieuwkoop found that grafting ventrolateral vegetal blastomeres to animal caps induced ventrolateral mesendoderm, while only the most dorsal vegetal blastomeres induced the dorsal organizer and dorsal mesendoderm. The Nieuwkoop center at the mid- to late-blastula stage in amphibians therefore constitutes a dorsal mesendoderm-inducing activity (reviewed in Gerhart, 1999; Kimelman, 2006). Molecularly, the Nieuwkoop center constitutes a region of Wnt/β-catenin and Nodal signaling activity.

The zebrafish YSL is the source of Nodal ligands and also exhibits Wnt/β-catenin signaling activity. In particular, although β-catenin first localizes to nuclei of prospective dorsal marginal blastomeres by the 128- to 256- cell stage, it subsequently localizes to nuclei of the dorsal YSL when the YSL forms at the 512-cell stage (Dougan et al., 2003; Nojima et al., 2010; Schneider et al., 1996). Furthermore, the dorsal YSL and dorsal blastomeres both transcribe *dharma*, and by the mid-blastula stage (sphere, 4 hpf) *dharma* expression becomes restricted to the dorsal YSL (Chen & Kimelman, 2000; Koos & Ho, 1998; Yamanaka et al., 1998). Thus, the dorsal YSL is a source of mesendoderm-inducing signals and expresses an organizer-inducing transcription factor, suggesting that the dorsal YSL is similar to the amphibian Nieuwkoop center. However, in zebrafish the dorsal determinants promote nuclear β-catenin localization in prospective dorsal blastomeres before the YSL forms at the 512-cell stage. Thus, while organizer induction in zebrafish requires Wnt/β-catenin and Nodal signaling activity as in *Xenopus*, the YSL is the primary source of Nodal ligands, while Wnt/β-catenin activity is shared between the YSL and the dorsal blastomeres.

In grafting experiments, the zebrafish YSL has inductive abilities similar to the Nieuwkoop center. For example, grafting a YSL and attached yolk cell to the animal pole of a host embryo induces an ectopic ring of expression of the pan-mesodermal marker *tbxta* and a smaller ectopic domain of the organizer marker *gsc*, showing that the YSL can induce dorsal mesoderm (Koshida et al., 1998; Mizuno et al., 1999). Similarly, grafting a zebrafish animal cap to a YSL and attached yolk cell also induces *gsc* and *tbxta* in the animal cap (Ober & Schulte-Merker, 1999). While these experiments cannot distinguish between whether the YSL itself or a factor from elsewhere within the yolk cell induces the mesendoderm, the induced mesendoderm in the host is in direct contact with the donor YSL (Mizuno et al., 1999). Overall, these elegant grafting experiments suggest that the YSL shares similar inductive abilities to the amphibian Nieuwkoop center.

Loss-of-function experiments support the primary role of the YSL as a source of Nodal signals. For example, injection of *ndr1* and *ndr2* morpholinos into the YSL blocks notochord induction (Fan et al., 2007; Hong et al., 2011), showing that Nodal signals from the YSL are required for the induction of a dorsal mesodermal derivative. Furthermore, injection of RNase into the YSL, which ablates all RNA expression in the YSL only, completely ablates ventrolateral expression of *tbxta* (Chen & Kimelman, 2000). Interestingly, however, injecting RNase into the YSL does not affect dorsal expression of either *gsc* or *tbxta*, suggesting that some dorsal mesoderm can form in the absence of YSL transcription. Nuclear β-catenin in the prospective dorsal blastomeres likely maintains *gsc* and *tbxta* expression in the dorsal margin when RNA is depleted from the YSL (Dougan et al., 2003; Nojima et al., 2010; Schneider et al., 1996). Taken together, these observations suggest that the YSL has limited similarities to the amphibian Nieuwkoop center: it is a source of Nodal signals that induce the mesendoderm, and along with the yolk it has the ability to induce the organizer in grafting experiments.

It is interesting that the YSL, which does not contribute to any adult structure of the zebrafish, has a critical role as the source of Nodal ligands to induce the mesendoderm. Further work is needed to understand the evolution and molecular function of the YSL. In the last decade, a microarray study of transcripts isolated from the YSL identified new genes expressed in this structure (Hong et al., 2010), one of which, *mxtx2*, was identified as an upstream regulator of Nodal transcription in the YSL (Hong et al., 2011). Identifying further YSL-specific genes could help to reveal additional molecular players underlying the inductive abilities of the YSL.

## 4. Cell signaling underlying the organizer's dorsalizing activities

### 4.1 Dorsoventral patterning is controlled by ventrolaterally expressed BMPs and secreted BMP antagonists from the organizer

#### *4.1.1 Ventrolaterally expressed Bmp2b/Bmp7a heterodimers ventralize the embryo*

Bmp2b/7a secreted heterodimers act as a morphogen to specify ventrolateral fates in the early zebrafish embryo. Initially, both *bmp2b* and *bmp7a* are ubiquitously expressed in the early blastoderm, but during mid-blastula stages (dome, 4.3 hpf) *bmp2b* and *bmp7a* expression is repressed by Dharma in dorsal marginal cells (Leung, Bischof, et al., 2003a; Nikaido et al., 1997; Schmid et al., 2000). *bmp2b* and *bmp7a* expression becomes further confined to ventral domains of the embryo during late blastula stages (Nikaido et al., 1997; Schmid et al., 2000). During mid- to late blastula stages, FGF signaling directly represses *bmp* gene expression (Fürthauer et al., 2004; Londin et al., 2005) (Fig. 5A). During gastrulation, the BMP antagonists Chordin, Noggin, and Follistatin, which are secreted by the organizer, antagonize BMP signaling by direct binding to BMP ligands (Fig. 5B). These secreted antagonists also repress *bmp* gene expression through a feedback loop, as

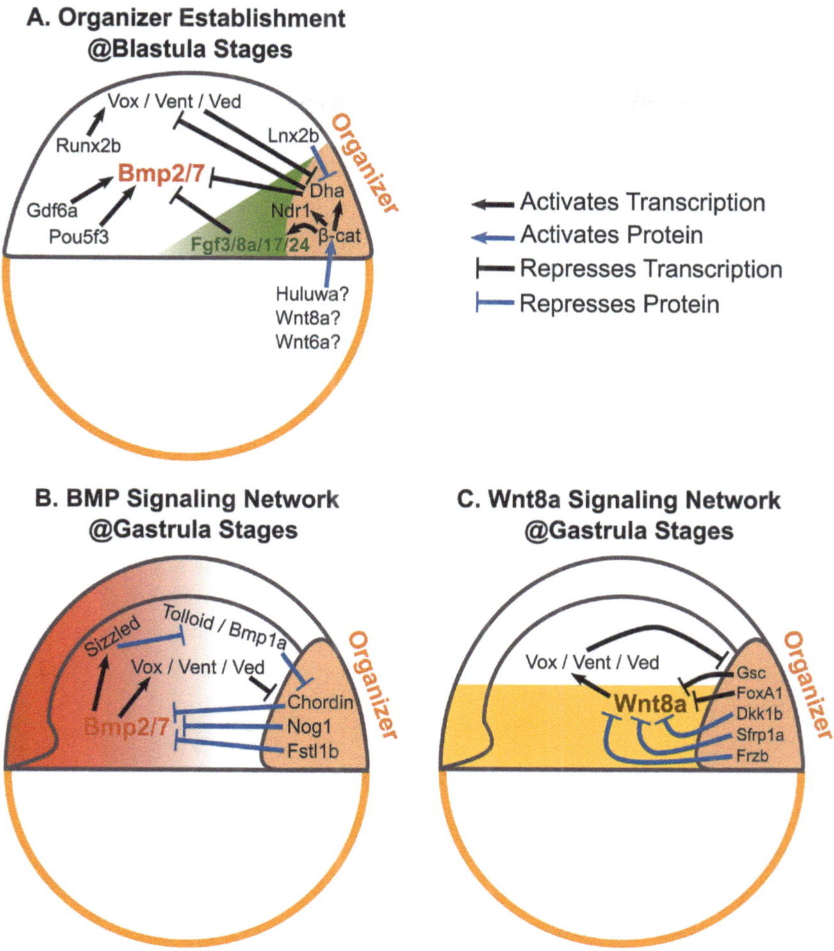


**Fig. 5** Organizer formation and function at blastula and gastrula stages in the zebrafish embryo. (A) Dorsal determinants (such as Huluwa, Wnt8a, and/or Wnt6a) promote nuclear localization of β-catenin in prospective dorsal blastomeres, promoting transcription of *dharma* (*dha*), *ndr1*, and *fgfs*. During blastula stages, transcription of broadly expressed BMP ligands *bmp2b* and *bmp7a* is repressed dorsally by Dharma and FGF signaling. Transcription of *vox*, *vent*, and *ved* is thought to be activated by Runx2b and promoted primarily by Wnt8a beginning at late blastula stages (not shown). (B) During gastrula stages, BMP heterodimers are bound dorsally by a sink of BMP antagonists, namely Chordin, Noggin 1 (Nog1), and Follistatin-like 1b (Fstl1b), which are secreted from the organizer. Meanwhile Chordin is cleaved by metalloproteases Tolloid and Bmp1a, which themselves are antagonized by Sizzled. BMP signaling also maintains transcription of *vox*, *vent*, and *ved* at mid-gastrula stages. (C) Signaling by Wnt8a in the ventrolateral margin represses the organizer via the transcription factors Vox, Vent, and Ved. Meanwhile, the organizer represses *wnt8a* expression in the dorsal midline via Gsc and FoxA3 and antagonizes Wnt8a signaling via the secreted factors Dkk1b, Sfrp1a, and Frzb.

BMP signaling maintains *bmp2b* and *bmp7a* expression during gastrulation (Nguyen et al., 1998; Schulte-Merker, Lee, McMahon, & Hammerschmidt, 1997).

BMP signaling ventralizes embryos, promoting non-neural ectoderm and ventrolateral mesoderm at the expense of neuroectoderm and dorsolateral mesoderm, respectively. The ventralizing activity of BMP signaling has been demonstrated by overexpressing BMP ligands, mutating or knocking down BMP inhibitors Chordin, Noggin 1, and Follistatin-like 1b, or overexpressing a constitutively active BMP receptor (Dal-Pra et al., 2006; Hammerschmidt, Serbedzija, & McMahon, 1996; Koshida et al., 2002; Nguyen et al., 1998; Nikaido, Tada, Takeda, Kuroiwa, & Ueno, 1999). Conversely, zygotic mutations in *bmp2b* or *bmp7a*, overexpression of Chordin, or overexpression of a dominant negative BMP receptor have the opposite effects: expanding neuroectoderm at the expense of non-neural ectoderm and expanding dorsolateral mesoderm at the expense of ventrolateral mesoderm (Miller-Bertoglio, Fisher, Sánchez, Mullins, & Halpern, 1997; Nguyen et al., 1998; Nikaido et al., 1999; Schmid et al., 2000).

Like Nodal and other TGFβ family members, BMP dimeric ligands signal by binding a heterotetrameric receptor complex consisting of two Type I and two Type II serine-threonine kinase receptors. Upon binding ligand, the Type II receptors phosphorylate and activate the kinases of the Type I receptors, which phosphorylate Smad1/5/8, the transcriptional effector that activates BMP target genes (reviewed in Heldin & Moustakas, 2016). During zebrafish dorsoventral patterning, a heterodimeric Bmp2/7 ligand exclusively signals, and this signaling occurs through a receptor complex containing two distinct type I receptors, Bmpr1 and Acvr1l (Little & Mullins, 2009; Schmid et al., 2000; Tajer et al., 2021). Unexpectedly, it has recently been shown that the Bmp2/7 signal is solely transmitted through the Acvr1l kinase activity. While Acvr1l kinase activity is essential for signaling, surprisingly Bmpr1 kinase activity is not, as a kinase dead Bmpr1 receptor can rescue Bmpr1 deficient embryos (Tajer et al., 2021).

Although both Bmp2 and Bmp7 homodimers form and are secreted in zebrafish, they cannot signal except when overexpressed (Little & Mullins, 2009; Tajer et al., 2021). It has recently been demonstrated that the increased potency of Bmp2/7 heterodimers in signaling is not caused by preferential binding of BMP homodimers to the secreted BMP antagonists Chordin, Noggin, and Follistatin (Tajer et al., 2021). Interestingly, overexpressed Bmp2 homodimers also signal through both Bmpr1 and Acvr1l (Tajer et al., 2021), although *in vitro* BMP2 fails to bind ACVR1 (Heinecke et al., 2009; Saremba et al., 2008). Instead, BMP7 normally binds and signals through ACVR1. The potency of Bmp2/7 heterodimers may stem from their unique ability to bind the two different type I receptors Acvr1l and Bmpr1, which may synergize in some manner in transducing the BMP signal (Tajer et al., 2021).

Bmp2/7 heterodimers act as a morphogen to pattern the embryonic dorsoventral axis. These heterodimers establish a ventral-to-dorsal morphogen gradient of BMP signaling, which has been quantitatively visualized using immunofluorescence of phosphorylated Smad5 protein (p-Smad5) (Pomreinke et al., 2017; Zinski et al., 2017). How the BMP morphogen signaling gradient is interpreted to activate target genes and specify cell fates has recently been investigated using multiple analyses (Greenfeld, Lin, & Mullins, 2021; Rogers, ElGamacy, Jordan, & Müller, 2020). In one study, direct BMP target genes were identified using RNA-seq, and their spatial expression profiles were determined using single-cell RNA-seq and Seurat analysis (Greenfeld et al., 2021). This enabled the identification of distinct spatial domains of BMP target gene expression, which correlated with different BMP signaling levels as determined by p-Smad5 immunofluorescence (Greenfeld et al., 2021). By comparing how target gene expression changes in wild-type and *chordin* mutant embryos, it was shown that the expression of four BMP target genes is controlled by 3 distinct threshold levels of BMP signaling rather than the slope of the BMP gradient (Greenfeld et al., 2021). Duration of BMP signaling was excluded as a mechanism of gradient interpretation, based on the ability of a short (<30 min) pulse of BMP2/7 to activate expression of all 3 target gene domains (Greenfeld et al., 2021). Furthermore, a prolonged (2 h) but low amplitude dose of BMP2/7 was unable to induce high threshold genes, eliminating this signal duration induction model for the target genes examined (Greenfeld et al., 2021).

In a second study, candidate BMP target genes were identified using RNA-seq and their temporal expression profiles were determined using Nanostring (Rogers et al., 2020). Using an optogenetically inducible BMP signaling system, the BMP target genes were tested for differences in temporal responsiveness to BMP signaling, rate of transcriptional activation, and the level of BMP signaling required for activation (Rogers et al., 2020). In this analysis, differences in BMP activation thresholds were not sufficient to explain the spatiotemporal expression patterns of all the BMP target genes examined (Rogers et al., 2020). Some of these BMP target genes are additionally controlled by combined input from BMP, Nodal, and FGF signaling pathways (Rogers et al., 2020). For example, ventral ectodermal genes are activated by BMP signaling but repressed at the ventral margin by Nodal and Fgf signaling, while other ventral marginally-expressed genes require BMP and Nodal/Fgf for their activation (Rogers et al., 2020). Further studies are needed to determine how these signaling pathways are integrated at the chromatin level to activate or repress transcription of target genes.

The initiation of *bmp2b* and *bmp7a* zygotic expression is controlled by maternal pathways (Fig. 5A). One such maternal factor is Pou5f3, a zebrafish homolog of mammalian Oct4. Pou5f3 acts broadly in initiating zygotic transcription along with Nanog and SoxB1 transcription factors, binding DNA to prime genes for activation (Lee et al., 2013; Leichsenring, Maes, Mossner, Driever, & Onichtchouk, 2013). Maternal-zygotic *pou5f3* mutant embryos display a dorsalized phenotype including reduced *bmp* gene expression (Reim & Brand, 2006). Furthermore, expression of *pou5f3* in ventral blastomeres is sufficient to rescue *bmp* expression in maternal-zygotic *pou5f3* mutant embryos, supporting a role for Pou5f3 in activating ventrolaterally-expressed *bmp* ligand gene expression (Reim & Brand, 2006). Notably, RNA polymerase II associates with Sox-Pou binding regions in the *bmp2b* promoter prior to zygotic genome activation, suggesting that *bmp2b* transcription is directly activated by Pou5f3 (Leichsenring et al., 2013).

The TGFβ family member Growth differentiation factor 6a (Gdf6a/Radar) is a second factor implicated upstream of *bmp* expression. *gdf6a* is maternally expressed and ubiquitously distributed until the late blastula stage (Goutel, Kishimoto, Schulte-Merker, & Rosa, 2000). Depletion of Gdf6a with a translation-blocking morpholino or overexpression of a dominant negative form of Gdf6a reduces *bmp2b* and *bmp7a* gene expression, while overexpression of Gdf6a upregulates *bmp* expression and reduces *gsc* expression in the organizer (Goutel et al., 2000; Sidi, Goutel, Peyrieras, & Rosa, 2003). Furthermore, the ventralizing effect of Gdf6a overexpression is blocked by a morpholino targeting the Type I BMP receptor Acvr1l, suggesting that Gdf6a acts through Acvr1l to initiate *bmp* expression (Sidi et al., 2003). While zygotic *gdf6a* mutants have been generated and do not display dorsoventral patterning defects (Valdivia et al., 2016), maternal mutants are needed to show whether the initiation of *bmp* expression depends on maternal Gdf6a.

#### 4.1.2 The organizer inhibits BMP signaling to dorsalize the embryo

The dorsal organizer regulates BMP signaling by secreting three BMP antagonists: Chordin, Noggin 1, and Follistatin-like 1b (Fig. 5B). Experiments *in vitro* have shown that these proteins bind directly to BMPs to antagonize them: Chordin and Noggin prevent BMPs from interacting with their receptors, while Follistatin antagonizes BMPs without preventing receptor binding (Iemura et al., 1998; Piccolo, Sasai, Lu, & De Robertis, 1996; Zimmerman, De Jesus-Escobar, & Harland, 1996). *chordin* mutant embryos exhibit moderate ventralization (Schulte-Merker et al., 1997), which is strongly enhanced by morpholino depletion of either Noggin 1 or Follistatin-like 1b (Dal-Pra et al., 2006). In embryos triply depleted of *chordin*, *noggin*, and *follistatin-like 1b*, non-neural ectoderm and ventral mesoderm expand at the expense of neuroectoderm and dorsal mesoderm, respectively (Dal-Pra et al., 2006). Thus, BMP inhibitors secreted from the dorsal organizer promote neuroectoderm and dorsal mesoderm, counteracting the ventralizing BMPs expressed ventrolaterally. However, depletion of Noggin 1 and Follistatin-like 1b together does not affect dorsoventral patterning, showing that Chordin is the primary inhibitor of BMP signaling secreted from the dorsal organizer (Dal-Pra et al., 2006). As discussed above, ectopic inhibition of BMP signaling can induce a partial but not a complete ectopic axis: in general, misexpression of Noggin or Chordin induces or expands neuroectoderm and dorsolateral mesoderm but never induces dorsal midline mesoderm (Dixon Fox & Bruce, 2009; Fürthauer, Thisse, & Thisse, 1999; Koshida et al., 1998; Tanaka et al., 2017; Varga, Maegawa, & Weinberg, 2011). Thus, inhibition of BMP signaling is necessary but not sufficient to induce a complete body axis.

Structural analysis of Chordin protein has provided an interesting model for how Chordin binds BMP in the extracellular space. Chordin protein consists of four von Willebrand factor Type C domains (VWFC1–4) (Troilo et al., 2014) that vary in their ability to bind different BMPs: BMP2 binds preferably to VWFC1 and 3, and BMP7 to VWFC1 and 4 (Zhang, Huang, Qiu, Nickel, & Sebald, 2007). Between VWFC1 and 2 lie four Chordin-specific domains of unknown function. The arrangement of these domains led to the long-running prediction that Chordin displays a horseshoe-shaped structure, with BMP binding domains at the tips of the horseshoe binding opposite sides of a BMP ligand (Larrain et al., 2000). This prediction was recently verified through a three-dimensional structure of human Chordin determined from electron microscopy and small-angle X-ray scattering (Troilo et al., 2014). These studies reveal that the VWFC1 and VWFC2–4 domains of Chordin reside on opposite tips of the horseshoe-shaped structure, enabling the VWFC domains to cooperatively bind BMP ligands (Larrain et al., 2000; Troilo et al., 2014).

Restricting the activity of Chordin ventrally are the metalloproteases Tolloid and Bmp1a, which cleave Chordin and prevent it from interacting with BMP ligand. These two metalloproteases cleave Chordin at two sites: the first site is located just C-terminal to VWFC1, and the second is between VWFC3–4 (Piccolo et al., 1997; Scott et al., 1999). Tolloid and Bmp1a function partially redundantly, and in their absence ventral cell fates fail to be specified and the embryo is strongly dorsalized (Jasuja et al., 2006; Tuazon, Wang, Andrade, Umulis, & Mullins, 2020). A ventrally-expressed competitive inhibitor, Sizzled, binds the active sites of Tolloid and Bmp1a, preventing them from cleaving Chordin (Lee, Ambrosio, Reversade, & De Robertis, 2006; Miller-Bertoglio et al., 1999; Muraoka et al., 2006; Yabe et al., 2003). The *sizzled* gene is a direct transcriptional target of BMP signaling, enabling negative feedback inhibition of BMP signaling (Greenfeld et al., 2021; Yabe et al., 2003). Defects in *sizzled* mutant embryos are first evident during mid-gastrula stages, and embryos exhibit a duplicated ventral tail fin at 1 day post fertilization (dpf), indicating a later function in patterning dorsoventral tissues (Hammerschmidt, Pelegri, et al., 1996; Martyn & Schulte-Merker, 2003; Miller-Bertoglio et al., 1999; Solnica-Krezel et al., 1996; Tuazon et al., 2020; Yabe et al., 2003). Consistent with a later role, *sizzled* mutants display a normal p-Smad5 gradient at early gastrula stages, indicating that it does not play a role in establishing the BMP signaling gradient in zebrafish (Miller-Bertoglio et al., 1999; Tuazon et al., 2020). As a negative feedback regulator, Sizzled has also been shown to play an important role in regulating robustness in dorsoventral patterning in changes to embryo size in *Xenopus*, effectively scaling Chordin proteolysis to maintain dorsoventral proportions in embryos of altered sizes (Inomata, Shibata, Haraguchi, & Sasai, 2013). As such, Sizzled may act at early gastrula stages to provide robustness to BMP signaling gradient formation due to environmental or biological perturbations.

BMP signaling is also regulated by three additional extracellular regulators: Twisted gastrulation 1a (Twsg1a), 1b (Twsg1b), and BMP binding endothelial regulator (Bmper). These factors can bind BMP ligands and Chordin to regulate BMP signaling (reviewed in Zinski et al., 2018). Surprisingly, in the zebrafish, both depletion and overexpression of Twisted gastrulation and Bmper cause dorsalized phenotypes, showing that these factors can both enhance and antagonize BMP signaling during zebrafish dorsoventral patterning (Little & Mullins, 2004; Rentzsch, Zhang, Kramer, Sebald, & Hammerschmidt, 2006; Xie & Fisher, 2005; Zhang et al., 2010). Further study using Twisted gastrulation and Bmper mutants is needed to understand how these factors regulate BMP signaling during gastrulation.

It has long been believed that Chordin secreted from the dorsal organizer rapidly diffuses ventrally and forms a dorsal-to-ventral gradient that generates an inverse ventral-to-dorsal BMP signaling gradient. In this so-called counter-gradient model, BMP ligands are rapidly bound by Chordin near their site of production, producing a BMP gradient that is opposite the Chordin gradient. The counter-gradient model is supported by data from *Drosophila* and *Xenopus* suggesting that Chordin is highly diffusible and forms a gradient (Blitz, Shimmi, Wunnenberg-Stapleton, O'Connor, & Cho, 2000; Plouhinec, Zakin, Moriyama, & De Robertis, 2013; Srinivasan, Rashka, & Bier, 2002). Critically, the counter-gradient model relies on long-range diffusion of Chordin. However, in zebrafish, recent mathematical modeling combined with computational simulations of BMP signaling gradient formation in wild-type and *chordin*, *sizzled*, and *tolloid;bmp1a* double mutant embryos suggest an alternative source-sink model (Tuazon et al., 2020; Zinski et al., 2017). This source-sink model is also supported by the short-range action of Chordin in a study in *Xenopus* (Blitz et al., 2000). In the source-sink model, Chordin is restricted to the dorsal side of the embryo and acts as a dorsal BMP-binding sink, while BMPs exhibit long-range diffusion from their ventrolateral source (Tuazon et al., 2020; Zinski et al., 2017). In a recent test to distinguish between the counter-gradient and source-sink models, membrane-bound tethered Chordin, which cannot diffuse, was expressed dorsally in zebrafish embryos depleted of endogenous Chordin, Tolloid, and Bmp1a (Tuazon et al., 2020). Remarkably, these triply depleted embryos, which are normally ventralized, were rescued by the dorsally expressed tethered Chordin (Tuazon et al., 2020). Since membrane-bound Chordin cannot diffuse to form a Chordin gradient, these experiments showed that a BMP signaling gradient is established by a source-sink rather than a counter-gradient mechanism (Tuazon et al., 2020). The metalloproteases Tolloid and Bmp1a play a key role confining Chordin dorsally to form this dorsal sink of BMP activity (Tuazon et al., 2020).

While BMP and Chordin appear to behave as a source-sink system in the early gastrula, it is unclear whether the BMP signaling gradient is maintained by the same mechanism at later stages of development. Notably, BMP signaling patterns the dorsoventral axis in progressive anterior-to-posterior temporal intervals throughout gastrulation (Hashiguchi & Mullins, 2013; Tucker, Mintzer, & Mullins, 2008). Beginning at late blastula stages, the cells of the embryo are drawn down over the yolk in coordinated morphogenetic cell movements known as epiboly, enclosing the entire yolk by the end of gastrulation (Kimmel et al., 1995). These movements cause the dorsal-most and ventral-most marginal regions, which are the lowest and highest domains of BMP signaling, respectively, to become progressively closer and eventually fuse (reviewed in Tuazon & Mullins, 2015). Sizzled, an inhibitor of Tolloid and Bmp1a, begins to function during these stages (Hammerschmidt, Pelegri, et al., 1996; Miller-Bertoglio et al., 1999) and may alter the BMP gradient forming mechanism during this period of massive changes in the topography of these signaling centers. Further studies are needed to understand how the BMP signaling gradient is regulated during these late stages of gastrulation.

Interestingly, the progressive anteroposterior activity of BMP signaling in dorsoventral patterning is exquisitely coordinated with FGF and Wnt signaling that function in anteroposterior patterning (Hashiguchi & Mullins, 2013; Tucker et al., 2008). One way that FGF and BMP signaling are coordinated is via the linker region of p-Smad5 (Fuentealba et al., 2007; Hashiguchi & Mullins, 2013). FGF signaling, acting through Mitogen Acting Protein Kinase (MAPK), promotes phosphorylation of p-Smad5 at the linker region, attenuating p-Smad5 activity in ventral and vegetal regions during mid-to-late gastrula stages. Blocking phosphorylation of the p-Smad5 linker region by MAPK causes the precocious specification of posterior ventral tissues, suggesting that phosphorylation of the linker region by FGF/MAPK signaling regulates in part the timing of dorsoventral patterning along the anteroposterior axis (Hashiguchi & Mullins, 2013). Further studies are needed to determine the other regulatory mechanisms modulating the timing of dorsoventral patterning by BMP signaling.

## 4.2 Zygotic Wnt8a/β-catenin signaling promotes ventrolateral and posterior fates, while opposing the organizer's dorsalizing and anteriorizing activity

Although the organizer must inhibit BMP signaling to promote dorsal fates as discussed above, the organizer must also inhibit a second zygotic signaling pathway to establish a complete body axis: zygotic Wnt8a/β-catenin signaling. For example, overexpressing BMP antagonists in one blastomere of cleavage-stage zebrafish embryos produces an incomplete secondary body axis containing neural tissue and somites but never a complete head or notochord (Dixon Fox & Bruce, 2009; Fürthauer et al., 1999). Instead, overexpression of both BMP and Wnt antagonists is needed to establish a complete head (Tanaka et al., 2017). In this section, we discuss how the organizer inhibits the zygotic posteriorizing signal Wnt8a to establish anterior structures including the head. We further discuss how this zygotic

Wnt8a signal, which represses the organizer and dorsal midline mesoderm, contrasts with the maternal Wnt/β-catenin signal that induces the organizer and dorsal midline mesoderm.

### 4.2.1 Zygotic Wnt8a promotes ventrolateral fates and represses dorsal midline mesoderm

Zygotic Wnt8a, which is expressed during late blastula and early gastrula stages around the ventrolateral margin, functions in ventralizing the embryo together with BMP signaling (Kelly, Greenstein, Erezyilmaz, & Moon, 1995; Seiliez, Thisse, & Thisse, 2006) (Fig. 5C). Because Wnt8a maintains *bmp* gene expression during gastrulation, BMP signaling becomes severely reduced in *wnt8a* mutants (Baker, Ramel, & Lekven, 2010). Hence, disentangling the roles of BMP and Wnt signaling is experimentally challenging. Elegant studies have demonstrated how BMP and Wnt8a signaling act independently and in concert to pattern the embryo, using *bmp2b* and *wnt8a* single and double mutants (Ramel et al., 2005) and embryos doubly depleted of Wnt8a and Chordin, which over-activates BMP signaling despite the loss of Wnt8a (Baker et al., 2010). These studies revealed that Wnt8a and BMP signaling together promote ventrolateral mesoderm, but do so differently: Wnt8a represses dorsal midline mesoderm while BMP signaling represses dorsolateral but not dorsal midline mesoderm (Lekven et al., 2001; Ramel et al., 2005).

While both BMP and Wnt8a signaling promote ventrolateral fates at the expense of dorsal fates, only Wnt8a is required to restrict the size of the dorsal organizer itself (Lekven et al., 2001; Ramel et al., 2005; Ramel & Lekven, 2004). In particular, *wnt8a* mutant embryos display expanded ventrolateral expression of *gsc*, *chd*, and *flh* at early gastrula stages (Ramel & Lekven, 2004). However, their expression at these stages is normal in *bmp2b* mutant embryos (Ramel & Lekven, 2004). Hence, Wnt8a signaling alone is sufficient to repress the dorsal midline mesoderm, including the dorsal organizer. On the other hand, BMP signaling can repress the organizer in certain experimental contexts, in particular when BMP signaling is over-activated or Wnt8a signaling is absent. First, *bmp2b;wnt8a* double mutants show further ventrolateral expansion of *gsc*, *chd*, and *flh* than *wnt8a* single mutants (Ramel et al., 2005). Thus, in the absence of Wnt8a, BMP signaling can repress the organizer. Second, over-activating BMP signaling in wild-type embryos or when *wnt8a* is depleted can repress dorsal midline mesoderm (Baker et al., 2010; Nikaido et al., 1999; Schmid et al., 2000). Thus, BMP signaling can repress the organizer, but it does not normally do so.

Interestingly, Wnt8a represses the organizer through the transcription factors *vox*, *vent*, and *ved*, which are also BMP target genes as discussed above (Gilardelli et al., 2004; Ramel et al., 2005; Ramel & Lekven, 2004). Similar to *wnt8a* mutants, *vox;vent* double mutants or embryos triply depleted of *vox*, *vent*, and *ved* display a strongly dorsalized phenotype and exhibit expanded dorsal organizers (Imai et al., 2001; Shimizu et al., 2002). While both Wnt8a and BMP signaling promote *vox*, *vent*, and *ved*, they do so differently: Wnt8a is the primary regulator of *vox* and *vent* transcription at late blastula and early gastrula stages, while BMP signaling helps to maintain *vox* and *vent* transcription beginning at mid-gastrula stages (Melby et al., 2000; Ramel & Lekven, 2004).

### 4.2.2 Wnt8a acts as a morphogen to posteriorize the neuroectoderm

In addition to promoting ventrolateral mesoderm, Wnt8a promotes posterior neuroectoderm at the expense of anterior neuroectoderm (Erter, Wilm, Basler, Wright, & Solnica-Krezel, 2001; Lekven et al., 2001). In this context, Wnt8a is thought to act as a morphogen, patterning the neuroectoderm in a dose-dependent manner (reviewed in Green, Whitener, Mohanty, & Lekven, 2015). In support of this model, neural markers in distinct anteroposterior domains in *Xenopus* respond to Wnt8a in a dose-dependent manner (Kiecker & Niehrs, 2001). Similarly, increasing the dosage of a morpholino to knock down Wnt8a in zebrafish leads to progressive loss of posterior neuroectoderm, also suggesting that Wnt8a posteriorizes the neuroectoderm in a dose-dependent manner (Erter et al., 2001). Recently, the gradient of Wnt8a activity was quantified in zebrafish utilizing a transgenic Wnt reporter line: as expected, this activity gradient peaks near the margin (more posterior domains) and declines near the animal pole (more anterior domains) (Akieda et al., 2019). This work demonstrated that the Wnt8a activity gradient is initially noisy but is refined by apoptosis of cells with inappropriate Wnt activity levels (Akieda et al., 2019).

Unlike BMPs, which are thought to diffuse freely through the extracellular space (Pomreinke et al., 2017; Zinski et al., 2017), Wnts are post-translationally lipid-modified and therefore may not diffuse efficiently through the aqueous extracellular space (Nusse & Clevers, 2017). One hypothesis, supported by recent evidence in the zebrafish, is that Wnt8a may be transported by long filopodia called cytonemes that allow Wnts to signal at a distance from their source via cell-cell contacts rather than via ligand diffusion (Luz et al., 2014; Mattes et al., 2018). On the other hand, in zebrafish and *Xenopus*, mutated versions of Wnt8a that cannot be lipid modified still retain some ability to signal, albeit at reduced levels compared to wild-type Wnt8a (Luz et al., 2014; Speer et al., 2019). Further experiments are needed to probe the mechanism by which Wnt8a is distributed to target cells and determine how cells interpret the Wnt8a gradient.

### 4.2.3 The organizer inhibits Wnt8a to promote anterior fates

To promote anterior structures, the organizer inhibits Wnt8a both transcriptionally and by expressing several secreted Wnt antagonists (Fig. 5C). First, the organizer expresses Gsc and Forkhead box A3 (Foxa3), which are transcription factors that inhibit *wnt8a* transcription in the prospective dorsal shield (Seiliez et al., 2006). It has not been shown whether these transcription factors directly repress *wnt8a* transcription in zebrafish, but Gsc does so in *Xenopus* (Yao & Kessler, 2001), and during zebrafish axis patterning Gsc broadly functions as a transcriptional repressor (Dixon Fox & Bruce, 2009). Together, Gsc and Foxa3 promote head formation and anteriorize the neuroectoderm by repressing *wnt8a* expression in the dorsal organizer (Seiliez et al., 2006).

Second, the organizer expresses multiple secreted Wnt antagonists. Secreted Wnt inhibitor Dickkopf inhibits Wnt signaling by direct binding to Frizzled co-receptor LRP (Bafico, Liu, Yaniv, Gazit, & Aaronson, 2001). *dickkopf 1b* (*dkk1b*) is expressed in the organizer shortly after the mid-blastula transition and marks the prospective prechordal plate (Hashimoto et al., 2000; Shinya et al., 2000). Its expression is initiated in the organizer by the maternal Wnt/β-catenin dorsal determinant signal and is maintained by FGF signaling and Chordin activity (Hashimoto et al., 2000; Shinya et al., 2000; Tanaka et al., 2017). Depletion of Dkk1b blocks head formation (Seiliez et al., 2006), while overexpression of Dkk1b anteriorizes the neuroectoderm and causes loss of posterior tissues (Hashimoto et al., 2000; Hino et al., 2018; Shinya et al., 2000; Yan et al., 2018), showing that Dkk1b opposes the posteriorizing activity of Wnt8a.

In addition to Dickkopf, two other secreted Wnt antagonists are dorsally expressed: these are *secreted frizzled-related protein 1a* (*sfrp1a*), which is expressed broadly dorsally, and *frizzled related protein* (*frzb*), which is expressed only in the organizer region (Tendeng & Houart, 2006; Wang et al., 1997). Overexpressing Frzb causes loss of posterior tissues, similar to Dkk1b overexpression (Shinya et al., 2000). However, neither embryos depleted of Sfrp1a nor embryos mutant for *frzb* had any noted defects in head induction (Holly, Widen, Famulski, & Waskiewicz, 2014; Rochard et al., 2016), suggesting that Dickkopf is the primary Wnt antagonist secreted by the organizer. The generation of triple mutant embryos of *dkk1b*, *sfrp1a*, and *frzb* is needed to determine whether Sfrp1a and Frzb cooperate with Dkk1b in head induction.

### 4.2.4 Contrasting roles of Wnt/β-catenin signaling in dorsoventral patterning

Interestingly, Wnt/β-catenin signaling has two contrasting roles in zebrafish axis patterning. First, the dorsal determinant stabilizes maternal β-catenin, allowing it to translocate to prospective dorsal nuclei to establish the organizer and promote dorsal fates shortly after the mid-blastula transition. β-catenin no longer localizes to dorsal blastomere nuclei by the onset of gastrulation (Schneider et al., 1996). Second, beginning at late blastula and early gastrula stages, zygotic Wnt8a promotes ventrolateral fates and represses the organizer. These two Wnt/β-catenin signaling pathways have contrasting effects on dorsoventral marker expression: for example, maternal loss of *β-catenin 2* blocks formation of the organizer and expression of the organizer marker *gsc* (Kelly et al., 2000), while loss of zygotic *wnt8a* causes ventrolateral expansion of *gsc* expression (Baker et al., 2010; Lekven et al., 2001; Ramel & Lekven, 2004). How do these two temporally distinct Wnt/β-catenin signals promote contrasting effects on axis patterning?

One possible explanation for these contrasting roles of Wnt/β-catenin signaling is that at mid-blastula and gastrula stages Wnt target genes may be differentially competent to respond to Wnt signaling. For example, co-regulators of Wnt signaling could be expressed during late blastula stages that disable dorsalizing Wnt target genes and enable ventralizing Wnt target genes. In support of this timing hypothesis, overexpression of secreted Wnt antagonist Dkk1b initially reduces the size of the organizer at mid-blastula stages (Hino et al., 2018) but ultimately produces a dorsalized phenotype at the end of gastrulation (Hashimoto et al., 2000; Hino et al., 2018; Shinya et al., 2000; Yan et al., 2018), reflecting a de-repression of organizer activity.

A similar de-repression of organizer activity occurs when the embryo is depleted of all β-catenin. The zebrafish genome encodes two β-catenin proteins (Bellipanni et al., 2006). While the early maternal Wnt/β-catenin signal that induces the organizer is mediated exclusively by β-catenin 2, the zygotic Wnt/β-catenin signal that promotes ventrolateral and posterior fates is mediated by both β-catenin 1 and β-catenin 2 (Bellipanni et al., 2006). Interestingly, while maternal loss of *β-catenin 2* alone strongly ventralizes embryos, simultaneous depletion of both β-catenins produces a distinct and unusual dorsalized phenotype called "ciuffo" (Bellipanni et al., 2006;

Varga et al., 2011; Varga, Maegawa, Bellipanni, & Weinberg, 2007). "Ciuffo" embryos display no dorsal organizer at late blastula stages, but at the beginning of gastrulation dorsal organizer genes (*gsc, noggin 1,* and *chordin,* but not *dharma*) are expressed in expanded domains compared to wild-type embryos (Bellipanni et al., 2006; Varga et al., 2007, 2011). Ultimately, "ciuffo" embryos exhibit a radially dorsalized phenotype with expanded dorsal neuroectoderm and somitic mesoderm. However, these embryos lack a notochord, a normal derivative of the dorsal organizer, indicating that a fully functional organizer is not formed (Bellipanni et al., 2006; Varga et al., 2007, 2011). Thus, similar to Dkk1b overexpression, the depletion of all β-catenin initially blocks dorsal organizer formation but ultimately causes a dorsalized phenotype, albeit lacking notochord.

It is a surprising result that organizer genes can be expressed, albeit with a delay, in the absence of both maternal β-catenins. One possible explanation for this result is that Nodal and FGF signaling at the margin may be sufficient to induce the organizer in the absence of repressive, zygotic Wnt8a activity. Notably, loss of maternal *β-catenin 2* blocks the initial dorsal expression of *ndr1*, *fgf3*, and *fgf8a* at mid-blastula stages but does not affect their later, normal circumferential expression at the margin by the onset of gastrulation (Maegawa et al., 2006; Tsang et al., 2004). Furthermore, the expanded *gsc* and *chd* expression in "ciuffo" embryos requires Nodal and FGF signaling (Varga et al., 2007). Thus, prolonged exposure to Nodal or FGF signaling could induce organizer genes circumferentially in the absence of both maternal β-catenins and Wnt signaling altogether. Although the persistent marginal Nodal expression is not sufficient to induce an organizer in embryos maternally mutant for *β-catenin 2* only, overexpression of Ndr1 can rescue maternal *β-catenin 2* mutants. Thus, marginal Nodal and FGF could induce organizer genes in the absence of repression by Wnt8a. Further experiments are needed to understand how organizer genes can be induced in the absence of all Wnt/β-catenin signaling in the embryo.

Other dorsoventrally graded signals could also play a role in how cells respond to Wnt signaling in different domains of the embryo. For example, recent results from *Xenopus* indicate that β-catenin recruitment to target genes is not sufficient to trigger target gene activation and instead that Wnt target genes also require co-activation by Nodal, BMP, or FGF signaling (Afouda et al., 2020; Nakamura, de Paiva Alves, Veenstra, & Hoppler, 2016). In particular, these results indicate that maternal Wnt target genes may require Nodal signaling to be activated (Afouda et al., 2020), while zygotic Wnt8a target genes require BMP or FGF signaling

(Nakamura et al., 2016). Thus, transcriptional co-activators or repressors acting at different time-points or spatial domains of the embryo may both help to distinguish the dorsalizing and ventralizing responses to Wnt signaling.

## 4.3 Restricting the size of the dorsal organizer

Repressors of the dorsal organizer prevent the organizer from expanding ventrolaterally, ensuring the correct balance of dorsoventral tissue specification. Unlike the molecules that antagonize organizer specification by inhibiting β-catenin activity (discussed in Section 3.2 above), the repressors discussed here are thought to act after organizer induction. As discussed above, Wnt8a represses the organizer ventrolaterally through Vox, Vent, and Ved. In this section, we discuss additional repressors of the dorsal organizer that have been identified through loss-of-function analyses. We classify these organizer repressors by two possible modes of function (Fig. 6).

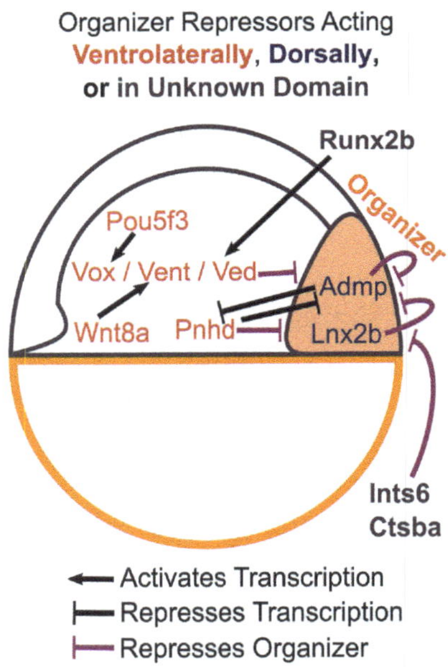


**Fig. 6** Restricting the size of the dorsal organizer in ventrolateral and dorsal domains. The organizer repressors Pinhead (Pnhd), Pou5f3, Wnt8a, and Vox/Vent/Ved all act ventrolaterally to prevent the organizer from expanding. Pou5f3 and Wnt8a both promote *vox/vent/ved* expression ventrolaterally. Meanwhile, Admp and Lnx2b act in dorsal domains to restrict the organizer's dorsal-inducing activity. Admp and Pinhead repress each other's expression. Finally, the domain of action of the organizer repressors Runx2b, Ints6, and Ctsba remains unknown, but Runx2b promotes *vox/vent/ved* expression, suggesting that it may act ventrolaterally.

First, like Wnt8a and Vox/Vent/Ved, some repressors act ventrolaterally to prevent organizer expansion, such as the POU domain transcription factor Pou5f3 and the BMP-like ligand Pinhead. Second, some repressors act dorsally to dampen the organizer's dorsal-inducing activity, such as the dorsally-expressed TGFβ family ligand Anti-dorsalizing morphogenetic protein (Admp) and Lnx2b, which destabilizes Dharma protein (discussed in Section 3.4.1 above). We begin with a ventrolaterally acting repressor, Pou5f3.

Maternal-zygotic mutants of transcription factor gene *pou5f3* (*spiel ohne grenzen*) display expanded dorsal organizers, showing that Pou5f3 represses the organizer (Reim & Brand, 2006). However, maternal-zygotic *pou5f3* mutants display normal *dharma* expression, suggesting that Pou5f3 represses the organizer after its induction (Reim & Brand, 2006). As discussed above, Pou5f3 plays a broad role in zygotic genome activation by binding the promoter regions of genes and priming them for transcription (Lee et al., 2013; Leichsenring et al., 2013). While *pou5f3* is ubiquitously expressed prior to the onset of gastrulation (Belting et al., 2001; Burgess, Reim, Chen, Hopkins, & Brand, 2002), expression of Pou5f3 in ventrolateral domains is sufficient to rescue maternal-zygotic *pou5f3* mutants, showing that Pou5f3 acts ventrolaterally to repress the organizer (Reim & Brand, 2006). Further experiments revealed that Pou5f3 acts as a transcriptional activator during zebrafish dorsoventral patterning, promoting expression of the organizer repressor gene *vox* and indirectly repressing *fgf8a* (Belting et al., 2011). Thus, Pou5f3 probably restricts the organizer by blocking dorsal gene expression in ventrolateral domains through Vox and other factors.

Unlike Pou5f3, the TGFβ family ligand Anti-dorsalizing morphogenetic protein (Admp) acts dorsally to repress the dorsal organizer. Unlike other BMP ligands, *admp* is only expressed dorsally (Lele, Nowak, & Hammerschmidt, 2001). Admp was originally implicated as a repressor of the dorsal organizer because morpholino knockdown of Admp causes organizer expansion, unlike *bmp2b* mutants (Lele et al., 2001; Ramel & Lekven, 2004; Willot et al., 2002). Interestingly, however, recently generated *admp* mutant embryos have a wild-type phenotype (Yan et al., 2019). The loss of *admp* produced no phenotype because another BMP-like ligand, *pinhead*, was upregulated in *admp* mutant embryos (Yan et al., 2019). Unlike *admp*, *pinhead* is expressed ventrolaterally, and Admp and Pinhead repress each other's expression (Yan et al., 2019). Double mutant *admp;pinhead* embryos exhibit an expanded dorsal organizer, indicating that Admp and Pinhead cooperate to restrict the dorsal organizer but act in opposite domains to do so (Yan et al., 2019).

Three additional organizer repressors in zebrafish have unknown domains of function. RUNX family transcription factor 2b (Runx2b) has been implicated as a repressor of the dorsal organizer based on morpholino knockdown studies (Flores, Lam, Crosier, & Crosier, 2008). RUNX transcription factors play multiple roles in development, including in hematopoiesis and development of the skeleton (Mevel, Draper, Lie, Kouskoff, & Lacaud, 2019). Zebrafish embryos depleted of Runx2b exhibit expanded organizers and a dorsalized phenotype (Flores et al., 2008). In this context, Runx2b acts as a direct transcriptional activator of *ved* (Flores et al., 2008). Depletion of Runx2b blocks *ved* transcription and delays but does not block *vox* and *vent* transcription (Flores et al., 2008). While *runx2b* is maternally and ubiquitously expressed, by late blastula stages *runx2b* expression is excluded from the dorsal organizer (Flores et al., 2008). Since Runx2b activates *ved* gene expression and it is ventrolaterally restricted by late blastula stages, it likely acts ventrolaterally to restrict the organizer. However, this has not been shown directly by mosaic rescue experiments. Generation of maternal *runx2b* mutants is needed to further investigate the role of Runx2b during zebrafish dorsoventral patterning.

A second repressor organizer repressor with an unknown domain of function is Integrator complex subunit 6 (Ints6) (Kapp, Abrams, Marlow, & Mullins, 2013). In maternal *ints6* mutant embryos, the organizer is expanded, leading to the induction of multiple ectopic body axes, a unique phenotype among the mutants with expanded organizers (Kapp et al., 2013). However, maternal *ints6* mutants display normal expression of *dharma* and a normal domain of nuclear β-catenin localization, demonstrating that Ints6 represses the organizer downstream of maternal Wnt/β-catenin-mediated organizer induction (Kapp et al., 2013). Overexpression of Ints6 has no effect on dorsoventral patterning, suggesting that Ints6 acts as a permissive factor in restricting the organizer (Kapp et al., 2013). The Integrator complex was originally discovered as an RNA polymerase II-associated complex that trims the 3′ ends of spliceosomal small nuclear RNAs (snRNAs) before they are incorporated into spliceosomes (Baillat et al., 2005). However, *ints6* maternal mutants can respond to both BMP and Wnt signaling, inconsistent with a broad splicing defect blocking many signaling pathway components (Kapp et al., 2013). More recent studies in other model systems have demonstrated that the Integrator complex participates in multiple aspects of gene transcription, including cleaving nascent mRNAs, activating enhancers, and promoting RNA polymerase II pause-release (Beckedorff et al., 2020; Elrod et al., 2019; Gardini et al., 2014; Lai, Gardini, Zhang, & Shiekhattar, 2015;

Tatomer et al., 2019). Further work is needed to determine the molecular mechanism by which Ints6 represses the dorsal organizer and prevents the formation of multiple body axes during dorsoventral patterning.

Finally, another surprising repressor of the dorsal organizer in zebrafish is the lysosomal endopeptidase Cathepsin Ba (Ctsba), identified via the recently characterized *split top* maternal-effect mutant in zebrafish (Langdon et al., 2016). Maternal *ctsba* mutants exhibit variable phenotypes including dorsalization and incomplete epiboly. In these mutants, both *dharma* and *ndr1* expression are normal, indicating that the organizer is induced normally. However, by mid-blastula stages, these mutants display expanded *gsc* expression, indicating that Ctsba represses the organizer after its induction, similar to Ints6 and Pou5f3. However, the molecular mechanism by which Ctsba represses the organizer remains unknown (Langdon et al., 2016).

## 4.4 Fibroblast Growth Factors promote organizer establishment and function

Fibroblast Growth Factors (FGFs) play numerous roles during zebrafish axial patterning, including promoting dorsal organizer fates (Fürthauer et al., 2004; Maegawa et al., 2006), posteriorizing the neuroectoderm (Fürthauer et al., 2004; Kudoh, Wilson, & Dawid, 2002; Varga et al., 2011), posteriorizing the mesoderm (Draper, Stock, & Kimmel, 2003), and promoting mesoderm at the expense of endoderm downstream of Nodal signaling (Mizoguchi, Izawa, Kuroiwa, & Kikuchi, 2006; Poulain, Fürthauer, Thisse, Thisse, & Lepage, 2006; van Boxtel et al., 2018). FGF ligands signal by binding FGF receptor (FGFR) tyrosine kinases, which then phosphorylate and activate intracellular targets, principally Mitogen Activated Protein Kinase (MAPK) (reviewed in Bottcher & Niehrs, 2005; Ornitz & Itoh, 2015). In this section, we focus on the function of FGFs in establishing the dorsal organizer and promoting dorsal fates.

The zebrafish genome encodes thirty-two putative FGF ligands (Ruzicka et al., 2019) (reviewed in Itoh, 2007). Among these, four so far have been implicated in zebrafish dorsoventral and anteroposterior patterning during gastrulation: Fgf3, Fgf8a, Fgf17, and Fgf24. Similar to the Nodal ligand Ndr1, these four FGFs are first expressed on the prospective dorsal side of the embryo shortly after the mid-blastula transition before spreading ventrolaterally, forming a dorsoventral expression gradient around the margin during mid-blastula stages (30% epiboly) (Cao et al., 2004; Fodor et al., 2013; Fürthauer et al., 1997, 2004; Fürthauer, Reifers, Brand, Thisse, &

Thisse, 2001; Raible & Brand, 2001; Reifers et al., 1998; Tsang et al., 2004). Consistent with a dorsal-to-ventral FGF signaling gradient, the expression of FGF target gene *sprouty4* (*spry4*) and the distribution of phosphorylated Erk (p-Erk), which is the intracellular signal transducer of FGF/MAPK signaling, both form dorsal-to-ventral gradients by the mid-blastula stage (Fürthauer et al., 2001, 2004; Poulain et al., 2006). While FGF signaling dorsalizes the embryo during mid-blastula stages (discussed below), it remains unclear whether FGF ligands act as morphogens along the dorsoventral axis, as FGF target genes expressed in distinct dorsoventral domains have not been identified.

During mid-blastula stages, FGF signaling induces the zebrafish dorsal organizer downstream of maternal Wnt/β-catenin signaling. This role of FGF signaling in axis induction was revealed by rescue experiments in maternal *β-catenin 2* mutant embryos, which lack a dorsal organizer but can be rescued by expressing β-catenin 2 (Kelly et al., 2000). However, if FGF signaling is blocked in maternal *β-catenin 2* mutant embryos using a chemical FGF inhibitor, a dominant negative FGF receptor, or the MAPK inhibitor Dusp6, the embryos fail to be rescued by expression of β-catenin 2 (Maegawa et al., 2006). Hence, FGF signaling is necessary downstream of β-catenin for zebrafish organizer induction. However, FGF signaling is not sufficient for complete organizer function in the absence of maternal *β-catenin 2*, as overexpressing Fgf3 or Fgf8a in *β-catenin 2* mutant embryos produces a body axis lacking a notochord (Maegawa et al., 2006; Tanaka et al., 2017). Dharma and Nodal signaling likely act in conjunction with FGF signaling to induce the organizer.

FGF signaling promotes dorsal fates in zebrafish by inhibiting both BMP and Wnt signaling. In particular, FGF signaling directly represses transcription of BMP ligand genes *bmp2b* and *bmp7a* during blastula stages (Fürthauer et al., 1997, 2004; Londin et al., 2005; Maegawa et al., 2006). During these stages, Chordin overexpression does not reduce *bmp* expression, suggesting that FGF signaling is primarily responsible for repressing *bmp* expression dorsally at blastula stages (Londin et al., 2005), in addition to repression by Dharma in the dorsal midline. Additionally, FGF signaling represses zygotic Wnt8a signaling by maintaining expression of the Wnt inhibitor *dkk1b* in the shield during gastrulation (Tanaka et al., 2017). Consistent with these functions, overexpressing Fgf3, Fgf8a, Fgf17, or Fgf24 can dorsalize the embryo (Cao et al., 2004; Fürthauer et al., 2004). These and other FGFs likely function redundantly in dorsoventral patterning. For example, embryos triply depleted of *fgf3*, *fgf8a*, and *fgf24* have no reported dorsoventral patterning defects (Picker et al., 2009), probably due to redundancy with *fgf17*. It is also possible that other FGFs reported to be expressed at blastula or early gastrula stages in zebrafish, such as *fgf4* (Yamauchi, Miyakawa, Miyake, & Itoh, 2009) and *fgf6a* (Thisse et al., 2004), may also function redundantly to these four FGF ligands. Simultaneous loss-of-function of four or more FGFs will be needed to concretely establish roles for these ligands in dorsoventral patterning. Alternatively, recently generated zebrafish mutants of the five FGFRs, *fgfr1a*, *fgfr1b*, *fgfr2*, *fgfr3*, and *fgfr4*, may shed more light on the role of FGF signaling during gastrulation (Leerberg, Hopton, & Draper, 2019).

Separate from its role in dorsoventral patterning, FGF signaling also plays an important role in mesoderm induction in zebrafish. By early gastrula stages in zebrafish, FGF signaling forms a marginal-to-animal activity gradient, peaking a few cell tiers from the margin and declining animally (van Boxtel et al., 2018). Nodal signaling inhibits FGF signaling nearest the margin by inducing transcription of FGF inhibitor *dual specificity phosphatase 4* (*dusp4*) (van Boxtel et al., 2018). High FGF levels a few cell tiers from the margin repress endoderm and promote mesoderm downstream of Nodal signaling, while lower FGF levels nearest the margin permit endoderm development (van Boxtel et al., 2018).

FGFs are thought to function as a morphogens along the anteroposterior axis (reviewed in Bokel & Brand, 2013). This is supported by the identification of FGF target genes expressed in distinct anteroposterior domains (Nowak, Machate, Yu, Gupta, & Brand, 2011). Studies utilizing a localized source of overexpressed Fgf8 in zebrafish indicate that an Fgf8 activity gradient forms by a source-sink mechanism in which the "sink" is controlled by clathrin-mediated endocytosis of the Fgf8 ligand-receptor complex (Rengarajan, Matzke, Reiner, Orian-Rousseau, & Scholpp, 2014; Scholpp & Brand, 2004; Yu et al., 2009). These findings support the notion that endocytic trafficking helps shape the FGF signaling gradient. Interestingly, delaying endocytosis with a dominant negative version of the ubiquitin ligase Cbl in zebrafish causes expansion of the dorsal organizer gene *gsc* (Nowak et al., 2011), suggesting that endocytosis also plays a role in dorsoventral patterning, possibly by regulating FGF signaling.

FGF signaling in the embryo is regulated by five feedback inhibitors: Interleukin 17 receptor D (Il17rd, also known as Sef), the Sprouty proteins Sprouty2 and Sprouty4, Dusp4 (discussed above), and Dusp6 (also known as Mkp3). Il17rd is a transmembrane protein that likely inhibits FGF signaling by direct binding to FGFRs (Tsang, Friesel, Kudoh, & Dawid, 2002). *il17rd* is initially expressed maternally and ubiquitously but eventually becomes localized to the cells of the margin (Fürthauer, Lin, Ang, Thisse, & Thisse, 2002; Tsang et al., 2002). Sprouty proteins are thought to act by inhibiting MAPK/ERK signaling downstream of FGF signaling, although their exact mechanism of action remains unclear (Guy, Jackson, Yusoff, & Chow, 2009; Mason, Morrison, Basson, & Licht, 2006; Ornitz & Itoh, 2015). Finally, Dusp6 inhibits FGF signaling by dephosphorylating active MAPKs (Tsang et al., 2004). In zebrafish embryos, *sprouty2*, *sprouty4*, *dusp4*, and *dusp6* are all initially expressed dorsally, similar to *fgf3*, *fgf8a*, and *fgf24*, but spread around the margin by the onset of gastrulation (Fürthauer et al., 2001, 2004; Tsang et al., 2004; van Boxtel et al., 2018). Based on morpholino depletion and gain-of-function studies, Il17rd, Sprouty2, and Sprouty4 all inhibit the induction of dorsal fates by FGF signaling (Fürthauer et al., 2001, 2002, 2004; Tsang et al., 2002). Zebrafish *sprouty4* nonsense mutants have been generated but do not exhibit overt dorsoventral patterning defects, possibly due to redundancy with *sprouty2* (Nguyen-Chi et al., 2012). While overexpression of dominant-negative Dusp6 dorsalized zebrafish embryos, suggesting that Dusp6 also inhibits dorsal fates (Tsang et al., 2004), recently generated zygotic *dusp6* mutants are phenotypically wild-type (Maurer & Sagerstrom, 2018). Generation of zebrafish embryos mutant for several of these FGF antagonists may thus be necessary to establish requirements for these antagonists during zebrafish dorsoventral patterning.

The ventralizing effect of Dusp and Sprouty proteins, which inhibit MAPK signaling, revealed that MAPK/Ras signaling functions downstream of FGF signaling during zebrafish dorsoventral patterning (reviewed in Tsang & Dawid, 2004). In support of this, overexpression of a dominant-negative Ras causes a ventralized phenotype, similar to FGF signaling loss-of-function phenotypes (Fürthauer et al., 2004). Furthermore, morpholino depletion of the MAPKs ERK1 or ERK2 downregulates or ablates expression of organizer marker *gsc* expression and reduces mesoderm formation, consistent with roles for FGF signaling in organizer activity and mesoderm specification (Krens et al., 2008; Krens, Corredor-Adámez, He, Snaar-Jagalska, & Spaink, 2008). While it is clear that MAPK/Ras acts downstream of FGF signaling during dorsoventral patterning, the MAPK/Ras signal transduction pathway acts in many biological processes. How FGF signaling may interact with other MAPK regulators during zebrafish development remains unclear.

## 5. Conclusions

Body axis formation in zebrafish is controlled by the dorsal organizer, which acts via coordinated Wnt/β-catenin, Nodal, BMP, and FGF signaling pathways. The organizer has the fascinating ability to induce a secondary body axis in grafting experiments. The zebrafish has provided an excellent system in which to study the organizer, as genetic analysis in this model has revealed many factors controlling organizer establishment and function. The organizer is established by dorsal determinants that activate β-catenin on the prospective dorsal side of the embryo, but the identity and mode of action of the dorsal determinants remains inconclusive. The dorsalizing response to β-catenin is only transient, as β-catenin acting downstream of zygotic Wnt8a antagonizes the organizer beginning at late blastula stages. Broadly expressed Bmp2/7 heterodimers and Wnt8a promote ventrolateral fate specification. Meanwhile, the organizer blocks *bmp* and *wnt8a* transcription dorsally and antagonizes BMP and Wnt8a signaling via secreted inhibitors such as Chordin and Dickkopf. Additionally, Vox/Vent/Ved and other factors restrict the size of the dorsal organizer, ensuring the proper balance of dorsoventral tissue formation. Finally, FGF signaling is required for organizer induction and promotes dorsal fates, but due to the large number of FGF ligands and receptors, studying FGF signaling using loss-of-function genetic analysis remains experimentally challenging. Overall, the interaction of these and other signaling pathways in both dorsoventral, anteroposterior, and mesendodermal patterning both spatially and temporally remains a topic of active study in the zebrafish. The genetic tools, live imaging, and accessible large, transparent embryos afforded by the zebrafish model will continue to unravel the mechanisms of embryonic patterning by Wnt, BMP, Nodal, and FGF signaling and the dorsal organizer.

## Acknowledgments

We thank Jose L. Pelliccia for helpful comments on the manuscript. Work in the Mullins laboratory is supported by NIH grants R35-GM131908 to M.C.M. and F31-HD106718 to W.D.J.

## References

Abrams, E. W., Fuentes, R., Marlow, F. L., Kobayashi, M., Zhang, H., Lu, S., et al. (2020). Molecular genetics of maternally-controlled cell divisions. *PLoS Genetics*, *16*, e1008652.

Afouda, B. A., Nakamura, Y., Shaw, S., Charney, R. M., Paraiso, K. D., Blitz, I. L., et al. (2020). Foxh1/nodal defines context-specific direct maternal Wnt/beta-catenin target gene regulation in early development. *iScience*, *23*, 101314.

Akieda, Y., Ogamino, S., Furuie, H., Ishitani, S., Akiyoshi, R., Nogami, J., et al. (2019). Cell competition corrects noisy Wnt morphogen gradients to achieve robust patterning in the zebrafish embryo. *Nature Communications*, *10*, 4710.

Bafico, A., Liu, G., Yaniv, A., Gazit, A., & Aaronson, S. A. (2001). Novel mechanism of Wnt signalling inhibition mediated by Dickkopf-1 interaction with LRP6/arrow. *Nature Cell Biology*, *3*, 683–686.

Baillat, D., Hakimi, M. A., Naar, A. M., Shilatifard, A., Cooch, N., & Shiekhattar, R. (2005). Integrator, a multiprotein mediator of small nuclear RNA processing, associates with the C-terminal repeat of RNA polymerase II. *Cell*, *123*, 265–276.

Baker, K. D., Ramel, M. C., & Lekven, A. C. (2010). A direct role for Wnt8 in ventrolateral mesoderm patterning. *Developmental Dynamics*, *239*, 2828–2836.

Barolo, S., & Posakony, J. W. (2002). Three habits of highly effective signaling pathways: Principles of transcriptional control by developmental cell signaling. *Genes & Development*, *16*, 1167–1181.

Beckedorff, F., Blumenthal, E., daSilva, L. F., Aoi, Y., Cingaram, P. R., Yue, J., et al. (2020). The human integrator complex facilitates transcriptional elongation by Endonucleolytic cleavage of nascent transcripts. *Cell Reports*, *32*, 107917.

Beddington, R. S. (1994). Induction of a second neural axis by the mouse node. *Development*, *120*, 613–620.

Bellipanni, G., Varga, M., Maegawa, S., Imai, Y., Kelly, C., Myers, A. P., et al. (2006). Essential and opposing roles of zebrafish beta-catenins in the formation of dorsal axial structures and neurectoderm. *Development*, *133*, 1299–1309.

Belting, H. G., Hauptmann, G., Meyer, D., Abdelilah-Seyfried, S., Chitnis, A., Eschbach, C., et al. (2001). Spiel ohne grenzen/pou2 is required during establishment of the zebrafish midbrain-hindbrain boundary organizer. *Development*, *128*, 4165–4176.

Belting, H. G., Wendik, B., Lunde, K., Leichsenring, M., Mossner, R., Driever, W., et al. (2011). Pou5f1 contributes to dorsoventral patterning by positive regulation of vox and modulation of fgf8a expression. *Developmental Biology*, *356*, 323–336.

Bennett, J. T., Stickney, H. L., Choi, W. Y., Ciruna, B., Talbot, W. S., & Schier, A. F. (2007). Maternal nodal and zebrafish embryogenesis. *Nature*, *450* (E1-2; discussion E2-4).

Bisgrove, B. W., Su, Y. C., & Yost, H. J. (2017). Maternal Gdf3 is an obligatory cofactor in nodal signaling for embryonic axis formation in zebrafish. *eLife*, *6*, e28534.

Blitz, I. L., Shimmi, O., Wunnenberg-Stapleton, K., O'Connor, M. B., & Cho, K. W. (2000). Is chordin a long-range- or short-range-acting factor? Roles for BMP1-related metalloproteases in chordin and BMP4 autofeedback loop regulation. *Developmental Biology*, *223*, 120–138.

Blumberg, B., Wright, C. V., De Robertis, E. M., & Cho, K. W. (1991). Organizer-specific homeobox genes in Xenopus laevis embryos. *Science*, *253*, 194–196.

Bokel, C., & Brand, M. (2013). Generation and interpretation of FGF morphogen gradients in vertebrates. *Current Opinion in Genetics & Development*, *23*, 415–422.

Bottcher, R. T., & Niehrs, C. (2005). Fibroblast growth factor signaling during early vertebrate development. *Endocrine Reviews*, *26*, 63–77.

Burgess, S., Reim, G., Chen, W., Hopkins, N., & Brand, M. (2002). The zebrafish spiel-ohne-grenzen (spg) gene encodes the POU domain protein Pou2 related to mammalian Oct4 and is essential for formation of the midbrain and hindbrain, and for pre-gastrula morphogenesis. *Development*, *129*, 905–916.

Campbell, P. D., Heim, A. E., Smith, M. Z., & Marlow, F. L. (2015). Kinesin-1 interacts with Bucky ball to form germ cells and is required to pattern the zebrafish body axis. *Development*, *142*, 2996–3008.

Cao, Y., Zhao, J., Sun, Z., Zhao, Z., Postlethwait, J., & Meng, A. (2004). fgf17b, a novel member of Fgf family, helps patterning zebrafish embryos. *Developmental Biology*, *271*, 130–143.

Carvalho, L., & Heisenberg, C. P. (2010). The yolk syncytial layer in early zebrafish development. *Trends in Cell Biology*, *20*, 586–592.

Cha, S. W., Tadjuidje, E., Tao, Q., Wylie, C., & Heasman, J. (2008). Wnt5a and Wnt11 interact in a maternal Dkk1-regulated fashion to activate both canonical and non-canonical signaling in Xenopus axis formation. *Development*, *135*, 3719–3729.

Chen, S., & Kimelman, D. (2000). The role of the yolk syncytial layer in germ layer patterning in zebrafish. *Development*, *127*, 4681–4689.

Chen, Y., & Schier, A. F. (2001). The zebrafish nodal signal squint functions as a morphogen. *Nature*, *411*, 607–610.

Chen, Y., & Schier, A. F. (2002). Lefty proteins are long-range inhibitors of squint-mediated nodal signaling. *Current Biology*, *12*, 2124–2128.

Ciruna, B., Weidinger, G., Knaut, H., Thisse, B., Thisse, C., Raz, E., et al. (2002). Production of maternal-zygotic mutant zebrafish by germ-line replacement. *Proceedings of the National Academy of Sciences of the United States of America*, *99*, 14919–14924.

Dal-Pra, S., Fürthauer, M., Van-Celst, J., Thisse, B., & Thisse, C. (2006). Noggin1 and Follistatin-like2 function redundantly to chordin to antagonize BMP activity. *Developmental Biology*, *298*, 514–526.

De Robertis, E. M. (2009). Spemann's organizer and the self-regulation of embryonic fields. *Mechanisms of Development*, *126*, 925–941.

Dixon Fox, M., & Bruce, A. E. (2009). Short- and long-range functions of Goosecoid in zebrafish axis formation are independent of chordin, noggin 1 and Follistatin-like 1b *Development*, *136*, 1675–1685.

Dougan, S. T., Warga, R. M., Kane, D. A., Schier, A. F., & Talbot, W. S. (2003). The role of the zebrafish nodal-related genes squint and cyclops in patterning of mesendoderm. *Development*, *130*, 1837–1851.

Draper, B. W., Stock, D. W., & Kimmel, C. B. (2003). Zebrafish fgf24 functions with fgf8 to promote posterior mesodermal development. *Development*, *130*, 4639–4654.

Dubrulle, J., Jordan, B. M., Akhmetova, L., Farrell, J. A., Kim, S. H., Solnica-Krezel, L., et al. (2015). Response to nodal morphogen gradient is determined by the kinetics of target gene induction. *eLife*, *4*, e05042.

Elrod, N. D., Henriques, T., Huang, K. L., Tatomer, D. C., Wilusz, J. E., Wagner, E. J., et al. (2019). The integrator complex attenuates promoter-proximal transcription at protein-coding genes. *Molecular Cell*, *76*(738–752), e737.

Erter, C. E., Solnica-Krezel, L., & Wright, C. V. (1998). Zebrafish nodal-related 2 encodes an early mesendodermal inducer signaling from the extraembryonic yolk syncytial layer. *Developmental Biology*, *204*, 361–372.

Erter, C. E., Wilm, T. P., Basler, N., Wright, C. V., & Solnica-Krezel, L. (2001). Wnt8 is required in lateral mesendodermal precursors for neural posteriorization in vivo. *Development*, *128*, 3571–3583.

Fan, X., Hagos, E. G., Xu, B., Sias, C., Kawakami, K., Burdine, R. D., et al. (2007). Nodal signals mediate interactions between the extra-embryonic and embryonic tissues in zebrafish. *Developmental Biology*, *310*, 363–378.

Fauny, J. D., Thisse, B., & Thisse, C. (2009). The entire zebrafish blastula-gastrula margin acts as an organizer dependent on the ratio of nodal to BMP activity. *Development*, *136*, 3811–3819.

Fekany-Lee, K., Gonzalez, E., Miller-Bertoglio, V., & Solnica-Krezel, L. (2000). The homeobox gene bozozok promotes anterior neuroectoderm formation in zebrafish through negative regulation of BMP2/4 and Wnt pathways. *Development*, *127*, 2333–2345.

Fekany, K., Yamanaka, Y., Leung, T., Sirotkin, H. I., Topczewski, J., Gates, M. A., et al. (1999). The zebrafish bozozok locus encodes dharma, a homeodomain protein essential for induction of gastrula organizer and dorsoanterior embryonic structures. *Development*, *126*, 1427–1438.

Feldman, B., Dougan, S. T., Schier, A. F., & Talbot, W. S. (2000). Nodal-related signals establish mesendodermal fate and trunk neural identity in zebrafish. *Current Biology*, *10*, 531–534.

Feldman, B., Gates, M. A., Egan, E. S., Dougan, S. T., Rennebeck, G., Sirotkin, H. I., et al. (1998). Zebrafish organizer development and germ-layer formation require nodal-related signals. *Nature*, *395*, 181–185.

Flores, M. V., Lam, E. Y., Crosier, K. E., & Crosier, P. S. (2008). Osteogenic transcription factor Runx2 is a maternal determinant of dorsoventral patterning in zebrafish. *Nature Cell Biology*, *10*, 346–352.

Fodor, E., Zsigmond, A., Horvath, B., Molnar, J., Nagy, I., Toth, G., et al. (2013). Full transcriptome analysis of early dorsoventral patterning in zebrafish. *PLoS One*, *8*, e70053.

Fuentealba, L. C., Eivers, E., Ikeda, A., Hurtado, C., Kuroda, H., Pera, E. M., et al. (2007). Integrating patterning signals: Wnt/GSK3 regulates the duration of the BMP/Smad1 signal. *Cell*, *131*, 980–993.

Fuentes, R., Tajer, B., Kobayashi, M., Pelliccia, J. L., Langdon, Y., Abrams, E. W., et al. (2020). The maternal coordinate system: Molecular-genetics of embryonic axis formation and patterning in the zebrafish. *Current Topics in Developmental Biology*, *140*, 341–389.

Fürthauer, M., Lin, W., Ang, S. L., Thisse, B., & Thisse, C. (2002). Sef is a feedback-induced antagonist of Ras/MAPK-mediated FGF signalling. *Nature Cell Biology*, *4*, 170–174.

Fürthauer, M., Reifers, F., Brand, M., Thisse, B., & Thisse, C. (2001). sprouty4 acts in vivo as a feedback-induced antagonist of FGF signaling in zebrafish. *Development*, *128*, 2175–2186.

Fürthauer, M., Thisse, C., & Thisse, B. (1997). A role for FGF-8 in the dorsoventral patterning of the zebrafish gastrula. *Development*, *124*, 4253–4264.

Fürthauer, M., Thisse, B., & Thisse, C. (1999). Three different noggin genes antagonize the activity of bone morphogenetic proteins in the zebrafish embryo. *Developmental Biology*, *214*, 181–196.

Fürthauer, M., Van Celst, J., Thisse, C., & Thisse, B. (2004). Fgf signalling controls the dorsoventral patterning of the zebrafish embryo. *Development*, *131*, 2853–2864.

Gardini, A., Baillat, D., Cesaroni, M., Hu, D., Marinis, J. M., Wagner, E. J., et al. (2014). Integrator regulates transcriptional initiation and pause release following activation. *Molecular Cell*, *56*, 128–139.

Ge, X., Grotjahn, D., Welch, E., Lyman-Gingerich, J., Holguin, C., Dimitrova, E., et al. (2014). Hecate/Grip2a acts to reorganize the cytoskeleton in the symmetry-breaking event of embryonic axis induction. *PLoS Genetics*, *10*, e1004422.

Gerhart, J. (1999). Pieter Nieuwkoop's contributions to the understanding of meso-endoderm induction and neural induction in chordate development. *The International Journal of Developmental Biology*, *43*, 605–613.

Gerhart, J. (2010). Enzymes, embryos, and ancestors. *Annual Review of Cell and Developmental Biology*, *26*, 1–20.

Gilardelli, C. N., Pozzoli, O., Sordino, P., Matassi, G., & Cotelli, F. (2004). Functional and hierarchical interactions among zebrafish vox/vent homeobox genes. *Developmental Dynamics*, *230*, 494–508.

Glinka, A., Wu, W., Delius, H., Monaghan, A. P., Blumenstock, C., & Niehrs, C. (1998). Dickkopf-1 is a member of a new family of secreted proteins and functions in head induction. *Nature*, *391*, 357–362.

Gore, A. V., Maegawa, S., Cheong, A., Gilligan, P. C., Weinberg, E. S., & Sampath, K. (2005). The zebrafish dorsal axis is apparent at the four-cell stage. *Nature*, *438*, 1030–1035.

Goudarzi, M., Berg, K., Pieper, L. M., & Schier, A. F. (2019). Individual long non-coding RNAs have no overt functions in zebrafish embryogenesis, viability and fertility. *eLife, 8*, e40815.
Goutel, C., Kishimoto, Y., Schulte-Merker, S., & Rosa, F. (2000). The ventralizing activity of radar, a maternally expressed bone morphogenetic protein, reveals complex bone morphogenetic protein interactions controlling dorso-ventral patterning in zebrafish. *Mechanisms of Development, 99*, 15–27.
Green, D., Whitener, A. E., Mohanty, S., & Lekven, A. C. (2015). Vertebrate nervous system posteriorization: Grading the function of Wnt signaling. *Developmental Dynamics, 244*, 507–512.
Green, J. B., & Sharpe, J. (2015). Positional information and reaction-diffusion: Two big ideas in developmental biology combine. *Development, 142*, 1203–1211.
Greenfeld, H., Lin, J., & Mullins, M. C. (2021). The BMP signaling gradient is interpreted through concentration thresholds in dorsal-ventral axial patterning. *PLoS Biology, 19*, e3001059.
Gritsman, K., Talbot, W. S., & Schier, A. F. (2000). Nodal signaling patterns the organizer. *Development, 127*, 921–932.
Gritsman, K., Zhang, J., Cheng, S., Heckscher, E., Talbot, W. S., & Schier, A. F. (1999). The EGF-CFC protein one-eyed pinhead is essential for nodal signaling. *Cell, 97*, 121–132.
Guglielmi, L., Heliot, C., Kumar, S., Alexandrov, Y., Gori, I., Papaleonidopoulou, F., et al. (2021). Smad4 controls signaling robustness and morphogenesis by differentially contributing to the nodal and BMP pathways. *Nature Communications, 12*, 6374.
Guy, G. R., Jackson, R. A., Yusoff, P., & Chow, S. Y. (2009). Sprouty proteins: Modified modulators, matchmakers or missing links? *The Journal of Endocrinology, 203*, 191–202.
Hagos, E. G., & Dougan, S. T. (2007). Time-dependent patterning of the mesoderm and endoderm by nodal signals in zebrafish. *BMC Developmental Biology, 7*, 22.
Halpern, M. E., Thisse, C., Ho, R. K., Thisse, B., Riggleman, B., Trevarrow, B., et al. (1995). Cell-autonomous shift from axial to paraxial mesodermal development in zebrafish floating head mutants. *Development, 121*, 4257–4264.
Hammerschmidt, M., Pelegri, F., Mullins, M. C., Kane, D. A., van Eeden, F. J., Granato, M., et al. (1996). Dino and Mercedes, two genes regulating dorsal development in the zebrafish embryo. *Development, 123*, 95–102.
Hammerschmidt, M., Serbedzija, G. N., & McMahon, A. P. (1996). Genetic analysis of dorsoventral pattern formation in the zebrafish: Requirement of a BMP-like ventralizing activity and its dorsal repressor. *Genes & Development, 10*, 2452–2461.
Harland, R., & Gerhart, J. (1997). Formation and function of Spemann's organizer. *Annual Review of Cell and Developmental Biology, 13*, 611–667.
Harvey, S. A., & Smith, J. C. (2009). Visualisation and quantification of morphogen gradient formation in the zebrafish. *PLoS Biology, 7*, e1000101.
Hashiguchi, M., & Mullins, M. C. (2013). Anteroposterior and dorsoventral patterning are coordinated by an identical patterning clock. *Development, 140*, 1970–1980.
Hashimoto, H., Itoh, M., Yamanaka, Y., Yamashita, S., Shimizu, T., Solnica-Krezel, L., et al. (2000). Zebrafish Dkk1 functions in forebrain specification and axial mesendoderm formation. *Developmental Biology, 217*, 138–152.
He, M., Zhang, R., Jiao, S., Zhang, F., Ye, D., Wang, H., et al. (2020). Nanog safeguards early embryogenesis against global activation of maternal beta-catenin activity by interfering with TCF factors. *PLoS Biology, 18*, e3000561.
Heinecke, K., Seher, A., Schmitz, W., Mueller, T. D., Sebald, W., & Nickel, J. (2009). Receptor oligomerization and beyond: A case study in bone morphogenetic proteins. *BMC Biology, 7*, 59.
Heisenberg, C. P., Houart, C., Take-Uchi, M., Rauch, G. J., Young, N., Coutinho, P., et al. (2001). A mutation in the Gsk3-binding domain of zebrafish Masterblind/Axin1 leads to a fate transformation of telencephalon and eyes to diencephalon. *Genes & Development, 15*, 1427–1434.

Helde, K. A., & Grunwald, D. J. (1993). The DVR-1 (Vg1) transcript of zebrafish is maternally supplied and distributed throughout the embryo. *Developmental Biology, 159*, 418–426.

Heldin, C. H., & Moustakas, A. (2016). Signaling receptors for TGF-beta family members. *Cold Spring Harbor Perspectives in Biology, 8*.

Hemmati-Brivanlou, A., Kelly, O. G., & Melton, D. A. (1994). Follistatin, an antagonist of activin, is expressed in the Spemann organizer and displays direct neuralizing activity. *Cell, 77*, 283–295.

Hino, H., Nakanishi, A., Seki, R., Aoki, T., Yamaha, E., Kawahara, A., et al. (2018). Roles of maternal wnt8a transcripts in axis formation in zebrafish. *Developmental Biology, 434*, 96–107.

Holly, V. L., Widen, S. A., Famulski, J. K., & Waskiewicz, A. J. (2014). Sfrp1a and Sfrp5 function as positive regulators of Wnt and BMP signaling during early retinal development. *Developmental Biology, 388*, 192–204.

Hong, S. K., Jang, M. K., Brown, J. L., McBride, A. A., & Feldman, B. (2011). Embryonic mesoderm and endoderm induction requires the actions of non-embryonic nodal-related ligands and Mxtx2. *Development, 138*, 787–795.

Hong, S. K., Levin, C. S., Brown, J. L., Wan, H., Sherman, B. T., Huang, D. W., et al. (2010). Pre-gastrula expression of zebrafish extraembryonic genes. *BMC Developmental Biology, 10*, 42.

Houston, D. W. (2012). Cortical rotation and messenger RNA localization in Xenopus axis formation. *Wiley Interdisciplinary Reviews: Developmental Biology, 1*, 371–388.

Iemura, S., Yamamoto, T. S., Takagi, C., Uchiyama, H., Natsume, T., Shimasaki, S., et al. (1998). Direct binding of follistatin to a complex of bone-morphogenetic protein and its receptor inhibits ventral and epidermal cell fates in early Xenopus embryo. *Proceedings of the National Academy of Sciences of the United States of America, 95*, 9337–9342.

Imai, Y., Gates, M. A., Melby, A. E., Kimelman, D., Schier, A. F., & Talbot, W. S. (2001). The homeobox genes vox and vent are redundant repressors of dorsal fates in zebrafish. *Development, 128*, 2407–2420.

Inomata, H., Shibata, T., Haraguchi, T., & Sasai, Y. (2013). Scaling of dorsal-ventral patterning by embryo size-dependent degradation of Spemann's organizer signals. *Cell, 153*, 1296–1311.

Itoh, N. (2007). The Fgf families in humans, mice, and zebrafish: Their evolutional processes and roles in development, metabolism, and disease. *Biological & Pharmaceutical Bulletin, 30*, 1819–1825.

Jasuja, R., Voss, N., Ge, G., Hoffman, G. G., Lyman-Gingerich, J., Pelegri, F., et al. (2006). bmp1 and mini fin are functionally redundant in regulating formation of the zebrafish dorsoventral axis. *Mechanisms of Development, 123*, 548–558.

Jesuthasan, S., & Strähle, U. (1997). Dynamic microtubules and specification of the zebrafish embryonic axis. *Current Biology, 7*, 31–42.

Kapp, L. D., Abrams, E. W., Marlow, F. L., & Mullins, M. C. (2013). The integrator complex subunit 6 (Ints6) confines the dorsal organizer in vertebrate embryogenesis. *PLoS Genetics, 9*, e1003822.

Kawahara, A., Wilm, T., Solnica-Krezel, L., & Dawid, I. B. (2000). Antagonistic role of vega1 and bozozok/dharma homeobox genes in organizer formation. *Proceedings of the National Academy of Sciences of the United States of America, 97*, 12121–12126.

Kelly, C., Chin, A. J., Leatherman, J. L., Kozlowski, D. J., & Weinberg, E. S. (2000). Maternally controlled (beta)-catenin-mediated signaling is required for organizer formation in the zebrafish. *Development, 127*, 3899–3911.

Kelly, G. M., Erezyilmaz, D. F., & Moon, R. T. (1995). Induction of a secondary embryonic axis in zebrafish occurs following the overexpression of beta-catenin. *Mechanisms of Development, 53*, 261–273.

Kelly, G. M., Greenstein, P., Erezyilmaz, D. F., & Moon, R. T. (1995). Zebrafish wnt8 and wnt8b share a common activity but are involved in distinct developmental pathways. *Development*, *121*, 1787–1799.
Kiecker, C., & Niehrs, C. (2001). A morphogen gradient of Wnt/beta-catenin signalling regulates anteroposterior neural patterning in Xenopus. *Development*, *128*, 4189–4201.
Kimelman, D. (2006). Mesoderm induction: From caps to chips. *Nature Reviews. Genetics*, *7*, 360–372.
Kimmel, C. B., Ballard, W. W., Kimmel, S. R., Ullmann, B., & Schilling, T. F. (1995). Stages of embryonic development of the zebrafish. *Developmental Dynamics*, *203*, 253–310.
Kimmel, C. B., & Law, R. D. (1985). Cell lineage of zebrafish blastomeres. II. Formation of the yolk syncytial layer. *Developmental Biology*, *108*, 86–93.
Kimmel, C. B., Warga, R. M., & Schilling, T. F. (1990). Origin and organization of the zebrafish fate map. *Development*, *108*, 581–594.
Klein, P. S., & Melton, D. A. (1996). A molecular mechanism for the effect of lithium on development. *Proceedings of the National Academy of Sciences of the United States of America*, *93*, 8455–8459.
Koos, D. S., & Ho, R. K. (1998). The nieuwkoid gene characterizes and mediates a Nieuwkoop-center-like activity in the zebrafish. *Current Biology*, *8*, 1199–1206.
Koos, D. S., & Ho, R. K. (1999). The nieuwkoid/dharma homeobox gene is essential for bmp2b repression in the zebrafish pregastrula. *Developmental Biology*, *215*, 190–207.
Koshida, S., Shinya, M., Mizuno, T., Kuroiwa, A., & Takeda, H. (1998). Initial anteroposterior pattern of the zebrafish central nervous system is determined by differential competence of the epiblast. *Development*, *125*, 1957–1966.
Koshida, S., Shinya, M., Nikaido, M., Ueno, N., Schulte-Merker, S., Kuroiwa, A., et al. (2002). Inhibition of BMP activity by the FGF signal promotes posterior neural development in zebrafish. *Developmental Biology*, *244*, 9–20.
Krens, S. F., Corredor-Adámez, M., He, S., Snaar-Jagalska, B. E., & Spaink, H. P. (2008). ERK1 and ERK2 MAPK are key regulators of distinct gene sets in zebrafish embryogenesis. *BMC Genomics*, *9*, 196.
Krens, S. F., He, S., Lamers, G. E., Meijer, A. H., Bakkers, J., Schmidt, T., et al. (2008). Distinct functions for ERK1 and ERK2 in cell migration processes during zebrafish gastrulation. *Developmental Biology*, *319*, 370–383.
Kudoh, T., Wilson, S. W., & Dawid, I. B. (2002). Distinct roles for Fgf, Wnt and retinoic acid in posteriorizing the neural ectoderm. *Development*, *129*, 4335–4346.
Lai, F., Gardini, A., Zhang, A., & Shiekhattar, R. (2015). Integrator mediates the biogenesis of enhancer RNAs. *Nature*, *525*, 399–403.
Langdon, Y. G., Fuentes, R., Zhang, H., Abrams, E. W., Marlow, F. L., & Mullins, M. C. (2016). Split top: A maternal cathepsin B that regulates dorsoventral patterning and morphogenesis. *Development*, *143*, 1016–1028.
Larrain, J., Bachiller, D., Lu, B., Agius, E., Piccolo, S., & De Robertis, E. M. (2000). BMP-binding modules in chordin: A model for signalling regulation in the extracellular space. *Development*, *127*, 821–830.
Lee, H. X., Ambrosio, A. L., Reversade, B., & De Robertis, E. M. (2006). Embryonic dorsal-ventral signaling: Secreted frizzled-related proteins as inhibitors of tolloid proteinases. *Cell*, *124*, 147–159.
Lee, M. T., Bonneau, A. R., Takacs, C. M., Bazzini, A. A., DiVito, K. R., Fleming, E. S., et al. (2013). Nanog, Pou5f1 and SoxB1 activate zygotic gene expression during the maternal-to-zygotic transition. *Nature*, *503*, 360–364.
Leerberg, D. M., Hopton, R. E., & Draper, B. W. (2019). Fibroblast growth factor receptors function redundantly during zebrafish embryonic development. *Genetics*, *212*, 1301–1319.

Leichsenring, M., Maes, J., Mossner, R., Driever, W., & Onichtchouk, D. (2013). Pou5f1 transcription factor controls zygotic gene activation in vertebrates. *Science, 341*, 1005–1009.

Lekven, A. C., Thorpe, C. J., Waxman, J. S., & Moon, R. T. (2001). Zebrafish wnt8 encodes two wnt8 proteins on a bicistronic transcript and is required for mesoderm and neurectoderm patterning. *Developmental Cell, 1*, 103–114.

Lele, Z., Nowak, M., & Hammerschmidt, M. (2001). Zebrafish admp is required to restrict the size of the organizer and to promote posterior and ventral development. *Developmental Dynamics, 222*, 681–687.

Leung, T., Bischof, J., Soll, I., Niessing, D., Zhang, D., Ma, J., et al. (2003). Bozozok directly represses bmp2b transcription and mediates the earliest dorsoventral asymmetry of bmp2b expression in zebrafish. *Development, 130*, 3639–3649.

Leung, T., Soll, I., Arnold, S. J., Kemler, R., & Driever, W. (2003). Direct binding of Lef1 to sites in the boz promoter may mediate pre-midblastula-transition activation of boz expression. *Developmental Dynamics, 228*, 424–432.

Li-Villarreal, N., Forbes, M. M., Loza, A. J., Chen, J., Ma, T., Helde, K., et al. (2015). Dachsous1b cadherin regulates actin and microtubule cytoskeleton during early zebrafish embryogenesis. *Development, 142*, 2704–2718.

Li, Y., Li, Q., Long, Y., & Cui, Z. (2011). Lzts2 regulates embryonic cell movements and dorsoventral patterning through interaction with and export of nuclear beta-catenin in zebrafish. *The Journal of Biological Chemistry, 286*, 45116–45130.

Lim, S., Kumari, P., Gilligan, P., Quach, H. N., Mathavan, S., & Sampath, K. (2012). Dorsal activity of maternal squint is mediated by a non-coding function of the RNA. *Development, 139*, 2903–2915.

Little, S. C., & Mullins, M. C. (2004). Twisted gastrulation promotes BMP signaling in zebrafish dorsal-ventral axial patterning. *Development, 131*, 5825–5835.

Little, S. C., & Mullins, M. C. (2009). Bone morphogenetic protein heterodimers assemble heteromeric type I receptor complexes to pattern the dorsoventral axis. *Nature Cell Biology, 11*, 637–643.

Londin, E. R., Niemiec, J., & Sirotkin, H. I. (2005). Chordin, FGF signaling, and mesodermal factors cooperate in zebrafish neural induction. *Developmental Biology, 279*, 1–19.

Long, S., Ahmad, N., & Rebagliati, M. (2003). The zebrafish nodal-related gene southpaw is required for visceral and diencephalic left-right asymmetry. *Development, 130*, 2303–2316.

Lord, N. D., Carte, A. N., Abitua, P. B., & Schier, A. F. (2021). The pattern of nodal morphogen signaling is shaped by co-receptor expression. *eLife, 10*, e54894.

Lu, F. I., Thisse, C., & Thisse, B. (2011). Identification and mechanism of regulation of the zebrafish dorsal determinant. *Proceedings of the National Academy of Sciences of the United States of America, 108*, 15876–15880.

Luz, M., Spannl-Müller, S., Özhan, G., Kagermeier-Schenk, B., Rhinn, M., Weidinger, G., et al. (2014). Dynamic association with donor cell filopodia and lipid-modification are essential features of Wnt8a during patterning of the zebrafish neuroectoderm. *PLoS One, 9*, e84922.

Lyman Gingerich, J., Westfall, T. A., Slusarski, D. C., & Pelegri, F. (2005). Hecate, a zebrafish maternal effect gene, affects dorsal organizer induction and intracellular calcium transient frequency. *Developmental Biology, 286*, 427–439.

Maegawa, S., Varga, M., & Weinberg, E. S. (2006). FGF signaling is required for β-catenin-mediated induction of the zebrafish organizer. *Development, 133*, 3265–3276.

Malhotra, D., Shin, J., Solnica-Krezel, L., & Raz, E. (2018). Spatio-temporal regulation of concurrent developmental processes by generic signaling downstream of chemokine receptors. *eLife, 7*, e33574.

Martyn, U., & Schulte-Merker, S. (2003). The ventralized ogon mutant phenotype is caused by a mutation in the zebrafish homologue of sizzled, a secreted frizzled-related protein. *Developmental Biology, 260*, 58–67.
Mason, J. M., Morrison, D. J., Basson, M. A., & Licht, J. D. (2006). Sprouty proteins: Multifaceted negative-feedback regulators of receptor tyrosine kinase signaling. *Trends in Cell Biology, 16*, 45–54.
Mattes, B., Dang, Y., Greicius, G., Kaufmann, L. T., Prunsche, B., Rosenbauer, J., et al. (2018). Wnt/PCP controls spreading of Wnt/beta-catenin signals by cytonemes in vertebrates. *eLife*, 7, e36953.
Maurer, J. M., & Sagerstrom, C. G. (2018). A parental requirement for dual-specificity phosphatase 6 in zebrafish. *BMC Developmental Biology, 18*, 6.
Mei, W., Lee, K. W., Marlow, F. L., Miller, A. L., & Mullins, M. C. (2009). hnRNP I is required to generate the Ca2+ signal that causes egg activation in zebrafish. *Development, 136*, 3007–3017.
Melby, A. E., Beach, C., Mullins, M., & Kimelman, D. (2000). Patterning the early zebrafish by the opposing actions of bozozok and vox/vent. *Developmental Biology, 224*, 275–285.
Melby, A. E., Warga, R. M., & Kimmel, C. B. (1996). Specification of cell fates at the dorsal margin of the zebrafish gastrula. *Development, 122*, 2225–2237.
Mevel, R., Draper, J. E., Lie, A. L. M., Kouskoff, V., & Lacaud, G. (2019). RUNX transcription factors: Orchestrators of development. *Development, 146*.
Miller-Bertoglio, V., Carmany-Rampey, A., Fürthauer, M., Gonzalez, E. M., Thisse, C., Thisse, B., et al. (1999). Maternal and zygotic activity of the zebrafish ogon locus antagonizes BMP signaling. *Developmental Biology, 214*, 72–86.
Miller-Bertoglio, V. E., Fisher, S., Sánchez, A., Mullins, M. C., & Halpern, M. E. (1997). Differential regulation of chordin expression domains in mutant zebrafish. *Developmental Biology, 192*, 537–550.
Mintzer, K. A., Lee, M. A., Runke, G., Trout, J., Whitman, M., & Mullins, M. C. (2001). Lost-a-fin encodes a type I BMP receptor, Alk8, acting maternally and zygotically in dorsoventral pattern formation. *Development, 128*, 859–869.
Mizoguchi, T., Izawa, T., Kuroiwa, A., & Kikuchi, Y. (2006). Fgf signaling negatively regulates nodal-dependent endoderm induction in zebrafish. *Developmental Biology, 300*, 612–622.
Mizuno, T., Yamaha, E., Kuroiwa, A., & Takeda, H. (1999). Removal of vegetal yolk causes dorsal deficencies and impairs dorsal-inducing ability of the yolk cell in zebrafish. *Mechanisms of Development, 81*, 51–63.
Mlodzik, M. (2016). The Dishevelled protein family: Still rather a mystery after over 20 years of molecular studies. *Current Topics in Developmental Biology, 117*, 75–91.
Mo, S., Wang, L., Li, Q., Li, J., Li, Y., Thannickal, V. J., et al. (2010). Caveolin-1 regulates dorsoventral patterning through direct interaction with beta-catenin in zebrafish. *Developmental Biology, 344*, 210–223.
Montague, T. G., & Schier, A. F. (2017). Vg1-nodal heterodimers are the endogenous inducers of mesendoderm. *eLife, 6*, e28183.
Moriyama, Y., & De Robertis, E. M. (2018). Embryonic regeneration by relocalization of the Spemann organizer during twinning in Xenopus. *Proceedings of the National Academy of Sciences of the United States of America, 115*, E4815–E4822.
Müller, P., Rogers, K. W., Jordan, B. M., Lee, J. S., Robson, D., Ramanathan, S., et al. (2012). Differential diffusivity of nodal and lefty underlies a reaction-diffusion patterning system. *Science, 336*, 721–724.
Mullins, M. C., Hammerschmidt, M., Kane, D. A., Odenthal, J., Brand, M., van Eeden, F. J., et al. (1996). Genes establishing dorsoventral pattern formation in the zebrafish embryo: The ventral specifying genes. *Development, 123*, 81–93.

Muraoka, O., Shimizu, T., Yabe, T., Nojima, H., Bae, Y. K., Hashimoto, H., et al. (2006). Sizzled controls dorso-ventral polarity by repressing cleavage of the chordin protein. *Nature Cell Biology*, *8*, 329–338.

Nakamura, Y., de Paiva Alves, E., Veenstra, G. J., & Hoppler, S. (2016). Tissue- and stage-specific Wnt target gene expression is controlled subsequent to beta-catenin recruitment to cis-regulatory modules. *Development*, *143*, 1914–1925.

Nguyen-Chi, M. E., Bryson-Richardson, R., Sonntag, C., Hall, T. E., Gibson, A., Sztal, T., et al. (2012). Morphogenesis and cell fate determination within the adaxial cell equivalence group of the zebrafish myotome. *PLoS Genetics*, *8*, e1003014.

Nguyen, V. H., Schmid, B., Trout, J., Connors, S. A., Ekker, M., & Mullins, M. C. (1998). Ventral and lateral regions of the zebrafish gastrula, including the neural crest progenitors, are established by a bmp2b/swirl pathway of genes. *Developmental Biology*, *199*, 93–110.

Nieuwkoop, P. D. (1969). The formation of the mesoderm in urodelean amphibians : I. induction by the endoderm. *Wilhelm Roux' Archiv für Entwicklungsmechanik der Organismen*, *162*, 341–373.

Nikaido, M., Tada, M., Saji, T., & Ueno, N. (1997). Conservation of BMP signaling in zebrafish mesoderm patterning. *Mechanisms of Development*, *61*, 75–88.

Nikaido, M., Tada, M., Takeda, H., Kuroiwa, A., & Ueno, N. (1999). In vivo analysis using variants of zebrafish BMPR-IA: Range of action and involvement of BMP in ectoderm patterning. *Development*, *126*, 181–190.

Nojima, H., Rothhamel, S., Shimizu, T., Kim, C. H., Yonemura, S., Marlow, F. L., et al. (2010). Syntabulin, a motor protein linker, controls dorsal determination. *Development*, *137*, 923–933.

Nojima, H., Shimizu, T., Kim, C. H., Yabe, T., Bae, Y. K., Muraoka, O., et al. (2004). Genetic evidence for involvement of maternally derived Wnt canonical signaling in dorsal determination in zebrafish. *Mechanisms of Development*, *121*, 371–386.

Nowak, M., Machate, A., Yu, S. R., Gupta, M., & Brand, M. (2011). Interpretation of the FGF8 morphogen gradient is regulated by endocytic trafficking. *Nature Cell Biology*, *13*, 153–158.

Nusse, R., & Clevers, H. (2017). Wnt/beta-catenin signaling, disease, and emerging therapeutic modalities. *Cell*, *169*, 985–999.

Ober, E. A., & Schulte-Merker, S. (1999). Signals from the yolk cell induce mesoderm, neuroectoderm, the trunk organizer, and the notochord in zebrafish. *Developmental Biology*, *215*, 167–181.

Ornitz, D. M., & Itoh, N. (2015). The fibroblast growth factor signaling pathway. *Wiley Interdisciplinary Reviews: Developmental Biology*, *4*, 215–266.

Ozair, M. Z., Kintner, C., & Brivanlou, A. H. (2013). Neural induction and early patterning in vertebrates. *Wiley Interdisciplinary Reviews: Developmental Biology*, *2*, 479–498.

Pelliccia, J. L., Jindal, G. A., & Burdine, R. D. (2017). Gdf3 is required for robust nodal signaling during germ layer formation and left-right patterning. *eLife*, *6*, e28635.

Piccolo, S., Agius, E., Lu, B., Goodman, S., Dale, L., & De Robertis, E. M. (1997). Cleavage of chordin by Xolloid metalloprotease suggests a role for proteolytic processing in the regulation of Spemann organizer activity. *Cell*, *91*, 407–416.

Piccolo, S., Sasai, Y., Lu, B., & De Robertis, E. M. (1996). Dorsoventral patterning in Xenopus: Inhibition of ventral signals by direct binding of chordin to BMP-4. *Cell*, *86*, 589–598.

Picker, A., Cavodeassi, F., Machate, A., Bernauer, S., Hans, S., Abe, G., et al. (2009). Dynamic coupling of pattern formation and morphogenesis in the developing vertebrate retina. *PLoS Biology*, 7, e1000214.

Plouhinec, J. L., Zakin, L., Moriyama, Y., & De Robertis, E. M. (2013). Chordin forms a self-organizing morphogen gradient in the extracellular space between ectoderm and

mesoderm in the Xenopus embryo. *Proceedings of the National Academy of Sciences of the United States of America, 110*, 20372–20379.
Pomreinke, A. P., Soh, G. H., Rogers, K. W., Bergmann, J. K., Blassle, A. J., & Müller, P. (2017). Dynamics of BMP signaling and distribution during zebrafish dorsal-ventral patterning. *eLife, 6*, e25861.
Poulain, M., Fürthauer, M., Thisse, B., Thisse, C., & Lepage, T. (2006). Zebrafish endoderm formation is regulated by combinatorial nodal, FGF and BMP signalling. *Development, 133*, 2189–2200.
Raible, F., & Brand, M. (2001). Tight transcriptional control of the ETS domain factors erm and Pea3 by Fgf signaling during early zebrafish development. *Mechanisms of Development, 107*, 105–117.
Ramel, M. C., Buckles, G. R., Baker, K. D., & Lekven, A. C. (2005). WNT8 and BMP2B co-regulate non-axial mesoderm patterning during zebrafish gastrulation. *Developmental Biology, 287*, 237–248.
Ramel, M. C., & Lekven, A. C. (2004). Repression of the vertebrate organizer by Wnt8 is mediated by vent and vox. *Development, 131*, 3991–4000.
Rebagliati, M. R., Toyama, R., Fricke, C., Haffter, P., & Dawid, I. B. (1998). Zebrafish nodal-related genes are implicated in axial patterning and establishing left-right asymmetry. *Developmental Biology, 199*, 261–272.
Reid, C. D., Zhang, Y., Sheets, M. D., & Kessler, D. S. (2012). Transcriptional integration of Wnt and nodal pathways in establishment of the Spemann organizer. *Developmental Biology, 368*, 231–241.
Reifers, F., Bohli, H., Walsh, E. C., Crossley, P. H., Stainier, D. Y., & Brand, M. (1998). Fgf8 is mutated in zebrafish acerebellar (ace) mutants and is required for maintenance of midbrain-hindbrain boundary development and somitogenesis. *Development, 125*, 2381–2395.
Reim, G., & Brand, M. (2006). Maternal control of vertebrate dorsoventral axis formation and epiboly by the POU domain protein Spg/Pou2/Oct4. *Development, 133*, 2757–2770.
Rengarajan, C., Matzke, A., Reiner, L., Orian-Rousseau, V., & Scholpp, S. (2014). Endocytosis of Fgf8 is a double-stage process and regulates spreading and signaling. *PLoS One, 9*, e86373.
Rentzsch, F., Zhang, J., Kramer, C., Sebald, W., & Hammerschmidt, M. (2006). Crossveinless 2 is an essential positive feedback regulator of bmp signaling during zebrafish gastrulation. *Development, 133*, 801–811.
Ro, H., & Dawid, I. B. (2009). Organizer restriction through modulation of Bozozok stability by the E3 ubiquitin ligase Lnx-like. *Nature Cell Biology, 11*, 1121–1127.
Rochard, L., Monica, S. D., Ling, I. T., Kong, Y., Roberson, S., Harland, R., et al. (2016). Roles of Wnt pathway genes wls, wnt9a, wnt5b, frzb and gpc4 in regulating convergent-extension during zebrafish palate morphogenesis. *Development, 143*, 2541–2547.
Rogers, K. W., ElGamacy, M., Jordan, B. M., & Müller, P. (2020). Optogenetic investigation of BMP target gene expression diversity. *eLife, 9*, e58641.
Rogers, K. W., Lord, N. D., Gagnon, J. A., Pauli, A., Zimmerman, S., Aksel, D. C., et al. (2017). Nodal patterning without lefty inhibitory feedback is functional but fragile. *eLife, 6*, e28785.
Roszko, I., Sawada, A., & Solnica-Krezel, L. (2009). Regulation of convergence and extension movements during vertebrate gastrulation by the Wnt/PCP pathway. *Seminars in Cell & Developmental Biology, 20*, 986–997.
Ruzicka, L., Howe, D. G., Ramachandran, S., Toro, S., Van Slyke, C. E., Bradford, Y. M., et al. (2019). The Zebrafish Information Network: New support for non-coding genes,

richer Gene Ontology annotations and the Alliance of Genome Resources. *Nucleic Acids Research*, *47*, D867–D873.

Ryu, S. L., Fujii, R., Yamanaka, Y., Shimizu, T., Yabe, T., Hirata, T., et al. (2001). Regulation of dharma/bozozok by the Wnt pathway. *Developmental Biology*, *231*, 397–409.

Sako, K., Pradhan, S. J., Barone, V., Inglés-Prieto, A., Müller, P., Ruprecht, V., et al. (2016). Optogenetic control of nodal signaling reveals a temporal pattern of nodal signaling regulating cell fate specification during gastrulation. *Cell Reports*, *16*, 866–877.

Sampath, K., Rubinstein, A. L., Cheng, A. M., Liang, J. O., Fekany, K., Solnica-Krezel, L., et al. (1998). Induction of the zebrafish ventral brain and floorplate requires cyclops/nodal signalling. *Nature*, *395*, 185–189.

Saremba, S., Nickel, J., Seher, A., Kotzsch, A., Sebald, W., & Mueller, T. D. (2008). Type I receptor binding of bone morphogenetic protein 6 is dependent on N-glycosylation of the ligand. *The FEBS Journal*, *275*, 172–183.

Sasai, Y., Lu, B., Steinbeisser, H., Geissert, D., Gont, L. K., & De Robertis, E. M. (1994). Xenopus chordin: A novel dorsalizing factor activated by organizer-specific homeobox genes. *Cell*, *79*, 779–790.

Saúde, L., Woolley, K., Martin, P., Driever, W., & Stemple, D. L. (2000). Axis-inducing activities and cell fates of the zebrafish organizer. *Development*, *127*, 3407–3417.

Saxén, L., & Toivonen, S. (1962). *Primary embryonic induction*. Prentice-Hall, London, Englewood Cliffs, NJ: Logos Press.

Schmid, B., Fürthauer, M., Connors, S. A., Trout, J., Thisse, B., Thisse, C., et al. (2000). Equivalent genetic roles for bmp7/snailhouse and bmp2b/swirl in dorsoventral pattern formation. *Development*, *127*, 957–967.

Schneider, P. N., Slusarski, D. C., & Houston, D. W. (2012). Differential role of Axin RGS domain function in Wnt signaling during anteroposterior patterning and maternal axis formation. *PLoS One*, *7*, e44096.

Schneider, S., Steinbeisser, H., Warga, R. M., & Hausen, P. (1996). Beta-catenin translocation into nuclei demarcates the dorsalizing centers in frog and fish embryos. *Mechanisms of Development*, *57*, 191–198.

Scholpp, S., & Brand, M. (2004). Endocytosis controls spreading and effective signaling range of Fgf8 protein. *Current Biology*, *14*, 1834–1841.

Schulte-Merker, S., Hammerschmidt, M., Beuchle, D., Cho, K. W., De Robertis, E. M., & Nüsslein-Volhard, C. (1994). Expression of zebrafish goosecoid and no tail gene products in wild-type and mutant no tail embryos. *Development*, *120*, 843–852.

Schulte-Merker, S., Lee, K. J., McMahon, A. P., & Hammerschmidt, M. (1997). The zebrafish organizer requires chordino. *Nature*, *387*, 862–863.

Scott, I. C., Blitz, I. L., Pappano, W. N., Imamura, Y., Clark, T. G., Steiglitz, B. M., et al. (1999). Mammalian BMP-1/Tolloid-related metalloproteinases, including novel family member mammalian Tolloid-like 2, have differential enzymatic activities and distributions of expression relevant to patterning and skeletogenesis. *Developmental Biology*, *213*, 283–300.

Seiliez, I., Thisse, B., & Thisse, C. (2006). FoxA3 and goosecoid promote anterior neural fate through inhibition of Wnt8a activity before the onset of gastrulation. *Developmental Biology*, *290*, 152–163.

Shao, M., Lin, Y., Liu, Z., Zhang, Y., Wang, L., Liu, C., et al. (2012). GSK-3 activity is critical for the orientation of the cortical microtubules and the dorsoventral axis determination in zebrafish embryos. *PLoS One*, *7*, e36655.

Shih, J., & Fraser, S. E. (1996). Characterizing the zebrafish organizer: Microsurgical analysis at the early-shield stage. *Development*, *122*, 1313–1322.

Shimizu, T., Yamanaka, Y., Nojima, H., Yabe, T., Hibi, M., & Hirano, T. (2002). A novel repressor-type homeobox gene, ved, is involved in dharma/bozozok-mediated dorsal organizer formation in zebrafish. *Mechanisms of Development*, *118*, 125–138.

Shimizu, T., Yamanaka, Y., Ryu, S. L., Hashimoto, H., Yabe, T., Hirata, T., et al. (2000). Cooperative roles of Bozozok/dharma and nodal-related proteins in the formation of the dorsal organizer in zebrafish. *Mechanisms of Development*, *91*, 293–303.

Shinya, M., Eschbach, C., Clark, M., Lehrach, H., & Furutani-Seiki, M. (2000). Zebrafish Dkk1, induced by the pre-MBT Wnt signaling, is secreted from the prechordal plate and patterns the anterior neural plate. *Mechanisms of Development*, *98*, 3–17.

Sidi, S., Goutel, C., Peyrieras, N., & Rosa, F. M. (2003). Maternal induction of ventral fate by zebrafish radar. *Proceedings of the National Academy of Sciences of the United States of America*, *100*, 3315–3320.

Sirotkin, H. I., Dougan, S. T., Schier, A. F., & Talbot, W. S. (2000). Bozozok and squint act in parallel to specify dorsal mesoderm and anterior neuroectoderm in zebrafish. *Development*, *127*, 2583–2592.

Smith, W. C., & Harland, R. M. (1992). Expression cloning of noggin, a new dorsalizing factor localized to the Spemann organizer in Xenopus embryos. *Cell*, *70*, 829–840.

Soh, G. H., Pomreinke, A. P., & Müller, P. (2020). Integration of nodal and BMP signaling by mutual signaling effector antagonism. *Cell Reports*, *31*, 107487.

Solnica-Krezel, L., Stemple, D. L., Mountcastle-Shah, E., Rangini, Z., Neuhauss, S. C., Malicki, J., et al. (1996). Mutations affecting cell fates and cellular rearrangements during gastrulation in zebrafish. *Development*, *123*, 67–80.

Speer, K. F., Sommer, A., Tajer, B., Mullins, M. C., Klein, P. S., & Lemmon, M. A. (2019). Non-acylated Wnts can promote signaling. *Cell Reports*, *26*(875–883), e875.

Spemann, H. (1962). *Embryonic development and induction*. New York: Hafner Pub. Co.

Spemann, H., & Mangold, H. (2001). Induction of embryonic primordia by implantation of organizers from a different species. 1923. *The International Journal of Developmental Biology*, *45*, 13–38.

Srinivasan, S., Rashka, K. E., & Bier, E. (2002). Creation of a Sog morphogen gradient in the Drosophila embryo. *Developmental Cell*, *2*, 91–101.

Stachel, S. E., Grunwald, D. J., & Myers, P. Z. (1993). Lithium perturbation and goosecoid expression identify a dorsal specification pathway in the pregastrula zebrafish. *Development*, *117*, 1261–1274.

Tajer, B., Dutko, J. A., Little, S. C., & Mullins, M. C. (2021). BMP heterodimers signal via distinct type I receptor class functions. *Proceedings of the National Academy of Sciences of the United States of America*, 118.

Talbot, W. S., Trevarrow, B., Halpern, M. E., Melby, A. E., Farr, G., Postlethwait, J. H., et al. (1995). A homeobox gene essential for zebrafish notochord development. *Nature*, *378*, 150–157.

Tanaka, S., Hosokawa, H., Weinberg, E. S., & Maegawa, S. (2017). Chordin and dickkopf-1b are essential for the formation of head structures through activation of the FGF signaling pathway in zebrafish. *Developmental Biology*, *424*, 189–197.

Tatomer, D. C., Elrod, N. D., Liang, D., Xiao, M. S., Jiang, J. Z., Jonathan, M., et al. (2019). The integrator complex cleaves nascent mRNAs to attenuate transcription. *Genes & Development*, *33*, 1525–1538.

Tendeng, C., & Houart, C. (2006). Cloning and embryonic expression of five distinct sfrp genes in the zebrafish Danio rerio. *Gene Expression Patterns*, *6*, 761–771.

Thisse, B., Heyer, V., Lux, A., Alunni, V., Degrave, A., Seiliez, I., et al. (2004). Spatial and temporal expression of the zebrafish genome by large-scale in situ hybridization screening. *Methods in Cell Biology*, 77, 505–519.

Thisse, B., Wright, C. V., & Thisse, C. (2000). Activin- and nodal-related factors control antero-posterior patterning of the zebrafish embryo. *Nature*, *403*, 425–428.

Tiso, N., Filippi, A., Pauls, S., Bortolussi, M., & Argenton, F. (2002). BMP signalling regulates anteroposterior endoderm patterning in zebrafish. *Mechanisms of Development*, *118*, 29–37.

Tran, L. D., Hino, H., Quach, H., Lim, S., Shindo, A., Mimori-Kiyosue, Y., et al. (2012). Dynamic microtubules at the vegetal cortex predict the embryonic axis in zebrafish. *Development, 139*, 3644–3652.

Troilo, H., Zuk, A. V., Tunnicliffe, R. B., Wohl, A. P., Berry, R., Collins, R. F., et al. (2014). Nanoscale structure of the BMP antagonist chordin supports cooperative BMP binding. *Proceedings of the National Academy of Sciences of the United States of America, 111*, 13063–13068.

Tsang, M., & Dawid, I. B. (2004). Promotion and attenuation of FGF signaling through the Ras-MAPK pathway. *Sci STKE, 2004*. pe17.

Tsang, M., Friesel, R., Kudoh, T., & Dawid, I. B. (2002). Identification of Sef, a novel modulator of FGF signalling. *Nature Cell Biology, 4*, 165–169.

Tsang, M., Maegawa, S., Kiang, A., Habas, R., Weinberg, E., & Dawid, I. B. (2004). A role for MKP3 in axial patterning of the zebrafish embryo. *Development, 131*, 2769–2779.

Tuazon, F. B., & Mullins, M. C. (2015). Temporally coordinated signals progressively pattern the anteroposterior and dorsoventral body axes. *Seminars in Cell & Developmental Biology, 42*, 118–133.

Tuazon, F. B., Wang, X., Andrade, J. L., Umulis, D., & Mullins, M. C. (2020). Proteolytic restriction of chordin range underlies BMP gradient formation. *Cell Reports, 32*, 108039.

Tucker, J. A., Mintzer, K. A., & Mullins, M. C. (2008). The BMP signaling gradient patterns dorsoventral tissues in a temporally progressive manner along the anteroposterior axis. *Developmental Cell, 14*, 108–119.

Valdivia, L. E., Lamb, D. B., Horner, W., Wierzbicki, C., Tafessu, A., Williams, A. M., et al. (2016). Antagonism between Gdf6a and retinoic acid pathways controls timing of retinal neurogenesis and growth of the eye in zebrafish. *Development, 143*, 1087–1098.

van Boxtel, A. L., Chesebro, J. E., Heliot, C., Ramel, M. C., Stone, R. K., & Hill, C. S. (2015). A temporal window for signal activation dictates the dimensions of a nodal signaling domain. *Developmental Cell, 35*, 175–185.

van Boxtel, A. L., Economou, A. D., Heliot, C., & Hill, C. S. (2018). Long-range signaling activation and local inhibition separate the mesoderm and endoderm lineages. *Developmental Cell, 44*(179–191), e175.

Varga, M., Maegawa, S., Bellipanni, G., & Weinberg, E. S. (2007). Chordin expression, mediated by nodal and FGF signaling, is restricted by redundant function of two beta-catenins in the zebrafish embryo. *Mechanisms of Development, 124*, 775–791.

Varga, M., Maegawa, S., & Weinberg, E. S. (2011). Correct anteroposterior patterning of the zebrafish neurectoderm in the absence of the early dorsal organizer. *BMC Developmental Biology, 11*, 26.

Verhey, K. J., Kaul, N., & Soppina, V. (2011). Kinesin assembly and movement in cells. *Annual Review of Biophysics, 40*, 267–288.

Vopalensky, P., Pralow, S., & Vastenhouw, N. L. (2018). Reduced expression of the nodal co-receptor Oep causes loss of mesendodermal competence in zebrafish. *Development, 145*.

Waddington, C. H. (1932). Experiments on the development of Chick and Duck embryos, cultivated in vitro. Philosophical transactions of the Royal Society of London. *Series B, Containing Papers of a Biological Character, 221*, 179–230.

Wang, S., Krinks, M., Lin, K., Luyten, F. P., & Moos, M., Jr. (1997). Frzb, a secreted protein expressed in the Spemann organizer, binds and inhibits Wnt-8. *Cell, 88*, 757–766.

Warga, R. M., & Nüsslein-Volhard, C. (1999). Origin and development of the zebrafish endoderm. *Development, 126*, 827–838.

Williams, M. L. K., & Solnica-Krezel, L. (2020). Cellular and molecular mechanisms of convergence and extension in zebrafish. *Current Topics in Developmental Biology, 136*, 377–407.

Willot, V., Mathieu, J., Lu, Y., Schmid, B., Sidi, S., Yan, Y. L., et al. (2002). Cooperative action of ADMP- and BMP-mediated pathways in regulating cell fates in the zebrafish gastrula. *Developmental Biology, 241*, 59–78.

Wolpert, L. (1969). Positional information and the spatial pattern of cellular differentiation. *Journal of Theoretical Biology, 25*, 1–47.

Wolpert, L. (2011). Positional information and patterning revisited. *Journal of Theoretical Biology*, *269*, 359–365.

Wu, S. Y., Shin, J., Sepich, D. S., & Solnica-Krezel, L. (2012). Chemokine GPCR signaling inhibits beta-catenin during zebrafish axis formation. *PLoS Biology*, *10*, e1001403.

Xie, J., & Fisher, S. (2005). Twisted gastrulation enhances BMP signaling through chordin dependent and independent mechanisms. *Development*, *132*, 383–391.

Xie, X. W., Liu, J. X., Hu, B., & Xiao, W. (2011). Zebrafish foxo3b negatively regulates canonical Wnt signaling to affect early embryogenesis. *PLoS One*, *6*, e24469.

Xing, Y. Y., Cheng, X. N., Li, Y. L., Zhang, C., Saquet, A., Liu, Y. Y., et al. (2018). Mutational analysis of dishevelled genes in zebrafish reveals distinct functions in embryonic patterning and gastrulation cell movements. *PLoS Genetics*, *14*, e1007551.

Xiong, B., Rui, Y., Zhang, M., Shi, K., Jia, S., Tian, T., et al. (2006). Tob1 controls dorsal development of zebrafish embryos by antagonizing maternal beta-catenin transcriptional activity. *Developmental Cell*, *11*, 225–238.

Yabe, T., Shimizu, T., Muraoka, O., Bae, Y. K., Hirata, T., Nojima, H., et al. (2003). Ogon/secreted frizzled functions as a negative feedback regulator of bmp signaling. *Development*, *130*, 2705–2716.

Yamanaka, Y., Mizuno, T., Sasai, Y., Kishi, M., Takeda, H., Kim, C. H., et al. (1998). A novel homeobox gene, dharma, can induce the organizer in a non-cell-autonomous manner. *Genes & Development*, *12*, 2345–2353.

Yamauchi, H., Miyakawa, N., Miyake, A., & Itoh, N. (2009). Fgf4 is required for left-right patterning of visceral organs in zebrafish. *Developmental Biology*, *332*, 177–185.

Yan, L., Chen, J., Zhu, X., Sun, J., Wu, X., Shen, W., et al. (2018). Maternal Huluwa dictates the embryonic body axis through beta-catenin in vertebrates. *Science*, *362*.

Yan, Y., Ning, G., Li, L., Liu, J., Yang, S., Cao, Y., et al. (2019). The BMP ligand pinhead together with Admp supports the robustness of embryonic patterning. *Science Advances*, *5*, eaau6455.

Yao, J., & Kessler, D. S. (2001). Goosecoid promotes head organizer activity by direct repression of Xwnt8 in Spemann's organizer. *Development*, *128*, 2975–2987.

Yu, S. R., Burkhardt, M., Nowak, M., Ries, J., Petrasek, Z., Scholpp, S., et al. (2009). Fgf8 morphogen gradient forms by a source-sink mechanism with freely diffusing molecules. *Nature*, *461*, 533–536.

Zhang, J., Talbot, W. S., & Schier, A. F. (1998). Positional cloning identifies zebrafish one-eyed pinhead as a permissive EGF-related ligand required during gastrulation. *Cell*, *92*, 241–251.

Zhang, J. L., Huang, Y., Qiu, L. Y., Nickel, J., & Sebald, W. (2007). von Willebrand factor type C domain-containing proteins regulate bone morphogenetic protein signaling through different recognition mechanisms. *The Journal of Biological Chemistry*, *282*, 20002–20014.

Zhang, J. L., Patterson, L. J., Qiu, L. Y., Graziussi, D., Sebald, W., & Hammerschmidt, M. (2010). Binding between Crossveinless-2 and chordin von Willebrand factor type C domains promotes BMP signaling by blocking chordin activity. *PLoS One*, *5*, e12846.

Zimmerman, L. B., De Jesus-Escobar, J. M., & Harland, R. M. (1996). The Spemann organizer signal noggin binds and inactivates bone morphogenetic protein 4. *Cell*, *86*, 599–606.

Zinski, J., Bu, Y., Wang, X., Dou, W., Umulis, D., & Mullins, M. C. (2017). Systems biology derived source-sink mechanism of BMP gradient formation. *eLife*, *6*, e22199.

Zinski, J., Tajer, B., & Mullins, M. C. (2018). TGF-beta family signaling in early vertebrate development. *Cold Spring Harbor Perspectives in Biology*, *10*.

Zoltewicz, J. S., & Gerhart, J. C. (1997). The Spemann organizer of Xenopus is patterned along its anteroposterior axis at the earliest gastrula stage. *Developmental Biology*, *192*, 482–491.

CHAPTER SIX

# Local BMP signaling: A sensor for synaptic activity that balances synapse growth and function

**Rosario Vicidomini and Mihaela Serpe***
Neurosciences and Cellular and Structural Biology Division, *Eunice Kennedy Shiver* National Institute of Child Health and Human Development, National Institutes of Health, Bethesda, MD, United States
*Corresponding author: e-mail address: mihaela.serpe@nih.gov

## Contents



## Abstract

Synapse development is coordinated by intercellular communication between the pre- and postsynaptic compartments, and by neuronal activity itself. In flies as in vertebrates, neuronal activity induces input-specific changes in the synaptic strength so that the entire circuit maintains stable function in the face of many challenges, including changes in synapse number and strength. But how do neurons sense synapse activity? In several studies carried out using the *Drosophila* neuromuscular junction (NMJ), we demonstrated that local BMP signaling provides an exquisite sensor for synapse activity. Here we review the main features of this exquisite sensor and discuss its functioning beyond monitoring the synapse activity but rather as a key controller that operates in coordination with other BMP signaling pathways to balance synapse growth, maturation and function.

*Current Topics in Developmental Biology*, Volume 150
ISSN 0070-2153
https://doi.org/10.1016/bs.ctdb.2022.04.001

# 1. BMP signaling pathways

Bone morphogenetic proteins (BMPs) are a functionally diverse group of potent secreted molecules belonging to the TGF-β superfamily of growth and differentiation signaling factors. Originally identified in bone extracts as critical inducers of bone deposition (Urist, 1965), BMPs are now recognized to regulate a variety of cellular processes from cell fate specification, renewal and maintenance of stem cell populations to axis determination, organogenesis and tissue patterning (Hogan, 1996). BMPs also play crucial roles in the development of the nervous system, including neuroectoderm induction, neural stem cells and neural crest cells specification and patterning (Bond, Bhalala, & Kessler, 2012). Dysregulation of BMP signaling is associated with many developmental abnormalities and disease states highlighting the need for tight control of this pathway.

BMPs are secreted as pro-proteins with large pro-domains that are removed by furin-type enzymes to generate biologically active ligands (Cui, Jean, Thomas, & Christian, 1998). The pro-domains are required for the proper trafficking and folding of these molecules and formation of several disulfide bonds that (i) stabilize the characteristic cysteine-knot configuration of individual secreted monomers and (ii) covalently link biologically active BMP homo- and hetero-dimers (Degnin, Jean, Thomas, & Christian, 2004; Goldman et al., 2006). The BMP signaling cascade is initiated by binding of active dimeric ligands to a multi-component signaling complex composed of at least two different types of transmembrane Ser/Thr kinases, type I and type II receptors (Ehrlich, Gutman, Knaus, & Henis, 2012; Massague, 1990). Within this complex, Type II receptor is a constitutive kinase that phosphorylates and activates Type I receptor. In the canonical BMP pathway, activated Type I receptor binds to and phosphorylates the intracellular R-Smad effectors (Smad 1, 5, or 8 in vertebrates, and Mad in *Drosophila*) (Feng & Derynck, 2005; Schmierer & Hill, 2007). Phosphorylated R-Smads (pSmads) have a propensity to form trimeric complexes that favor their dissociation from the receptors (Kawabata, Inoue, Hanyu, Imamura, & Miyazono, 1998). Cytosolic pSmads associate with highly related co-Smads and translocate into the nucleus where, in conjunction with additional transcription factors, they activate or repress transcription of target genes (Hill, 2016; Zhang, Feng, We, & Derynck, 1996).

Activated BMPRs can also signal independently of Smads through non-canonical pathways that include mitogen-activated protein kinase (MAPK),

LIM kinase, phosphatidylinositol 3-kinase/Akt (PI3K/Akt), and Rho-like small GTPases (Derynck & Zhang, 2003; Foletta et al., 2003; Moustakas & Heldin, 2005; Zhang, 2009).

More recently, pSmad accumulation at cell membranes has been reported in at least two instances: (i) at tight junctions during neural tube closure (Eom, Amarnath, Fogel, & Agarwala, 2011), and (ii) at *Drosophila* NMJ (Dudu et al., 2006; Smith, Machamer, Kim, Hays, & Marques, 2012). During neural tube closure, pSmad1/5/8 binds to apical polarity complexes and functions as a bridge to mediate stabilization of BMP/BMPR complexes at tight junctions (Eom et al., 2011); prolonged BMP blockade disrupts tight junctions and disrupts epithelial organization (Eom, Amarnath, Fogel, & Agarwala, 2012). At the fly NMJ, pMad accumulates at the active zones, which are highly specialized membrane regions where synaptic vesicles dock to release the neurotransmitter. Each presynaptic active zone forms an asymmetric tight junction with a postsynaptic density, where neurotransmitter receptors concentrate. In both cases, the Smad-dependent junctional complexes do not participate in any transcriptional regulation activities. Instead, these local BMP signaling complexes appear engaged in interactions that control the integrity and function of the respective cellular junctions.

In epithelial cells, immunofluorescent staining of pSmads reveals very strong nuclear signals, overpowering any junctional pSmads signals. In contrast, at the fly NMJ, junctional pMad localizes at synaptic terminals whereas nuclear pMad accumulates in MN nuclei located in the ventral ganglion, the fly counterpart of the spinal cord. The spatial separation between nuclear and synaptic pMad was crucial to uncovering the function of this Smad-dependent local/junctional BMP signaling pathway in monitoring and modulating the synaptic junction (Sulkowski et al., 2016; Sulkowski, Kim, & Serpe, 2014).

The synaptic BMP signaling is genetically distinct from the other canonical and non-canonical BMP pathways that the fly motor neurons (MNs) must receive and integrate during development (Fig. 1). The NMJ, the synapse between a motor neuron (MN) terminal and a muscle fiber, is established during embryogenesis (Keshishian, Broadie, Chiba, & Bate, 1996). Once it hatches from the egg case, a first instar larva experiences rapid growth expanding its larval muscle ~160-fold during time it takes until it reaches the third instar larval stage, the last stage of larval development. Canonical BMP signaling pathways coordinate the growing muscles with the MNs, adjusting the NMJs size and neurotransmitter release to ensure

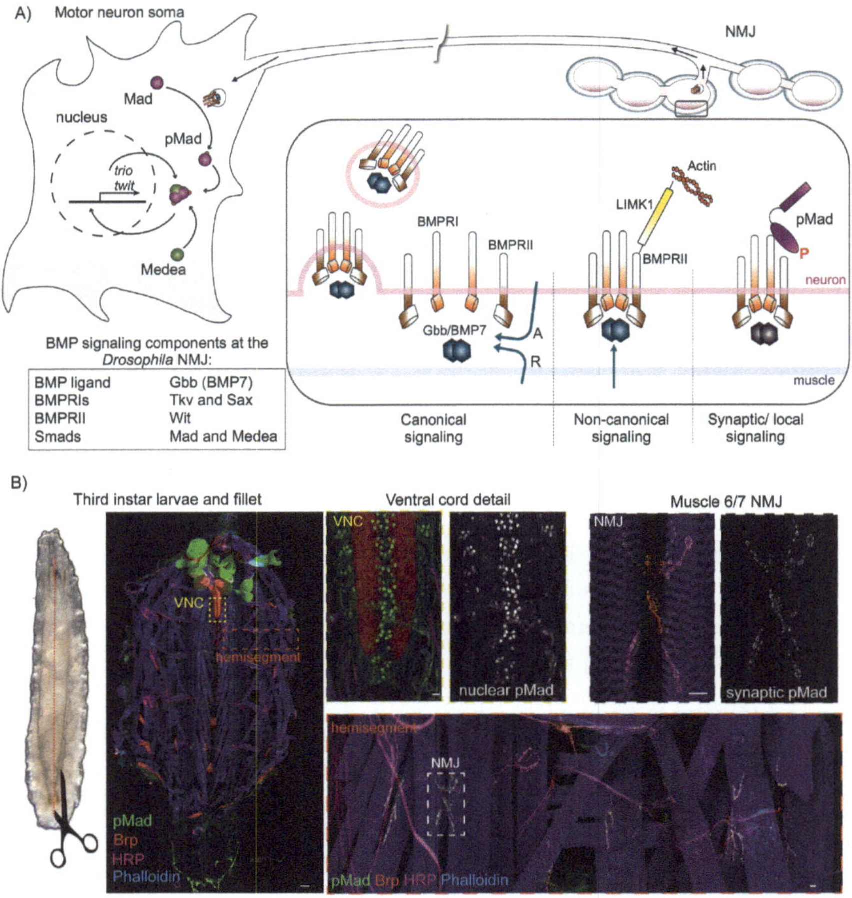


**Fig. 1** Fly motor neurons receive and integrate multiple BMP signaling pathways. (A) Diagram of various BMP pathways signaling to fly motor neurons. At synaptic terminals, Gbb/ BMP7 binding to BMP receptors on the motor neuron membrane triggers canonical and non-canonical BMP signaling. Gbb is secreted from both muscle (retrograde signaling—marked here "R") and motor neurons (autocrine signaling—labeled "A"). During canonical BMP signaling, the high-order BMP/BMPR (Gbb/Wit/Tkv/Sax) complexes are endocytosed and transported to the motor neuron soma where they phosphorylate Mad to regulate various transcriptional programs required for NMJ growth and function. During non-canonical signaling, Wit, the BMPRII, signals through LIMK1 to regulate synapse stability. This pathway does not involve Mad or Medea. In addition, the BMPRs Wit, Tkv and Sax, but not Gbb, enable the synaptic/local BMP signaling. (B) Dissection of a third instar larva along the dorsal side exposes the highly stereotyped body wall muscles labeled with phalloidin (blue) and imaged by confocal microscopy. The anti-horseradish peroxidase antibodies (HRP, magenta) label neuronal membranes. Bruchpilot (Brp, red) is a synaptic scaffold which marks the active zones. In response to canonical BMP signaling, pMad (green) accumulates in the motor neuron nuclei within the ventral nerve cord (VNC), the fly equivalent of mammalian spinal cord. In addition, pMad also accumulates in motor neuron synaptic terminals (right upper panels—muscle 6/7 NMJ). Synaptic/junctional pMad forms discrete puncta that co-localize with the active zone scaffold, Brp. Scale bars: 100 μm (larval fillet), 10 μm (others).

adequate depolarization of the muscle membrane and therefore proper muscle contraction and locomotion. These signaling pathways are triggered by the BMP7 homolog, Glass bottom boat (Gbb), secreted from the muscle (retrograde signaling) (McCabe et al., 2003) or from the motor neurons (autocrine signaling) (James et al., 2014) and require two type-I BMP receptors (BMPRI), Tkv and Sax, one type-II BMP receptor, Wit, and the downstream effectors Mad and Medea (Aberle et al., 2002; Marques et al., 2003; McCabe et al., 2004, 2003; Rawson, Lee, Kennedy, & Selleck, 2003). Loss of either Gbb, Sax, Tkv, Wit, Mad or Medea results in small NMJs with reduced number of synaptic boutons that are unable to depolarize the muscle fiber to normal levels. These mutant NMJs have reduced evoked excitatory junction potentials (EJPs) but fairly normal spontaneous miniature potentials (mEJPs, or minis in short). Beside canonical BMP signaling, a non-canonical BMP signaling mediated through BMP7/Gbb and BMPRII/Wit connects the presynaptic structures with the cytoskeleton to ensure structural stability (Eaton & Davis, 2005; Piccioli & Littleton, 2014). This pathway does not involve Mad or Medea. Finally, synaptic BMP signaling does not require Gbb, but relies on type-I and type-II BMPRs and Mad.

In the next sections we will review molecular mechanisms of synapse assembly and maturation then discuss the role of synaptic BMP signaling as a sensor of synapse activity and as a modulator of synapse maturation and plasticity. In the past, we and others have referred to this this pathway as non-canonical BMP signaling. However, this term is technically incorrect since non-canonical BMP signaling is defined as Smad-independent signaling. Albeit not implicated in transcriptional control, Smads are at the center of this local signaling modality, prompting us to revise our nomenclature and refer to this pathway as "local BMP signaling," or in the case of the fly NMJ "synaptic BMP signaling."

## 2. Synapse assembly and recruitment of neurotransmitter receptors at the fly neuromuscular junction

The NMJ is one of the most studied synapses in both vertebrate and invertebrate systems, primarily because of size and accessibility toward histological assays and electrophysiological recordings. In particular, the fly NMJ has been a favorite system to study synapse development and function since 1978, when Jan and Jan performed the first recordings at the larval NMJ and discovered that, like all insects and crustaceans, flies use the amino acid L-Glutamate as neurotransmitter at their NMJ (Jan & Jan, 1976a, 1976b). Subsequent studies revealed that *Drosophila* NMJ is similar in composition and physiology to mammalian central glutamatergic synapses and could serve as a powerful genetic system to analyze and model defects in the structural and physiological plasticity of glutamatergic synapses, which are associated with a variety of human pathologies from learning, memory deficits to autism.

Synapse assembly begins during embryogenesis, as soon as MN axons complete their navigation into the muscles field and arrive at their postsynaptic targets. The MN axon collapses its growth cone into a terminal with varicosities called synaptic boutons which begin to fill with neurotransmitter-packed synaptic vesicles. On the postsynaptic muscle, prior to the MN arrival, the ionotropic glutamate receptors (iGluRs) form small, nascent clusters distributed in the vicinity of future synaptic sites. The arrival of the MN triggers formation of large synaptic iGluR aggregates and promotes expression of more iGluRs which enable synapse maturation and growth. Lack of iGluRs clustering effectively halts the formation of the synapse at a specific time point, after the motor neurons have completed the navigation to their muscle target and the future synapse site has already been prefigured/pre-patterned (Burden, 1998; Kim, Bao, Bonanno, Zhang, & Serpe, 2012; Schmid et al., 2006). A mature synapse contains 40–60 individual iGluR complexes organized by a myriad of postsynaptic proteins; together they form a compact, electron-dense structure, called postsynaptic density (PSD). Each PSD juxtaposes an active zone, with a characteristic presynaptic T-bar structure, where glutamate-filled synaptic vesicles dock prior to release (Fig. 2). One bouton contains up to 20 individual synapses. During growth, the MN terminals add more synaptic boutons along one or more branches that innervate the target muscles and assume "beads on a string" morphologies. One "beads on a string" structure is commonly referred to as an NMJ, but each of these "NMJs" actually include hundreds of individual synapses. It is important to emphasize that synapse assembly and NMJ development are very different processes which are coordinated during development but are controlled by largely different signaling pathways. Several excellent reviews go into more details on NMJ development (Harris & Littleton, 2015; Menon, Carrillo, & Zinn, 2013).

In flies as in humans, synapse strength and plasticity is determined by the interplay between different postsynaptic receptor subtypes with different channel properties. For the scope of this review, we will focus on the two types of postsynaptic iGluR complexes, called type-A and type-B receptors,

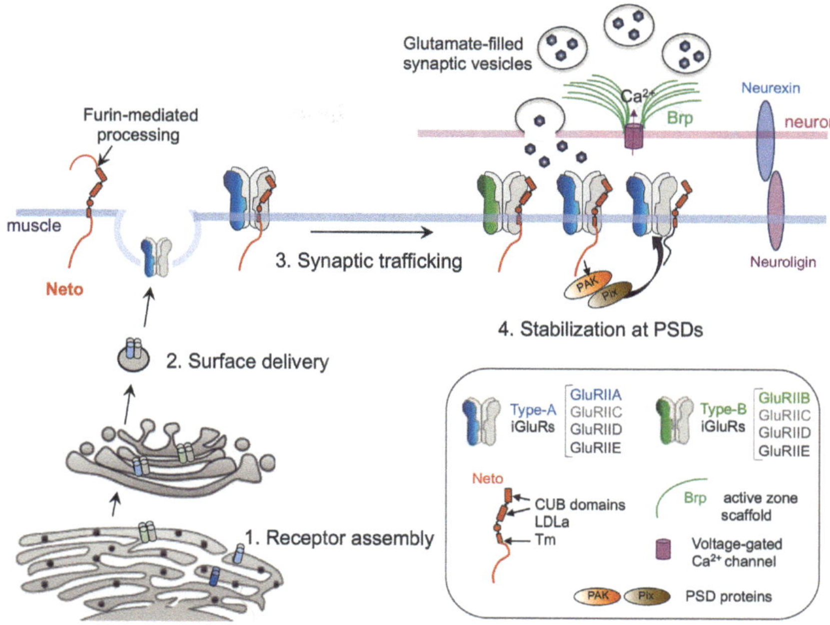


**Fig. 2** Synaptic recruitment marks the end of a long journey for iGluRs. The hetero-tetrameric iGluR NMJ complexes must assemble in ER before they could be delivered to muscle surface. Once on the membrane, iGluRs form complexes with Neto and together bind to scaffolds and motors which ensure trafficking to postsynaptic densities, juxtaposing the active zones. Synaptic iGluRs form large aggregates that are further stabilized through interactions with postsynaptic density components. Additional trans-synaptic interactions (such as Neurexin-Neuroligin) keep active zones and postsynaptic densities in register, further minimizing the distance that the neurotransmitter must travel to reach the postsynaptic receptors.

which are present at relevant NMJ PSDs. Both receptors form $Ca^{2+}$ permeable ligand-gated channels that open upon binding glutamate. However, this is an oversimplification of the system as this synapse is also modulated by presynaptic iGluRs, which function as auto-receptors, and by metabotropic glutamate receptors, which are G-protein-coupled receptors with modulatory roles at larval NMJ (Bogdanik et al., 2004; Kiragasi, Wondolowski, Li, & Dickman, 2017).

Type-A and type-B receptor channels consist of four different subunits: three shared subunits GluRIIC, -IID and -IIE, and either GluRIIA (type-A receptors) or GluRIIB (type-B) (DiAntonio, Petersen, Heckmann, & Goodman, 1999; Featherstone et al., 2005; Marrus, Portman, Allen, Moffat, & DiAntonio, 2004; Petersen, Fetter, Noordermeer, Goodman, & DiAntonio, 1997; Qin et al., 2005). Phylogenetic analysis indicates that all these subunits are closely related to the vertebrate kainate receptors, though the *Drosophila* receptors have strikingly different ligand binding profiles (Han, Dharkar, Mayer, & Serpe, 2015; Li et al., 2016). Like vertebrate kainate receptors, type-A and type-B receptors are modulated by an auxiliary subunit from the Neto (Neuropilin and Tolloid-like) family of proteins (Kim et al., 2012). These highly conserved auxiliary proteins modulate the gating properties of kainate-type glutamate receptors from worms to humans (Ng et al., 2009; Tomita & Castillo, 2012; Wang et al., 2012; Zhang et al., 2009). Neto proteins are single pass transmembrane proteins with two extracellular CUB (for complement C1r/C1s, UEGF, BMP-1) domains followed by an LDLa (low-density lipoprotein receptor domain class A) motif, a transmembrane domain and fairly divergent intracellular portion. At the fly NMJ, the postsynaptic iGluRs absolutely depend on Neto for their function and recruitment at synaptic sites indicating that Neto functions as an obligatory non-channel subunit for the postsynaptic iGluR complexes (Han et al., 2015; Kim et al., 2012, 2015).

The type-A and type-B receptors have similar single channel amplitude, but their kinetics are strikingly different, with type-B receptors desensitizing 10 times faster than the type-A (DiAntonio et al., 1999). Type-B channels can desensitize even before reaching the open state (Heckmann & Dudel, 1997), producing a much-reduced amplitude of the synaptic response comparing with the type-A channels (DiAntonio et al., 1999; Petersen et al., 1997). Because of this difference in channel properties, the dose of type-A vs type-B at individual synapses is a critical determinant of quantal size (or minis amplitude, mEJPs—the amplitude of the postsynaptic response to the spontaneous fusion of a single synaptic vesicle). Also, GluRIIA and GluRIIB compete with each other for the limiting shared iGluR subunits (GluRIIC, -IID and IIE) and the obligatory Neto. Muscle overexpression of GluRIIA triggers reduction of synaptic GluRIIB-containing receptors and a net increase in the quantal size, whereas overexpression of GluRIIB reduces the levels of synaptic GluRIIA/type-A receptors as well as the quantal size (Marrus & DiAntonio, 2004). Since different mechanisms modulate the synaptic recruitment and the activity of type-A and type-B receptors, flies can adjust synapse strength by differentially regulating these two receptor channels.

The life journey of glutamate receptors starts in ER, where all ionotropic glutamate receptors presumably assemble as tetramers followed by traffic to the cell surface, then to the synapse and stabilization at synaptic sites. Reconstitution studies in *Xenopus* oocytes indicate that only the NMJ iGluR complexes with four distinct receptor subunits are delivered to the cell surface (Han et al., 2015). This explains why in mutants lacking any of the shared subunits (GluRIIC, -IID or -IIE) or GluRIIA and GluRIIB together, none of the other iGluR subunits localizes to nascent synapses; such mutants never form NMJ synapses and die as late embryos, unable to move and hatch into the larval stages (DiAntonio et al., 1999; Featherstone et al., 2005; Marrus & DiAntonio, 2004; Qin et al., 2005). The surface delivery of either GluRIIA/C/D/E and GluRIIB/C/D/E reconstituted tetramer in heterologous systems is quite inefficient suggesting that additional chaperone(s) may facilitate the surface expression of these receptors *in vivo*. Neto has only a modest role in promoting the surface expression of both receptor channels (Han et al., 2015). Instead, Neto plays key roles during the subsequent steps in synapse assembly and function, including regulation of receptors recruitment, synaptic stabilization and function (Han et al., 2015; Kim et al., 2012, 2015; Ramos, Igiesuorobo, Wang, & Serpe, 2015). Genetics and live imaging studies show that Neto forms clusters at nascent NMJs at the time when iGluRs begin to accumulate and cluster (Kim et al., 2012). Moreover, Neto and iGluRs depend on each other for trafficking and stabilization at synapses: in the absence of Neto, the iGluR subunits fail to accumulate at the nascent NMJ, and, conversely, in the absence of iGluRs, Neto is not recruited either. Similar to mutants lacking iGluRs, *neto* mutants never form functional NMJs and die as completely paralyzed embryos. *neto* and *iGluR* mutants have also distinguishable phenotypes, as iGluR clusters are entirely absent from the muscle membrane in the *iGluR* mutants, whereas small, nascent receptor clusters can be detected in the vicinity of the prepatterned synaptic site (Kim & Serpe, 2013). Together these phenotypes indicate that Neto engages the iGluRs on the muscle membrane and together they traffic and are stabilized at the nascent synapse. Once at the synapse, Neto further modulates the receptors properties and, as will be discussed below, provides a dynamic scaffold for recruitment of PSD proteins and stabilization of selective receptor subtypes (Han et al., 2015; Kim et al., 2012, 2015; Ramos et al., 2015).

In flies as in vertebrates, MNs arrival of at the target muscles trigger clustering of postsynaptic neurotransmitter receptors and formation of large, stable receptor aggregates (Broadie & Bate, 1993; Chen & Featherstone, 2005; McMahan, 1990). Vertebrate MNs accomplish this task partly through the secretion of Agrin, a large glycoprotein implicated in the aggregation of postsynaptic (acetylcholine) receptors (Kim et al., 2008; Reist, Werle, & McMahan, 1992; Zhang et al., 2008). To date, no Agrin-like functions have been discovered at the *Drosophila* NMJ. Dynamic studies indicate that, upon MNs arrival at nascent synapses, fly NMJ iGluRs are incorporated into stably aggregates, with very little if any turnover. In contrast, most PSD components show significant mobility (Rasse et al., 2005). How iGluRs/Neto complexes are stably incorporated into large synaptic aggregates remains an open question. One key player appears to be the obligatory auxiliary subunit Neto which must be activated before being capable to mediate iGluRs clustering (Kim et al., 2015). Neto activities are restricted by an inhibitory prodomain which must be removed by Furin-mediated proteolysis. When the prodomain cleavage is blocked, Neto is properly targeted to the muscle membrane and engages the iGluR complexes *in vivo* but fails to enable the incorporation of iGluRs in stable synaptic clusters. The extracellular N-terminal CUB1 domain of Neto appears to be crucial for iGluR clustering: Neto mutants lacking the CUB1 domain, or its $Ca^{2+}$-binding capabilities, traffic to the synapses but cannot support formation of iGluR stable clusters and NMJ function (R. Vicidomini et al., unpublished data). Thus, Neto-enabled extracellular interactions provide a "receptors clustering capacity" function which is essential for synaptogenesis. Although the molecular nature of the iGluR clustering mechanism remains to be determined, the results from our laboratory indicate that this process requires an exposed, $Ca^{2+}$-binding, CUB1 domain.

Why is clustering of the receptors a critical step at the onset of synaptogenesis? There are at least two kinds of arguments for it: From a physiology perspective, the efficacy of the synaptic transmission decreases with the square distance between the site of neurotransmitters release and the site of their reception. Building a tightly packed postsynaptic receptor field that perfectly juxtaposes the release site is therefore paramount for the efficacy of the synaptic function. From a cell biology point of view, aggregates of membrane proteins could remain at the cell membrane and evade endocytosis if they are larger than the opening of the clathrin-coated pits. By clustering and building a large receptor aggregate, the nascent synapse may delay receptors internalization until postsynaptic scaffolds could be recruited and organized to enable further stabilization of postsynaptic receptors. It's not enough for postsynaptic neurotransmitter receptors to properly traffic to synaptic sites; aggregating these receptors at synaptic locations may hold them in place and initiate the cascade of events that stabilize the receptors at synaptic sites. In flies as in mammals, the postsynaptic densities function as both synaptic scaffolds and signaling hubs, ensuring the proper density for the neurotransmitter receptors, as well as their composition (in this case, the ratio of type-A: type-B receptors) (Albin & Davis, 2004; Lee & Schwarz, 2016; Liebl & Featherstone, 2008; Morimoto, Nobechi, Komatsu, Miyakawa, & Nose, 2009; Parnas, Haghighi, Fetter, Kim, & Goodman, 2001; Ramos et al., 2015). Later on, these mature postsynaptic structures are thought to function as grids where removal/ turnover of receptors leave empty "slots" and new receptors could slide right in with minimal disruption for the overall synaptic structure and physiology (Nicoll, 2017).

Trans-synaptic signaling and several layers of trans-synaptic interactions mediate the coordinated assembly of presynaptic and postsynaptic structures and control the proper alignment of the active zones with the postsynaptic receptor fields (Banovic et al., 2010; Dalva, McClelland, & Kayser, 2007; Giagtzoglou, Ly, & Bellen, 2009; Li, Ashley, Budnik, & Bhat, 2007; Mosca, Hong, Dani, Favaloro, & Luo, 2012; Ramesh et al., 2021; Thomas & Sigrist, 2012). Closer to the PSDs, presynaptic Nrx and postsynaptic Nlgs are engaged in dynamic interactions that control the sequential recruitment of type-A and type-B receptors to a growing PSD (Owald et al., 2012) and limit the size of the postsynaptic receptor fields (Banovic et al., 2010; Xing et al., 2014). A second layer of trans-synaptic interactions involves integrins and teneurins and anchors the synaptic specializations to the pre- and postsynaptic skeletons ensuring synaptic stability (Koch et al., 2008; Mosca et al., 2012; Stephan et al., 2015; Wang, Han, Nguyen, Jarnik, & Serpe, 2018). Finally, successful establishment of the synapse triggers transcriptional signals in both pre- and post-synaptic compartments and initiates the synthesis and trafficking of new components to the growing synapses.

## 3. Distinct mechanisms recruit type-A and type-B glutamate receptors

Coordinated locomotion in fly larvae requires several types of neurons, including tonic (Ib) and phasic (Is) glutamatergic MNs, type II octopaminergic and type III peptidergic neurons. Each muscle is innervated by only a single Ib MN that is the primary driver for contraction (Newman et al., 2017). Each Is MN innervates a group of 7–8 muscle fibers and appears to provide coordinated control within each hemisegment as well as along segments. The Ib and Is MNs have different composition of iGluRs: more type-A receptors at Ib and more type-B receptors at Is synapses (Marrus & DiAntonio, 2004). This suggests that the two types of glutamatergic MNs employ different mechanisms for the recruitment of glutamate receptors at their synapses.

During development, type-A receptors are the first to arrive at a nascent synapse and usually form the core of an individual synapse. Type-B receptors arrive after type-A and tend to accumulate at the periphery of a PSD (Akbergenova, Cunningham, Zhang, Weiss, & Littleton, 2018; Rasse et al., 2005). Studies in many laboratories uncovered a rich variety of mechanisms that ultimately impact the synaptic iGluRs abundance and composition (Liebl & Featherstone, 2005; Liebl et al., 2006; Zhao et al., 2020). However, most of this regulation seems to be primarily directed toward modulation of type-A receptors, starting from the stability of the *GluRIIA* transcript and the GluRIIA protein to the synaptic trafficking and retention of the type-A glutamate receptor channels. In this section we will review some of these mechanisms and highlight the complexity of efforts to control the level and activity of type-A receptors.

The signaling cascade that initiates synaptogenesis and mediates contact-dependent postsynaptic glutamate receptor expression remains unknown. However, this pathway signals to muscle nuclei and triggers a large yet transient burst of iGluRs transcription (Ganesan, Karr, & Featherstone, 2011). Upon neuron arrival, iGluR transcripts abundance increases rapidly in the muscle, but within a few hours decreases and continues to fall throughout larval development.

Several effectors that modulate the iGluRs mRNA abundance have been identified. For example, Lola (longitudinal lacking), a BTB-Zn finger transcription factor, has been implicated in the transcription of the iGluR subunits GluRIIA, -IIB, IIC, as well as dPak (p21 activating kinase), a PSD component (Fukui et al., 2012). Lola is negatively regulated by neural activity, which may account for the limited production of iGluR transcripts as development progresses.

TGF-β signaling promotes transcription of both *GluRIIA* and *GluRIIB* mRNAs but not *GluRIIC* or other postsynaptic components (such as dPak and Dlg) (Kim & O'Connor, 2014). This pathway, activated by either Activin, secreted from MNs, or Maverick from the glia, requires the type-I receptor Babo, the type-II receptor Punt and the dSmad2/Smox effector (Kim & O'Connor, 2014; Sulkowski et al., 2016). In the absence of Smox or Babo, the quantal size is strongly reduced due to the low abundance of both postsynaptic receptors. At the same time, Mav and an activin-like ligand, Dawdle, activate the TGF-β signaling pathway in the muscle to promote the secretion of Gbb and therefore upregulate NMJ growth (Ellis, Parker, Cho, & Arora, 2010; Fuentes-Medel et al., 2012). TGF-β signaling also controls GluRIIA synaptic abundance through post-transcriptional mechanisms; muscle overexpression of a *GluRIIB-GFP* transgene in *Smox* mutants restored the GluRIIB synaptic accumulation, whereas overexpression of *GluRIIA-GFP* did not, with the net GluRIIA-GFP muscle levels remaining undetectable (Kim & O'Connor, 2014). Thus, additional TGF-β transcriptional target(s) in the larval muscle function to stabilize either the GluRIIA *mRNA* or the GluRIIA protein.

Negative regulators of *GluRIIA* and *GluRIIB mRNA* levels include Dicer-1, the endoribonuclease necessary for microRNA synthesis, and microRNAs such as *mir-284* (Karr et al., 2009). Removal of *dicer-1* or *mir-284* de-represses the *GluRIIA* and *GluRIIB* transcripts and induces production of excess extra-synaptic receptors. However, there is no increase in synaptic receptors, consistent with additional mechanisms that limit the synaptic accumulation of these receptors. The translational repressor Pumilio (Pum) and its co-repressor partner Nanos (Nos) act in opposition to each other to regulate glutamate receptor subunit composition and synaptic physiology (Menon, Andrews, Murthy, Gavis, & Zinn, 2009; Menon et al., 2004). Pum binds to the 3′ untranslated regions of *Nos* and *GluRIIA* transcripts repressing their translation, whereas Nos represses GluRIIB. Since GluRIIA and GluRIIB compete with each other for limited, shared iGluR subunits, Pum and Nos form a regulatory network that controls postsynaptic receptor composition. This network which also involves the translation factor eIF-4E (Menon et al., 2009), has additional presynaptic functions in the control of NMJ growth.

Recent studies revealed that the postsynaptic levels of GluRIIA protein are down-regulated by calpains, which are calcium-activated proteases (Metwally, Zhao, Li, Wang, & Zhang, 2019). Since iGluRs are $Ca^{2+}$-permeable channels, the idea that calcium negatively regulates postsynaptic GluRIIA abundance is very appealing. It is still unclear whether calpains directly cleave GluRIIA *in vivo*; *in vitro* studies map putative processing sites within an extracellular (ligand binding) domain of GluRIIA, whereas calpains are primarily $Ca^{2+}$-sensing cytoplasmic enzymes.

During development, individual synapses grow until they reach a maximum size and do not split; instead, the larvae initiate a *de novo* synapse (Rasse et al., 2005). The limited size of mature synapses presumably reflects a limited capacity for receptor clustering and may contribute to the competition between type-A and type-B receptors for synaptic stabilization. Indeed, as described above, muscle overexpression of GluRIIA leads to reduction in GluRIIB synaptic abundance and vice versa (DiAntonio et al., 1999; Petersen et al., 1997). A growing body of literature indicates that the synaptic recruitment of type-A receptors depends on multiple postsynaptic components and on the activity of type-A receptor channels themselves (Lee & Schwarz, 2016; Liebl & Featherstone, 2005, 2008; Ljaschenko, Ehmann, & Kittel, 2013; Morimoto et al., 2009; Parnas et al., 2001; Petzoldt et al., 2014; Ramos et al., 2015). In contrast, type-B receptors appear to be incorporated by default and appears limited only by individual synapse's capacity for clustering receptors in stable aggregates. In other words, it takes work to capture/stabilize type-A receptors in stable synaptic aggregates, whereas type-B receptors seem to be incorporated by default, when the synaptic recruitment of type-A receptors fails. Examples supporting this view are abundant throughout embryonic and larval stages of development, indicating a consistent strategy for the synaptic recruitment of NMJ iGluRs.

For example, during late embryogenesis stages, Coracle, the *Drosophila* homologue of the mammalian cytoskeletal protein 4.1, binds to the GluRIIA cytoplasmic domain and appears to anchor type-A receptors to the actin cytoskeleton, promoting their synaptic accumulation (Chen, Merino, Sigrist, & Featherstone, 2005). The p21-activated kinase (PAK) co-localizes with the iGluR complexes at PSDs and, in conjunction with the guanine nucleotide exchange factor Pix and the adaptor protein Dreadlocks, promotes the synaptic accumulation of GluRIIA-containing receptors (Albin & Davis, 2004; Parnas et al., 2001).

The ability to recruit and stabilize type-A synaptic receptors appears to rely heavily on Neto-β, the predominant Neto isoform in the postsynaptic muscle (Ramos et al., 2015). Neto-β associates with iGluRs but also provides a dynamic postsynaptic scaffold through its large cytoplasmic domain rich in putative protein interaction motifs and docking sites. Animals lacking parts of Neto-β intracellular domain fail to recruit dPAK (Ramos et al., 2015) and dPix (Vicidomini et al, unpublished) at synaptic sites and have significantly reduced postsynaptic structures. These mutants have normal muscle levels of *GluRIIA* and *GluRIIB* transcripts but fail to concentrate type-A receptors and instead show increased synaptic accumulation of type-B iGluRs. Complete removal of Neto-β further decreases the ability of these synapses to recruit GluRIIB-containing receptors. These phenotypes support the view that at individual synapses Neto-mediated extracellular interactions limit the iGluRs clustering capacity, whereas Neto-dependent intracellular interactions provide a dynamic scaffold that controls PSD assembly and composition.

Importantly, not all type-A receptors can be incorporated in stable clusters. Studies on receptor dynamics and regulations indicate that inactivated and/or channel impaired GluRIIA-containing receptors can traffic to the synapses but do not accumulate there (Petzoldt et al., 2014). Postsynaptic PKA and CaMKII activities reduced the synaptic GluRIIA levels (Davis, DiAntonio, Petersen, & Goodman, 1998; Morimoto et al., 2009; Sulkowski et al., 2014), presumably by inactivating the GluRIIA-containing receptor channels. Thus, type-A receptors must be active for proper synaptic accumulation. This emerging sequence of events and regulatory networks will likely be further explored as additional studies and genetic screens uncover new putative modulators of the synaptic recruitment of type-A and type-B receptors. It is important to note that Ib synapses contain more GluRIIA-containing receptors, whereas Is synapses contain more GluRIIB (Marrus et al., 2004). Some of the experimental settings where more GluRIIB has been observed at the NMJ may be due to increased number of Is boutons relative to Ib. The relationship between Ib and Is terminals is a hot topic today and is carefully examined in several laboratories (Aponte-Santiago, Ormerod, Akbergenova, & Littleton, 2020; Wang, Lobb-Rabe, Ashley, Anand, & Carrillo, 2021).

Why would synaptic stabilization of type-A receptors require so much control and complex regulation? The reasons behind this preferential control may reflect several very practical considerations. First, type-A receptors have the potential to allow more ion flow than type-B receptors in response to neurotransmitter release. Type-A receptors have desensitization rates an order of magnitude lower than type-B (DiAntonio et al., 1999); upon binding of glutamate, they will stay open a lot longer than the type-B receptors, evoking substantially stronger postsynaptic depolarization. Also, type-B receptors seem to be utilized only during development and are not expressed in the adult abdominal muscle (Diao et al., 2015; Li et al., 2022). During evolution, flies had little incentives to invest in building regulatory networks for recruiting the type-B receptors. Instead, a tight control of the recruitment and stabilization of type-A receptors coupled with a "default" incorporation type-B receptors may have been a more sensible strategy. Should a larva need increased synaptic strength, the immediate solution would be to incorporate more type-A synaptic receptors and increase the GluRIIA/GluRIIB ratio of synaptic receptors. This is exactly what has been observed at individual synapses upon (optogenetic) stimulation; conversely, input-specific low stimulation drove type-A receptor out of the corresponding PSDs (Akbergenova et al., 2018). Developmental studies over a longer time frame yielded similar results: the levels of synaptic type-A receptors increased in response to increased locomotor activity and decreased in less mobile animals (Sigrist, Reiff, Thiel, Steinert, & Schuster, 2003). This positive feedback appears to (i) initially promote the incorporation type-A receptors at the core of new synapses during growth (Schmid et al., 2008) then (ii) adjust the levels of type-A receptors as a function of synapse activity during maturation. The absence of type-A receptors triggers a different kind of (negative) feedback mechanism characterized by a compensatory increase in presynaptic neurotransmitter release (reviewed in Davis & Muller, 2015; Frank, 2014).

Flies must be able to carefully monitor the type-A receptors synaptic distribution and function to efficiently adjust their postsynaptic accumulation and initiate positive or negative feedback mechanism. How do flies accomplish this task? In the next section we will review our findings that phosphorylated Mad (pMad) accumulates at the presynaptic active zone in a pattern that mirrors the postsynaptic type-A receptors activity (Sulkowski et al., 2016, 2014), indicating that synaptic pMad functions as a sensor of synapse activity.

## 4. pMad as a sensor of synapse activity

Detection of phosphorylated, BMP-activated Smads is routinely accomplished with phospho-specific antibodies raised against the C-terminal sequence -IS(pS)V(pS). In flies, a variety of phospho-specific antibodies capture pMad accumulation in MN nuclei, but also in discrete puncta at synaptic terminals. While many researchers tend to examine the two populations of pMad together, the nuclear and synaptic pools represent clearly different signaling inputs and have different developmental outcomes: the nuclear pMad accumulation is a result of canonical BMP signaling and will lead to transcriptional regulation of target genes, whereas synaptic pMad appears to be a bona fide sensor for synapse activity with no function in BMP-modulated transcriptional control (Smith et al., 2012; Sulkowski et al., 2014).

Several lines of evidence support a role for synaptic pMad as sensor of synapse activity. First, pMad signals are selectively lost at NMJs with reduced levels of postsynaptic iGluR/Neto complexes, such as observed in *neto* and

*GluRIIC* hypomorphs (Sulkowski et al., 2014). In contrast, nuclear pMad persists in MN nuclei, and expression of BMP target genes remains unaffected by alterations in synaptic pMad levels, indicating a specific impairment in pMad production/maintenance at synaptic terminals. Second, synaptic pMad mirrors the accumulation of postsynaptic type-A subtypes (Fig. 3). Third, synaptic pMad accumulation follows the activity and not the net levels of postsynaptic type-A receptors: a pulse of postsynaptic increase in PKA activity, which inhibits the activity of type-A receptors (Davis et al., 1998), decreases the synaptic pMad levels without changing the net levels of synaptic GluRIIA. Since pMad accumulation at synaptic terminals mirrors the activity and not the levels of type-A receptors, synaptic pMad appears to serve as an exquisite monitor for synapse activity.

Both synaptic and nuclear pMad accumulate in MNs: Expression of *Mad* transgenes in the MNs but not in the muscles restores both nuclear and synaptic pMad signals in *Mad* mutants (Sulkowski et al., 2016, 2014). However, the two pMad pools mark distinct pathways that are independently regulated, have different pathway components and different functions. Genetic manipulations of GluRIIA receptor levels in the muscle induce proportional changes in synaptic pMad and in quantal size (mEJPs amplitude) but have no effect on nuclear pMad. *GluRIIA* mutant animals have no synaptic pMad but have normal size NMJs indicating that synaptic pMad does not contribute to NMJ growth. Conversely, overexpression of

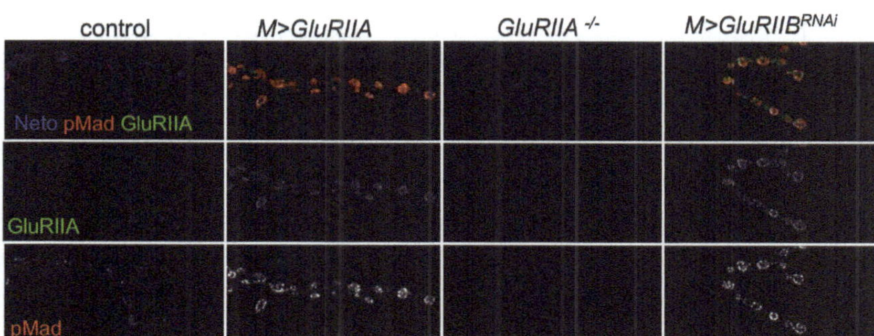


**Fig. 3** pMad mirrors the postsynaptic type-A receptors. Confocal images of NMJ boutons from third instar larvae of the indicated genotypes labeled for the obligatory auxiliary subunit Neto (blue), which marks both type-A and type-B glutamate receptors, GluRIIA (green), the glutamate receptor subunit specific for type-A receptors, and pMad (red). Synaptic pMad follows the distribution and intensity of GluRIIA-positive signals: It increases with the muscle overexpression of *GluRIIA* and becomes undetectable at *GluRIIA* mutant NMJs. Depletion of GluRIIB in the postsynaptic muscle triggers an increase of synaptic GluRIIA levels and therefore increased synaptic pMad.

Mad-GFP in MNs induces strong accumulation of nuclear pMad but has no effect on the synaptic pMad levels. Moreover, in the absence of GluRIIA, synaptic pMad is undetectable even though excess Mad-GFP accumulates at synaptic terminals. Therefore, nuclear pMad is completely independent of the GluRIIA status, whereas synaptic pMad absolutely mirrors the levels of active, GluRIIA-containing postsynaptic receptors. Overexpression of activated Sax and Tkv receptors in motoneurons increases the EJPs (which depends on nuclear pMad) without affecting quantal size, which indicates normal GluRIIA:GluRIIB levels, therefore normal synaptic pMad (Ball et al., 2010). Nuclear and synaptic pMad share some of the BMP pathway components, such as the BMPRIs Sax and Tkv, and the BMPRII Wit. Interestingly, Gbb is only required for nuclear pMad and is dispensable for synaptic pMad (Sulkowski et al., 2016). The presence of unique and also common pathway components demonstrates that synaptic pMad marks a genetically distinguishable BMP signaling pathway that is distinct from but coordinated with the other BMP pathways via shared components.

Super-resolution fluorescence microscopy studies indicate that the pMad-positive domains distribute into thin discs of ~700 nm diameter sandwiched in between the presynaptic active zones, the sites of neurotransmitter release and the postsynaptic iGluRs fields (Fig. 4). This is a very crowded presynaptic milieu critical for docking and priming readily releasable synaptic vesicles and for clustering and positioning voltage-gated $Ca^{2+}$ channels that control neurotransmitter release induced by action potentials (reviewed in Van Vactor & Sigrist, 2017; Zhai & Bellen, 2004). Why does a transcription factor like pMad accumulate at these synaptic specializations and what is it doing there? The size and shape of the pMad domains suggest that pMad associates with membrane-anchored complexes at the active zone. This is a very unusual distribution for Smad proteins, which are known to shuttle between cytoplasm and nucleus, depending on their phosphorylation status. The only enzyme known to phosphorylate Mad at the C terminus is Tkv, the BMPRI. Tkv can bind to and phosphorylate Mad only after it is phosphorylated by trans-activation upon the assembly of the BMP/BMPR signaling complexes. Since BMP signaling complexes are generally short lived, the synaptic pMad-positive domains likely represent pMad that, upon phosphorylation, remains associated with the activated BMP/BMPR complexes at presynaptic sites. Consistent with this model, synaptic pMad is lost upon knocking down BMPRs in the motor neurons (Sulkowski et al., 2016). This model also implies that active mechanisms must exist (i) to trap the active BMP/BMPR complexes at the active zones

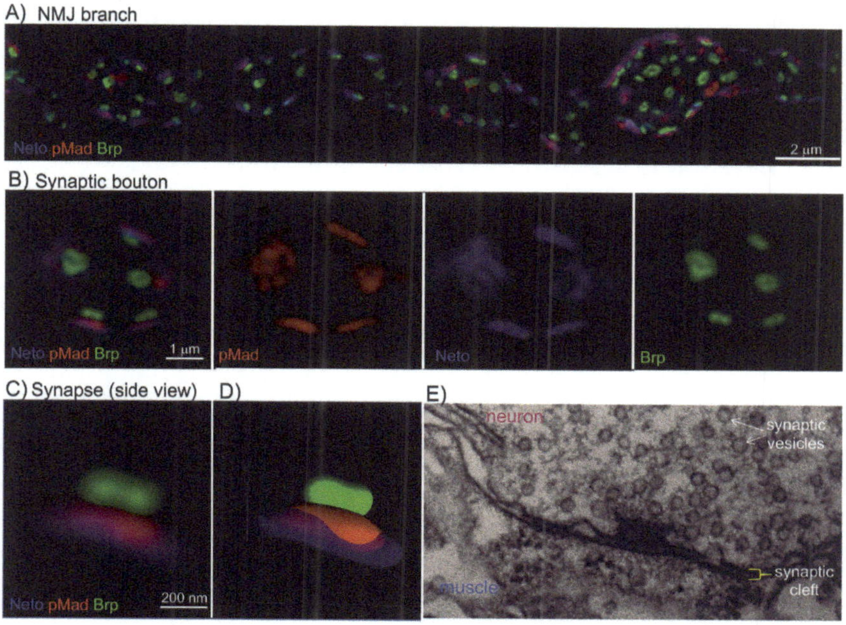


**Fig. 4** Synaptic pMad localizes at the active zone. (A, B) 3D structured illumination microscopy (3D-SIM) images of NMJ boutons from third instar larvae labeled for Brp-an active zone scaffold (green), pMad (red) and the obligatory auxiliary subunit Neto (blue). (C) High magnification view of a single synapse profile (from panel B). (D) Side view of a surface rendered volume of the synapse shown in panel (B). (E) Electron micrograph of a single synapse illustrating the characteristic T-bar structure juxtaposing the postsynaptic density. The anti-Brp monoclonal antibody recognizes the tip of the T-bar, hence the ring appearance. The anti-Neto antibodies recognize the extracellular CUB1 domain within the synaptic cleft and mark the postsynaptic densities. The pMad signals concentrate in between Brp and Neto, closer to Neto and form thin discs suggestive of a layer of pMad parallel to the presynaptic membrane.

and (ii) to keep the product of an enzymatic reaction, pMad, associated with its own enzyme, Tkv, within the BMP/BMPR signaling complexes. Some possible mechanisms will be discussed below.

More importantly, how is the status of postsynaptic receptors relayed across the synaptic cleft? Structural studies on glutamate receptors indicate that these receptors have unusually large extracellular domains that expands ~140Å within the 200Å synaptic cleft (He et al., 2021; Sobolevsky, Rosconi, & Gouaux, 2009). These domains undergo significant conformational changes during the receptor's gating cycle (Meyerson et al., 2016, 2014). The iGluR tetramers adopt a Y-shape structure organized in layers that include (i) the amino terminal domain (ATD), which plays a role in tetramer assembly, (ii) the ligand binding domain (LBD) which forms a clam shell shaped structure that closes to trap glutamate and (iii) the transmembrane domain (TMD), which forms the ion pore. Binding of glutamate triggers a corkscrew motion which shortens the Y tetramer and opens the ion pore (Meyerson et al., 2016). Cryo-electron microscopy studies revealed additional conformational states in which the upper arms of the Y-shaped tetramer splay apart to different extents (reviewed in Mayer, 2021). In addition, the native glutamate receptor complexes contain auxiliary subunits, such as Neto, that physically associate with the iGluRs and influence their biology (reviewed in Jackson & Nicoll, 2011; Tomita, 2010). Recent Cryo-EM studies revealed the large extent of Neto2 association with the vertebrate kainate receptor, GluK2 (He et al., 2021). Neto2 wraps around the receptor along an exposed receptor surface, with Neto's extracellular domains crosslinking different tetramer subunits within the ATD and LBD layers and the Neto's transmembrane domain associating tightly with the receptor TMD. This topology explains the role of Neto proteins in modulating channel gating (He et al., 2021). This topology also indicates that the CUB domains of Neto should be very sensitive to the channel state and should register any conformational changes within the receptor channel complex.

Expanding on this view, *Drosophila* Neto should easily distinguish between type-A and type-B postsynaptic glutamate receptors as well as their activity states. In addition, The CUB domains also bind BMPs or BMP modulators and have been implicated in a wide variety of extracellular protein interactions during development (Bork & Beckmann, 1993; Lee, Mendes, Plouhinec, & De Robertis, 2009). Neto may bind postsynaptic type-A receptors while at the same time use its BMP-binding CUB domains to reach out and engage presynaptic BMP signaling components, anchoring the presynaptic BMP/BMPR complexes at the active zones and promoting the synaptic pMad accumulation (Fig. 5). This model predicts that Neto enables trans-synaptic interactions only when bound to active postsynaptic type-A receptors. Depletion or inactivation of postsynaptic type-A receptors should trigger conformational changes that disrupts the Neto-dependent trans-synaptic interactions and restores the mobility of BMP/BMPR complexes, dampening the accumulation of synaptic pMad. By bridging both the postsynaptic iGluRs and the presynaptic BMP/BMPR complexes, Neto in a perfect position to sense the conformation of iGluR complexes and relay the status of these postsynaptic receptor channels to the presynaptic BMP signaling complexes.

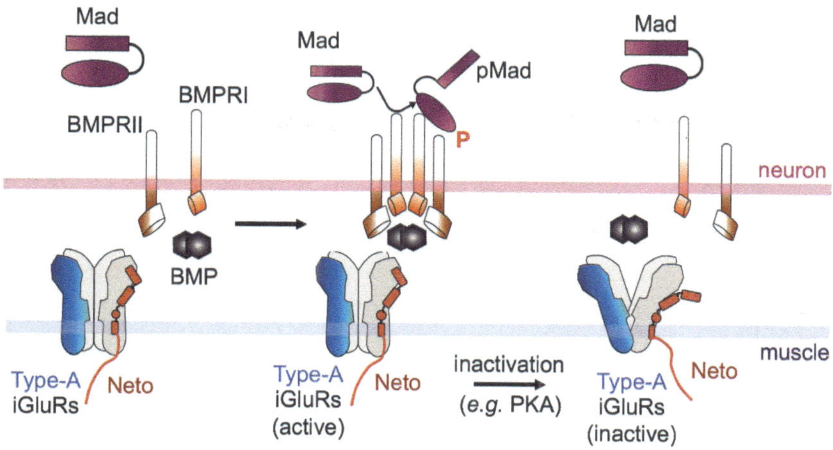


**Fig. 5** Model for how Neto-mediated trans-synaptic interactions relay the status of postsynaptic type-A receptors to presynaptic BMP/BMPR complexes. Neto and iGluRs traffic together at synaptic locations. Neto also has two extracellular BMP-interacting CUB domains that may localize BMP activities and anchor the presynaptic BMP/BMPR complexes at active zones via trans-synaptic interactions. These complexes phosphorylate Mad locally and induce pMad accumulation at the synaptic junction. Inactivation of type-A receptors induces conformational changes and dissociation of these trans-synaptic complexes.

The idea of Neto-centered trans-synaptic interactions raises the possibility that synaptic pMad may be a sensor of synapse activity with additional role(s) in influencing synapse composition and/or function. In the next section we will review a series of experiments that lead to the discovery of a positive feedback mechanism in which active type-A receptors induce accumulation of pMad at active zones which, in turn, promotes stabilization of type-A receptors at PSDs.

## 5. A positive feedback loop stabilizes glutamate receptor subtypes as a function of activity

The first hint of a positive feedback mechanism came from examining the composition of postsynaptic glutamate receptors in mutants with aberrant synaptic pMad levels. Interestingly, loss of synaptic pMad (in *impβ11*, *wit* and *mad* mutants) correlates with a decreased GluRIIA/GluRIIB ratio and reduced quantal size, while increased synaptic pMad (such as in *nrx* mutants) correlates with increased GluRIIA/GluRIIB ratio and increased quantal size (Sulkowski et al., 2016). In contrast, the presence of synaptic pMad even in a transcriptionally impaired BMP mutant (such as *gbb*) ensures relatively normal GluRIIA/GluRIIB ratio and quantal size. This tight correlation suggests a feedback mechanism whereby active postsynaptic GluRIIA receptors induce the accumulation of pMad at active zones, which in turn promotes the stabilization of GluRIIA receptors at postsynaptic sites.

But to demonstrate this feedback mechanism, one must disrupt the local pMad accumulation, without affecting the other BMP signaling pathways canonical. This precludes the use of any BMP signaling components or known BMP modulators, as any such manipulations will affect both local and transcriptional functions of BMP pathway. One possible solution came from post-translational modifications of Mad, which have the potential to influence the association between pMad and its own kinase, Tkv, and differentially disrupt the accumulation of synaptic pMad. Indeed, among mutants with increased synaptic pMad, previous studies identified the *nemo (nmo)* locus (Merino et al., 2009; Zeng, Rahnama, Wang, Sosu-Sedzorme, & Verheyen, 2007). Nmo is a MAPK-related kinase which phosphorylates Mad at S25 and promotes its nuclear export. Lack of S25 phosphorylation (in *nmo* mutants) decreases the nuclear pMad levels, due to increased nuclear export, but increases the synaptic pMad, presumably by promoting the pMad-BMPRs association at synaptic sites. Nmo does not appear to interfere with the ability of BMP/BMPR signaling complexes to phosphorylate Mad at its C-terminal residues. Intriguingly, neuronal overexpression of Tkv, but not Mad, rescues the normal levels of nuclear pMad in *nmo* mutants. This indicates that Tkv becomes limiting in the absence of Nmo and excess amount of neuronal Mad cannot compensate for limiting enzyme.

We predicted that overexpression of a Nemo-phosphomimetic Mad variant (S25D) in the MNs should not affect the nuclear pMad levels since excess $Mad^{S25D}$ should be efficiently exported from the MN nuclei. However, at active zones, excess $Mad^{S25D}$ should compete (via mass action) with endogenous Mad for BMPR-mediated phosphorylation. Only endogenous pMad should remain associated with BMP/BMPR complexes, whereas $pMad^{S25D}$ should likely fall off from the presynaptic BMP/BMPR complexes. The consequence of $Mad^{S25D}$ neuronal overexpression should be normal accumulation of nuclear pMad but diminished synaptic pMad. This is indeed what was experimentally observed: reduced synaptic pMad levels but no detectable deficits in the other BMP signaling pathways (Fig. 6) (Sulkowski et al., 2016). More importantly, selective loss of synaptic pMad was accompanied by reduced GluRIIA/GluRIIB ratio at synaptic sites and reduced quantal size. In contrast, overexpression of a

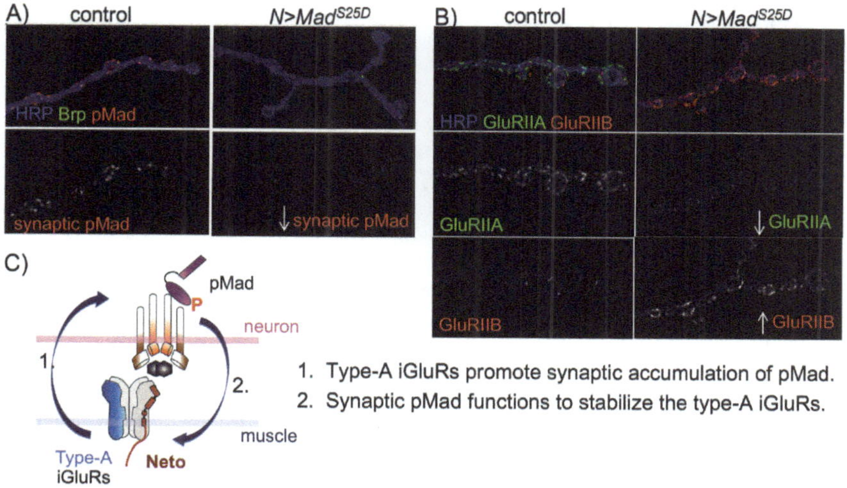


**Fig. 6** Disruption of presynaptic pMad reduces the levels of postsynaptic type-A receptors. (A, B) Confocal images of NMJ4 boutons from control and third instar larvae with a phosphomimetic Mad variant overexpressed in motor neurons ($N>Mad^{S25D}$). Neuronal expression of $Mad^{S25D}$ reduces the accumulation of synaptic pMad and GluRIIA (type-A receptors) and increases the GluRIIB synaptic accumulation. The anti-horseradish peroxidase (HRP-blue) labels neuronal membranes. (C) The positive feedback loop model

phospho-impaired $Mad^{S25A}$ variant has no influence on quantal size. These results indicate that diminished synaptic pMad accumulation in the motor neurons causes a direct reduction of postsynaptic type-A receptors. This is consistent with a positive feedback mechanism in which active type-A receptors induce accumulation of pMad at active zones which, in turn, promotes stabilization of type-A receptors at PSDs. Since GluRIIA and GluRIIB compete with each other for limiting components, reduced type-A receptors enable further synaptic stabilization of type-B receptors and induce a net change in the synaptic accumulation of iGluR subtypes toward more type-B receptors.

How do postsynaptic glutamate receptors modulate presynaptic pMad and in turn are stabilized by it? Trans-synaptic Neto-centered interactions that couple active postsynaptic type-A receptors with presynaptic BMP/BMPR complexes may provide the simplest explanation for this phenomenon (Fig. 7). Such trans-synaptic complexes could offer a versatile means for relaying iGluRs/Neto activity status to the presynaptic neuron via fast conformational modifications. At the same time, these trans-synaptic nanocolumns may function as "accumulation centers," holding the active type-A receptors at synaptic sites and facilitating interactions that stabilize the

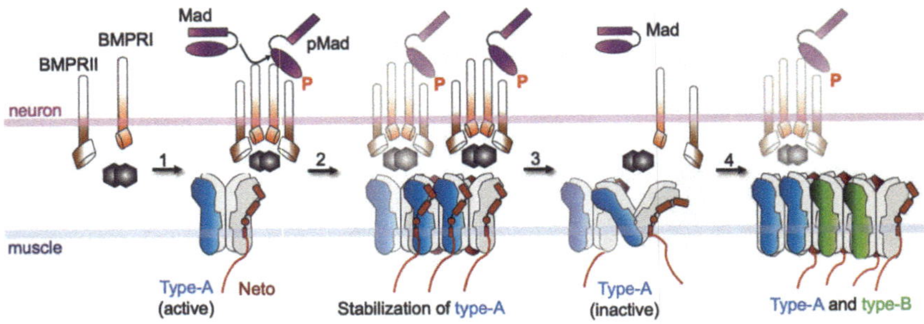


**Fig. 7** Model for type-A receptor stabilization via local BMP pathway. Neto in complexes with active type-A receptors localizes BMP activities, promoting the formation of presynaptic BMP/BMP receptor complexes (step 1). These complexes function as "accumulation centers" for stabilizing type-A receptors at nascent synapses (step 2). Dissociation of these local complexes terminates further incorporation of type-A receptors and allows for recruitment of type-B subtypes, which mark mature synapses (steps 3–4).

type-A receptors at PSDs. Inactivation of type-A receptors will induce conformational changes that disrupt the trans-synaptic complexes and stop further the incorporation of type-A receptors at PSDs. Also, type-A receptors arriving at the synapse in an inactive state will be unable to engage in trans-synaptic interactions and will evade incorporation in stable clusters, retaining their mobility. Importantly, only the type-A receptors require such positive feedback for their stable incorporation at PSDs. In contrast, the recruitment of (GluRIIB-containing) type-B receptors remains fairly constant during different developmental stages or in different synapse activity states and is limited by the competing GluRIIA and by the "clustering capacity" at individual synapses.

This positive feedback mechanism provides a molecular basis for thinking about several key steps during synapse assembly, maturation and plasticity. For example, this mechanism can explain the positive feedback that promotes incorporation of type-A receptors at new synapses, then restrains the type-A receptors accumulation during synapse maturation (Akbergenova et al., 2018; Schmid et al., 2008). Type-A receptors are the first to arrive at a nascent synapse and are incorporated in stable aggregates at the center of nascent synapses for as long as they are active. Inactivation of type-A receptors will pause their further synaptic stabilization and allow for the incorporation of type-B receptor channels, which mark more mature synapses. This sequence of events has been directly observed through dynamic studies, as discussed above (Schmid et al., 2008).

In addition, several indirect arguments demonstrated that channels properties and gating behavior influence the trafficking of GluRIIA-containing receptors and their stable incorporation at PSDs. First, photobleaching experiments showed that channels with reduced activity/charge transfer (such as the fast-desensitizing GluRIIA$^{E783A}$-containing mutant receptors) have increased mobility; these channels accumulate prematurely during early steps of synapse assembly but fail to be stabilized at synaptic locations and in time segregate away from wild-type GluRIIA-marked PSDs and from the Brp-marked active zones (Petzoldt et al., 2014). Second, the GluRIIA C-tail has been implicated in PKA- and calcium/calmodulin dependent protein kinase II (CaMKII)-dependent phosphorylation which reduces receptor activity and induces eviction from PSDs (Davis et al., 1998; Morimoto et al., 2009). Thus, postsynaptic signaling and calcium levels can immediately impact the activity of type-A receptors and therefore their ability to be incorporated in stable synaptic clusters.

This positive feedback mechanism can also explain the Hebbian mode of GluRIIA incorporation at PSDs (Ljaschenko et al., 2013). Indeed, local stimulation increases postsynaptic sensitivity by promoting synapse-specific recruitment/stabilization of type-A glutamate receptors. Conversely, GluRIIA-containing receptors are rapidly removed from synapses with reduced activity. Since the activity/ conformation status of type-A receptors can be immediately adjust in response to (calcium) signaling, any increase or decrease in synapse activity can be relayed to Neto-enabled trans-synaptic complexes which can adjust and fine-tune the incorporation of type-A receptors in stable aggregates.

Trans-synaptic nanocolumns that align active zone structures with post-synaptic receptor fields have been recently described at vertebrate and fly synapses. Studies that mapped vesicle fusion positions within individual vertebrate synapses indicate that action potential evoked fusion occurs within nanometer-scaled regions with higher local density of Rab3 interacting molecule (RIM), a protein that couples vesicle recruitment to release sites (Sudhof, 2012). In hippocampal neurons, these RIM nanoclusters align perfectly with high density patches of postsynaptic receptors and PSD95, a scaffold protein which stabilizes the receptors (Tang et al., 2016). Similarly, at the fly NMJ high probability release sites accumulate elevated levels of Cacophony (Cac), a voltage-gated $Ca^{2+}$ channel, Brp, Rim and Rim-binding protein complexes (Akbergenova et al., 2018; Van Vactor & Sigrist, 2017; Zhai & Bellen, 2004). Classic electrophysiology recordings revealed that glutamate receptors preferentially cluster at sites with high release probability (Marrus & DiAntonio, 2004). More recently, dynamic studies uncovered a strong positive correlation between the levels of postsynaptic type-A receptors and the active zones with high release probability (Akbergenova et al., 2018; Fouquet et al., 2009; Rasse et al., 2005). GluRIIA concentrates in "bright" PSD fields opposing high release probability sites, marked by high accumulation of Cac and Brp. In contrast, low release probability active zones have reduced presynaptic Brp and Cac intensities and more diffuse "dim" GluRIIA levels (Akbergenova et al., 2018; Gratz et al., 2019). Within individual synapses, line profiles drawn along various PSDs captured GluRIIA intense puncta concentrated at the center of the "bright" PSD, surrounded by a ring of GluRIIB signals. At "dim" PSDs, GluRIIB was more evenly distributed across the entire receptor field (Akbergenova et al., 2018).

The striking correlation between the intensity/activity of postsynaptic type-A receptors and the corresponding levels of both (i) presynaptic pMad and (ii) presynaptic Cac and Brp at individual active zones suggests that the Cac- and Brp-marked structures surrounded by pMad and the postsynaptic type-A receptors are organized in trans-synaptic nanocolumns. Such nanocolumns appear to be at the core of a positive feedback loop that enables recruitment of type-A receptors as a function of receptors activity. This model opens a lot of exciting possibilities as it provides a molecular framework for understanding how different subtypes of postsynaptic glutamate receptors are selectively recruited and stabilized at the onset of synaptogenesis. This model also brings additional insights into how pre- and post-synaptic structures develop coordinately. Future super-resolution and protein mobility studies should bring further clarity to the subsynaptic distribution of the individual protein components. Also, is the recruitment of presynaptic pMad coordinated with the recruitment of Cac and Brp and/or with the other presynaptic molecules implicated in neurotransmitter release? Does synaptic pMad mark a more general positive feedback mechanism that monitors postsynaptic activity status and coordinates and adjusts the recruitment of presynaptic active zone components? All these possibilities remain to be explored. It is important to note that this positive feedback mechanism is completely over-ruled when genetic or other challenges alter synapse activities; such perturbations trigger negative feedback mechanisms aimed at restoring the circuit steady-state function. For example, loss of postsynaptic GluRIIA activities induces a significant enhancement of Brp accumulation which mobilizes the readily released pool of synaptic vesicles and enables a compensatory increase of neurotransmitter release

(Davis & Muller, 2015; Frank, 2014). This is in sharp contrast with the normal settings for synapse development and function, where the positive feedback mechanism coordinates the recruitment of postsynaptic type-A receptors and effectively controls key steps of synapse assembly and maturation.

At the molecular level, this positive feedback also requires that synaptic pMad remains associated with the BMP/BMPR complexes at the active zones. The active zones localize to regions of subsynaptic membranes that experience abundant exocytosis of synaptic vesicles to release neurotransmitters; however, these specialized presynaptic membranes have no endocytosis. Therefore, the BMP/BMPRs confined within this membrane microdomain should be shielded from endocytosis and from further recycling or retrograde transport to the motor neuron soma. By remaining associated with its own kinase (the BMPRs), pMad may actively anchor the BMP/BMPR complexes at active zones, protecting them from endocytosis and retrograde transport. Tissue culture-based assays revealed that pMad indeed can accumulate at cell membranes through its association with the activated intracellular domain of the BMPRI, Tkv (Nguyen, Han, Newfeld, & Serpe, 2020). Genetics and molecular modeling studies indicate that this interaction is partly mediated through the enzyme-substrate/product (Tkv-Mad/pMad) binding interface; in addition, the Tkv-pMad interaction requires a highly conserved motif within the Smad-type proteins, the H2 helix. Genetic variants within the H2 helix have been uncovered in several patients with neuronal deficits or epithelial abnormalities suggesting that this motif may be critical for local BMP signaling and the integrity of the tight junctions throughout the animal kingdom (Nguyen et al., 2020). The H2 contribution may be direct, by shaping the Mad-Tkv interface, or indirect, via recruiting other protein(s) that may stabilize Mad-Tkv complexes at specialized cell junctions. In the chick neural tube, local BMP signaling controls apicobasal polarity partly by enabling the pSmad1/5/8-dependent association of BMP/BMPR signaling complexes with the partitioning defective (PAR) proteins, the PAR3-PAR5-aPKC complex, at the tight junctions (Eom et al., 2011). Reduction of junctional pSmad1/5/8 accumulation destabilizes the PAR complexes and disrupts the tight junctions. It is interesting to point out that the opposite phenomenon, the dissolution of the tight junctions and the epithelial-to-mesenchymal transition, requires TGF-β signaling and direct phosphorylation of another PAR protein, Par6, by the type-I TGF-β receptor (Ozdamar et al., 2005). Par6 mutants that can no longer be phosphorylated block TGF-β induced tight-junction dissolution. Likewise, loss-of-function disruptions of *Drosophila bazooka* (Par-3)-Par-6-aPKC complexes produces NMJs with increased levels of synaptic type-A receptors (therefore increased synaptic pMad levels) and significantly reduced number of boutons (a hallmark of reduced levels of nuclear pMad) (Ruiz-Canada et al., 2004). Together these studies suggest a dynamic interplay between various PAR complexes and the local pSmad accumulation in regulating specialized cellular junctions.

The stability of Tkv-pMad interactions may also be influenced by Scribble (Scrib), a basolateral determinant which instructs the polarized trafficking machinery and regulates the apical-basal protein distribution of epithelial cells (Bilder & Perrimon, 2000). In pupal wing epithelia, Scrib regulates Tkv localization to the basal side, where the secreted BMP ligands—which specify the posterior crossvein structure—accumulate (Gui, Huang, & Shimmi, 2016). It was proposed that Scrib facilitates Tkv internalization and thus optimizes signal transduction after the formation of BMP/BMPR complexes. Interestingly, Scrib appears to bind directly to Tkv and together they colocalize with pMad at the basal side of epithelial cells. Scrib is expressed in MNs and has been implicated in the regulation of vesicle release at the larval NMJ (Roche, Packard, Moeckel-Cole, & Budnik, 2002). It remains to be determined whether Scrib plays any role in the MNs in regulating Tkv/pMad interactions at the active zones or within the adjacent, endocytosis-able regions. Since many BMPRs are shared among different BMP signaling modalities, mechanisms that regulate their confinement to the endocytosis-free active zones vs distribution to other membrane domains of the synaptic boutons should have profound effects on the relative contribution of various types of BMP signaling and effectively sculpt synapse development.

## 6. Motor neurons coordinate multiple BMP signaling to balance NMJ growth with synapse maturation/stabilization

As discussed above (Fig. 1), several BMP signaling pathways shape the *Drosophila* NMJ growth and function. Muscle secreted BMP7/Gbb binds to BMPRs (Wit, Tkv, Sax) on the motor neurons terminals and signal retrogradely to control NMJ growth (reviewed in Marques & Zhang, 2006). Gbb secreted from MNs also signals in an autocrine manner to modulate neurotransmitter release. In both cases the high-order BMP/BMPR (Gbb/Wit/Tkv/Sax) complexes are endocytosed and transported to the MN soma where they phosphorylate Mad and regulate various transcriptional programs with distinct roles in the development and function of the NMJ. Besides these canonical signaling pathways, Gbb and Wit signal non-canonically, independently of Mad, through the effector protein LIM kinase 1 (LIMK1) to regulate synapse stability and to enable addition of new boutons with increased synaptic activity (Eaton & Davis, 2005; Piccioli & Littleton, 2014). LIMK1 regulates the presynaptic actin dynamics partly by controlling the activity of the actin depolymerizing protein Cofilin. LIMK1 also binds to the C-terminal domain of Wit, to a region required for synapse stability but dispensable for Mad-mediated transcriptional control. This implies that a pool of Wit must remain at the synaptic terminals to localize the LIMK1-dependent activities. Finally, the BMPRs Wit, Tkv and Sax, but not Gbb, are required for the synaptic BMP signaling (Sulkowski et al., 2016; Zhao et al., 2015), a pathway that confines the BMP/BMPR complexes to the active zones.

Motor neurons must coordinate all these multiple BMP pathways to build appropriate NMJs. Several lines of evidence indicate support this view. First, these genetically distinct pathways share selected components, some of them in limited and tightly controlled supplies. In both flies and mammals, posttranslational modifications such as phosphorylation and ubiquitination limit signaling activity and trigger degradation or deactivation of both Smads and receptors, keeping these signaling components in check (Alarcon et al., 2009; Dupont et al., 2009; Stinchfield et al., 2012; Zhu, Kavsak, Abdollah, Wrana, & Thomsen, 1999). Second, mutations that favor (or disfavor) one of these BMP signaling pathways appear do so at the expense of the other pathways. Finally, depending on the context, multiple modulators and regulatory feedbacks fine-tune the BMP signaling inputs/outputs and integrate this signaling network within the larger web of signaling cues that ultimately elicit specific cellular responses. Likewise, motor neurons employ complex modulatory mechanisms to coordinate various BMP signaling modalities and integrate them with additional signaling networks to control neuronal survival and circuit function. Several mechanisms that limit the net levels, subcellular distribution and regulation of various BMP pathway components are reviewed below. The range of phenotypes induced by genetics manipulations of relevant BMP pathways modulators emphasizes the remarkable interconnectedness among different BMP signaling pathways.

For example, many BMP pathway components, including Tkv, Wit and Sax, as well as Mad and Medea, are direct targets of *mir-124*, which limits BMP signaling (Sun et al., 2012). The phenotype of *mir-124* mutants resembles that of neuronal overexpression of Mad or activated Sax and Tkv receptors together, with increased basal neurotransmission (increased nuclear pMad) but normal quantal size (normal synaptic pMad). Mad protein levels are also downregulated by *brain tumor* (*brat*), a translational repressor that limits Mad translation (Shi et al., 2013). Interestingly, *brat* loss-of-function does not mirror the *Mad* gain-of-function phenotypes; instead, depletion of Brat moderately enhances nuclear pMad levels and induces (i) increased PSD size (with more iGluRs per PSD) and (ii) a ~10-fold increase of synaptic pMad. Consistently, *brat* mutant NMJs are overgrown and show normal basal neurotransmission and increased mini amplitudes/ quantal size. These phenotypical differences suggest that, besides Mad, Brat may repress additional NMJ modulators. Alternatively, since Brat expression appears limited to a subset of MNs, Brat may elicit different signaling outcomes via neuron-specific differential regulation of Mad. Moreover, Brat NMJ activities could be further modulated by the Pum-Nos translational repressor complex, which provides additional layers of coordinated regulation between the NMJ growth and the recruitment of postsynaptic glutamate receptors (Harris, Pargett, Sutcliffe, Umulis, & Ashe, 2011; Menon et al., 2009, 2004).

In MNs, canonical BMP signaling depends on endocytosis and endosomal trafficking (reviewed in Deshpande & Rodal, 2016). BMP/BMPR complexes, but not BMPRs alone, are endocytosed at synaptic terminals and routed to signaling endosomes that are transported retrogradely along the axon to the neuron soma (Smith et al., 2012). The retrograde transport requires Dynein, a motor complex which accomplishes most of the retrograde axonal transport, and a set of cargo-specific regulators that includes Vezatin-like (Vezl) (Spinner, Pinter, Drerup, & Herman, 2020). Studies on several endocytic regulators, including endophilin, dynamin, Dap160/intersectin indicate that endocytosis negatively regulates BMP signaling (Koh, Verstreken, & Bellen, 2004; Marie et al., 2004; O'Connor-Giles, Ho, & Ganetzky, 2008). Mutations in endocytic modulators show characteristic morphologies, with overgrown NMJs with supernumerary or satellite boutons, a mark of excessive BMP signaling. These mutants have elevated levels of both nuclear and synaptic pMad as endocytosis limits BMPRs overall availability. Following endocytosis, BMPRs are sorted and routed to three destinations: signaling endosomes, recycling endosomes and lysosomes. Proteins like Liquid facets (Lqf), the *Drosophila* epsin homolog, and the Sorting nexin 16 (Snx16), promote the routing of BMP/BMPRs to signaling endosomes (Rodal et al., 2011; Vanlandingham et al., 2013). Loss of these sorting components results in excessive BMP/BMPRs directed toward lysosomal degradation and/or recycling; this reduces the levels of BMP/BMPRs in signaling endosomes and attenuates canonical BMP signaling. Interestingly, *lqf* mutants have reduced nuclear pMad levels, but maintain significant levels of synaptic pMad, presumably because in these mutant backgrounds BMPR availability is specifically reduced within endosomal compartments but not at active zones (Vanlandingham et al., 2013). Conversely, mutations in endosome-localized proteins that interact with BMPRs and promote their lysosomal routing and degradation have excessive levels of BMPRs and consequently drastically enhanced BMP signaling in both MN nuclei and synaptic terminals (O'Connor-Giles et al., 2008; Sweeney & Davis, 2002; Tsang et al., 2009).

An additional endocytic mechanism, called micropinocytosis, contributes to Gbb-dependent BMPR internalization and degradation (Kim et al., 2019). The macropinosomes are actin-driven structures that require Abelson (Abl) tyrosine kinase, the Abl substrate and interactor Abi, and the Rac1 GTPase signaling. Mutations in any of these components (Abi, Abl or Rac1) produce satellite boutons morphologies characteristic of excess BMP signaling, accompanied by elevated nuclear and synaptic pMad levels. Tissue culture assays indicate that Gbb induces internalization of both Tvk and Wit via macropinocytosis. In addition, Tkv (but not Wit or Sax) is a substrate for Ube3A, an E3 ubiquitin ligase implicated in Angelman syndrome and autism (Li et al., 2016). Ube3A limits BMP signaling by binding and ubiquitinating Tkv for proteasomal degradation. This Ube3A function is conserved in mammalian cells, suggesting that the Ube3A-mediated downregulation of BMP signaling may explain the pathogenesis of Ube3A-associated Angelman syndrome and autism.

Mechanisms that regulate the levels and subcellular distribution of the BMP pathway components impact multiple BMP signaling pathways and may offer limited insights into their coordinated regulation. Additional BMP modulators must exist to allow the motor neuron to distinguish among different BMP signals and transduce specific pathway(s). For example, Gbb drives two different canonical BMP pathways in the MNs: Muscle-secreted Gbb signals retrogradely to control the NMJ growth, whereas neuronal Gbb signals in an autocrine manner to promote neurotransmitter release. The MNs seem to differentiate between the two different pools of Gbb via Crimpy (Cmpy), a neuronal Cysteine-rich transmembrane protein that binds to Gbb and delivers it to dense core vesicles for activity-dependent release (James & Broihier, 2011; James et al., 2014). In the absence of Cmpy, neuronal Gbb cannot sustain proper neurotransmitter release and instead induces excessive NMJ growth, augmenting the retrograde signaling. Thus, Cmpy effectively marks the pool of Gbb for autocrine signaling.

The non-canonical BMP signaling pathway appears to be selectively modulated by LIMK1 (as discussed above) and by the Fragile X mental retardation 1 (FMRP1), an RNA binding protein which causes fragile X syndrome (FXS), a common inherited form of intellectual disability and autism (Kashima et al., 2016; Zhang et al., 2001). FMRP1 binds to and downregulates the translation of two key molecules: (i) Futsch, the *Drosophila* homolog of the mammalian microtubule-associated protein MAP1B, a regulator of the microtubule cytoskeleton, and (ii) the Wit long isoform, which contains the LIMK1 binding domain. Loss of FMRP1 in larval MNs results in overgrown NMJs with elevated basal neurotransmission, presumably because of excessive (canonical and non-canonical) BMP signaling. However, *FMRP1* mutant NMJs have longer branches with increased number of enlarged synaptic boutons—a morphology distinctly different than the satellite boutons discussed above. The mini amplitude/quantal size is also normal at *FMRP1* mutant NMJs indicating that synaptic BMP signaling is largely unaffected. Importantly, the neuronal deficits observed in fly or mouse models of FXS can be rescued by lowering the Wit/BMPRII copy number or by pharmacological inhibition of LIMK1, suggesting that FXS pathologies are linked to elevated non-canonical BMP signaling.

Finally, synaptic BMP signaling is linked via positive feedback with active postsynaptic type-A glutamate receptors. Gbb is not part of this positive feedback, even though Gbb drives all other BMP signaling pathways and promotes BMPRs internalization, limiting their availability. With BMPRs in limited supply, the MNs cannot deploy all BMP signaling pathways at once. Instead, MNs must choose among several options: (i) endocytose and route BMP/BMPRs to signaling endosomes to grow the NMJs and ensure proper neurotransmission, (ii) connect Gbb-Wit-LIMK1 perisynaptic complexes with the cytoskeleton to provide structural stability for the NMJ, or (iii) trap the BMPRs at the active zones to stabilize active type-A receptors and thus initiate formation of new synapse and modulate their maturation and plasticity. Since local BMP signaling serves as a sensor for synapse activity, the neurons may use the local signaling complexes to monitor synapse activity then deploy the other BMP signaling pathways to coordinate NMJ growth with synapse maturation and stabilization. Such a balancing act contains all the features of a finely tuned machinery capable of constantly sampling the synapse activity status and implementing swift adjustments to optimize synapse strength and function.

## 7. Future challenges

The powerful fly NMJ model system was instrumental in identifying and characterizing different BMP pathways, uncovering a rich modulation of levels and availability for signaling for various pathway components, as well as the subcellular distribution, sorting and trafficking of various signaling complexes. However, our understanding of the composition and regulation of local BMP signaling complexes remains sketchy. Aside from the detection of pMad at the active zones, all the other signaling pathway components have been derived from genetics arguments. With new technologies constantly improving our analyses of multimolecular complexes and detection capabilities, the next immediate challenge will be to describe the composition of local BMP signaling complexes and localize these components at the active zones. Gbb is clearly not part of these complexes, but is there another ligand involved? Ligand binding decreases the mobility of BMPRs within the cell membrane and increases the stability of hetero-tetrameric BMPRs assembles (Marom, Heining, Knaus, & Henis, 2011), promoting BMPRs trans-phosphorylation and activation. Alternatively, the density and structural constraints within the active zone may limit the BMPRs mobility and effectively confine these complexes even in the absence of a ligand.

More intriguingly, how does pMad remain associated with its own kinase after the enzymatic reaction has been completed? The Tkv/Mad binding interface seems to be crucial for the stabilization of these complexes, but additional component(s) that alter complex dissociation rate should exist. Such components may function by binding and stabilizing the complexes directly, or they may introduce post-translational modifications in either the enzyme (Tkv) or the product (pMad) and indirectly influence their dissociation rate.

Our model also predicts that presynaptic BMP signaling complexes physically connect with postsynaptic iGluRs complexes via trans-synaptic nanocolumns. The iGluRs expand ~140Å within the 200Å synaptic cleft; the BMP-binding CUB domains of Neto expand even less, tucked just below the N-terminal domain of the iGluR complexes (as in the case of CUB1) or associated with the ligand binding domain (CUB2) (He et al., 2021). On the presynaptic site, the ectodomain of the BMPRs are ~55Å tall (Greenwald et al., 2003; Mace, Cutfield, & Cutfield, 2006). Even when BMP ligands bind to the BMPRs, these presynaptic complexes may not extend far enough to interact with the postsynaptic CUB-containing iGluRs/Neto complexes. Additional secreted molecule(s) may "bridge" these complexes and organize the proposed presynaptic BMPRs-postsynaptic iGluRs nanocolumns. In the future, screens with BioID methodologies, which label proximal endogenous proteins, may uncover both extracellular and intracellular scaffolds that configure and stabilize these nanocolumns.

As stressed above, the local BMP signaling pathway must be coordinated with the other BMP signaling modalities. But our current knowledge of transcriptional targets downstream of retrograde and autocrine BMP signaling is fairly limited. How do these BMP transcriptional targets function in adjusting the hard-wired transcriptional programs specific to each motor neuron lineage, so that each motor neuron output matches the status of the target muscle? The fast advances in RNA sequencing approaches should make such transcriptomics analyses accessible within the near future. Moreover, each muscle receives inputs from two different glutamatergic motor neurons, Ib and Is, with distinct synaptic structure, postsynaptic composition, neurotransmitter release properties and plasticity mechanisms. Recent studies reveal that the two neurons influence each other's structural and functional output: Ablation of one neuron causes selective NMJ expansion and increased neurotransmitter release in the remaining neuron (Aponte-Santiago et al., 2020; Wang et al., 2021). Gbb appears to have both shared and distinct functions in these two types of neurons, raising the possibility that BMP signaling may further enable the coordinated response of co-innervating neurons, balancing their synaptic plasticity. Elucidating neuron-specific BMP transcriptional targets and their relationship to local BMP signaling may uncover overarching principles and developmental strategies for the assembly and maturation of plastic synapses.

The complexity of BMP signaling reviewed here is reminiscent of neurotrophin-regulated signaling in vertebrate systems (reviewed in Reichardt, 2006). First identified as neuronal survival factors, neurotrophins are secreted as pro-proteins that must be processed to form mature ligands. Like BMPs, these active dimers form cysteine-knot structures and signal by binding to transmembrane kinase receptors and inducing their activation through trans-phosphorylation. Neurotrophin/receptor complexes are internalized and transported along axons to the cell soma (Thoenen & Barde, 1980). This canonical signaling pathway controls gene expression and promotes neuronal differentiation and growth. In addition, local neurotrophin signaling mediates activity-dependent synapse formation and maturation and promotes neurotransmitter release (reviewed in Park & Poo, 2013). While the molecular details of neurotrophin-induced local signaling remain largely unknown, the remarkable similarities between these signaling pathways suggest they may share central roles in the assembly and maturation of plastic synapses.

The local BMP signaling-based feedback mechanism reviewed here provides a molecular description of a key positive feedback loop that sculpts synapse assembly and maturation. At the same time, this positive feedback mechanism may serve as a template to elucidate other signaling networks utilized for building chemical synapses. Furthermore, a local role for BMP signaling in regulating cellular junctions has extremely broad implications, from stabilizing short lived intercellular interactions during patterning (Huang, Liu, & Kornberg, 2019) to supporting epithelia integrity and remodeling during development and homeostasis (Eom et al., 2012; Marmion, Jevtic, Springhorn, Pyrowolakis, & Yakoby, 2013). We envision this local modality of BMP signaling arose early during evolution to ensure the integrity of cellular junctions and evolved to fulfill additional functions at more specialized cellular junctions, the chemical synapses.

## Acknowledgments

R.V. and M.S. were supported by the Intramural Research Program of the *Eunice Kennedy Shriver* National Institute of Child Health and Human Development, National Institutes of Health, awards ZIA HD008914 and ZIA HD008869. We thank Ed Giniger and members of the Serpe laboratory for discussions and comments.

## References

Aberle, H., Haghighi, A. P., Fetter, R. D., McCabe, B. D., Magalhaes, T. R., & Goodman, C. S. (2002). wishful thinking encodes a BMP type II receptor that regulates synaptic growth in Drosophila. *Neuron, 33*, 545–558.

Akbergenova, Y., Cunningham, K. L., Zhang, Y. V., Weiss, S., & Littleton, J. T. (2018). Characterization of developmental and molecular factors underlying release heterogeneity at Drosophila synapses. *eLife, 7*, e38268.

Alarcon, C., Zaromytidou, A. I., Xi, Q., Gao, S., Yu, J., Fujisawa, S., et al. (2009). Nuclear CDKs drive Smad transcriptional activation and turnover in BMP and TGF-beta pathways. *Cell, 139*, 757–769.

Albin, S. D., & Davis, G. W. (2004). Coordinating structural and functional synapse development: Postsynaptic p21-activated kinase independently specifies glutamate receptor abundance and postsynaptic morphology. *The Journal of Neuroscience, 24*, 6871–6879.

Aponte-Santiago, N. A., Ormerod, K. G., Akbergenova, Y., & Littleton, J. T. (2020). Synaptic plasticity induced by differential manipulation of tonic and phasic motoneurons in Drosophila. *The Journal of Neuroscience*, *40*, 6270–6288.

Ball, R. W., Warren-Paquin, M., Tsurudome, K., Liao, E. H., Elazzouzi, F., Cavanagh, C., et al. (2010). Retrograde BMP signaling controls synaptic growth at the NMJ by regulating trio expression in motor neurons. *Neuron*, *66*, 536–549.

Banovic, D., Khorramshahi, O., Owald, D., Wichmann, C., Riedt, T., Fouquet, W., et al. (2010). Drosophila neuroligin 1 promotes growth and postsynaptic differentiation at glutamatergic neuromuscular junctions. *Neuron*, *66*, 724–738.

Bilder, D., & Perrimon, N. (2000). Localization of apical epithelial determinants by the basolateral PDZ protein Scribble. *Nature*, *403*, 676–680.

Bogdanik, L., Mohrmann, R., Ramaekers, A., Bockaert, J., Grau, Y., Broadie, K., et al. (2004). The Drosophila metabotropic glutamate receptor DmGluRA regulates activity-dependent synaptic facilitation and fine synaptic morphology. *The Journal of Neuroscience*, *24*, 9105–9116.

Bond, A. M., Bhalala, O. G., & Kessler, J. A. (2012). The dynamic role of bone morphogenetic proteins in neural stem cell fate and maturation. *Developmental Neurobiology*, *72*, 1068–1084.

Bork, P., & Beckmann, G. (1993). The CUB domain. A widespread module in developmentally regulated proteins. *Journal of Molecular Biology*, *231*, 539–545.

Broadie, K., & Bate, M. (1993). Innervation directs receptor synthesis and localization in Drosophila embryo synaptogenesis. *Nature*, *361*, 350–353.

Burden, S. J. (1998). The formation of neuromuscular synapses. *Genes & Development*, *12*, 133–148.

Chen, K., & Featherstone, D. E. (2005). Discs-large (DLG) is clustered by presynaptic innervation and regulates postsynaptic glutamate receptor subunit composition in Drosophila. *BMC Biology*, *3*, 1.

Chen, K., Merino, C., Sigrist, S. J., & Featherstone, D. E. (2005). The 4.1 protein coracle mediates subunit-selective anchoring of Drosophila glutamate receptors to the postsynaptic actin cytoskeleton. *The Journal of Neuroscience*, *25*, 6667–6675.

Cui, Y., Jean, F., Thomas, G., & Christian, J. L. (1998). BMP-4 is proteolytically activated by furin and/or PC6 during vertebrate embryonic development. *The EMBO Journal*, *17*, 4735–4743.

Dalva, M. B., McClelland, A. C., & Kayser, M. S. (2007). Cell adhesion molecules: Signalling functions at the synapse. *Nature Reviews. Neuroscience*, *8*, 206–220.

Davis, G. W., DiAntonio, A., Petersen, S. A., & Goodman, C. S. (1998). Postsynaptic PKA controls quantal size and reveals a retrograde signal that regulates presynaptic transmitter release in Drosophila. *Neuron*, *20*, 305–315.

Davis, G. W., & Muller, M. (2015). Homeostatic control of presynaptic neurotransmitter release. *Annual Review of Physiology*, *77*, 251–270.

Degnin, C., Jean, F., Thomas, G., & Christian, J. L. (2004). Cleavages within the prodomain direct intracellular trafficking and degradation of mature bone morphogenetic protein-4. *Molecular Biology of the Cell*, *15*, 5012–5020.

Derynck, R., & Zhang, Y. E. (2003). Smad-dependent and Smad-independent pathways in TGF-beta family signalling. *Nature*, *425*, 577–584.

Deshpande, M., & Rodal, A. A. (2016). The crossroads of synaptic growth signaling, membrane traffic and neurological disease: Insights from Drosophila. *Traffic*, *17*, 87–101.

DiAntonio, A., Petersen, S. A., Heckmann, M., & Goodman, C. S. (1999). Glutamate receptor expression regulates quantal size and quantal content at the Drosophila neuromuscular junction. *The Journal of Neuroscience*, *19*, 3023–3032.

Diao, F., Ironfield, H., Luan, H., Diao, F., Shropshire, W. C., Ewer, J., et al. (2015). Plug-and-play genetic access to drosophila cell types using exchangeable exon cassettes. *Cell Reports*, *10*, 1410–1421.

Dudu, V., Bittig, T., Entchev, E., Kicheva, A., Julicher, F., & Gonzalez-Gaitan, M. (2006). Postsynaptic mad signaling at the Drosophila neuromuscular junction. *Current Biology, 16*, 625–635.

Dupont, S., Mamidi, A., Cordenonsi, M., Montagner, M., Zacchigna, L., Adorno, M., et al. (2009). FAM/USP9x, a deubiquitinating enzyme essential for TGFbeta signaling, controls Smad4 monoubiquitination. *Cell, 136*, 123–135.

Eaton, B. A., & Davis, G. W. (2005). LIM Kinase1 controls synaptic stability downstream of the type II BMP receptor. *Neuron, 47*, 695–708.

Ehrlich, M., Gutman, O., Knaus, P., & Henis, Y. I. (2012). Oligomeric interactions of TGF-beta and BMP receptors. *FEBS Letters, 586*, 1885–1896.

Ellis, J. E., Parker, L., Cho, J., & Arora, K. (2010). Activin signaling functions upstream of Gbb to regulate synaptic growth at the Drosophila neuromuscular junction. *Developmental Biology, 342*, 121–133.

Eom, D. S., Amarnath, S., Fogel, J. L., & Agarwala, S. (2011). Bone morphogenetic proteins regulate neural tube closure by interacting with the apicobasal polarity pathway. *Development, 138*, 3179–3188.

Eom, D. S., Amarnath, S., Fogel, J. L., & Agarwala, S. (2012). Bone morphogenetic proteins regulate hinge point formation during neural tube closure by dynamic modulation of apicobasal polarity. *Birth Defects Research. Part A, Clinical and Molecular Teratology, 94*, 804–816.

Featherstone, D. E., Rushton, E., Rohrbough, J., Liebl, F., Karr, J., Sheng, Q., et al. (2005). An essential Drosophila glutamate receptor subunit that functions in both central neuropil and neuromuscular junction. *The Journal of Neuroscience, 25* 3199–3208.

Feng, X. H., & Derynck, R. (2005). Specificity and versatility in tgf-beta signaling through Smads. *Annual Review of Cell and Developmental Biology, 21*, 659–693.

Foletta, V. C., Lim, M. A., Soosairajah, J., Kelly, A. P., Stanley, E. G., Shannon, M., et al. (2003). Direct signaling by the BMP type II receptor via the cytoskeletal regulator LIMK1. *The Journal of Cell Biology, 162*, 1089–1098.

Fouquet, W., Owald, D., Wichmann, C., Mertel, S., Depner, H., Dyba, M., et al. (2009). Maturation of active zone assembly by Drosophila Bruchpilot. *The Journal of Cell Biology, 186*, 129–145.

Frank, C. A. (2014). Homeostatic plasticity at the Drosophila neuromuscular junction. *Neuropharmacology, 78*, 63–74.

Fuentes-Medel, Y., Ashley, J., Barria, R., Maloney, R., Freeman, M., & Budnik, V. (2012). Integration of a retrograde signal during synapse formation by glia-secreted TGF-beta Ligand. *Current Biology, 22*, 1831–1838.

Fukui, A., Inaki, M., Tonoe, G., Hamatani, H., Homma, M., Morimoto, T., et al. (2012). Lola regulates glutamate receptor expression at the Drosophila neuromuscular junction. *Biology Open, 1*, 362–375.

Ganesan, S., Karr, J. E., & Featherstone, D. E. (2011). Drosophila glutamate receptor mRNA expression and mRNP particles. *RNA Biology, 8*, 771–781.

Giagtzoglou, N., Ly, C. V., & Bellen, H. J. (2009). Cell adhesion, the backbone of the synapse: "Vertebrate" and "invertebrate" perspectives. *Cold Spring Harbor Perspectives in Biology, 1*, a003079.

Goldman, D. C., Hackenmiller, R., Nakayama, T., Sopory, S., Wong, C., Kulessa, H., et al. (2006). Mutation of an upstream cleavage site in the BMP4 prodomain leads to tissue-specific loss of activity. *Development, 133*, 1933–1942.

Gratz, S. J., Goel, P., Bruckner, J. J., Hernandez, R. X., Khateeb, K., Macleod, G. T., et al. (2019). Endogenous tagging reveals differential regulation of $Ca^{2+}$ channels at single AZs during presynaptic homeostatic potentiation and depression. *The Journal of Neuroscience, 39*(13), 2416–2429.

Greenwald, J., Groppe, J., Gray, P., Wiater, E., Kwiatkowski, W., Vale, W., et al. (2003). The BMP7/ActRII extracellular domain complex provides new insights into the cooperative nature of receptor assembly. *Molecular Cell*, *11*, 605–617.

Gui, J., Huang, Y., & Shimmi, O. (2016). Scribbled optimizes BMP signaling through its receptor internalization to the Rab5 endosome and promote robust epithelial morphogenesis. *PLoS Genetics*, *12*, e1006424.

Han, T. H., Dharkar, P., Mayer, M. L., & Serpe, M. (2015). Functional reconstitution of Drosophila melanogaster NMJ glutamate receptors. *Proceedings of the National Academy of Sciences of the United States of America*, *112*, 6182–6187.

Harris, K. P., & Littleton, J. T. (2015). Transmission, development, and plasticity of synapses. *Genetics*, *201*, 345–375.

Harris, R. E., Pargett, M., Sutcliffe, C., Umulis, D., & Ashe, H. L. (2011). Brat promotes stem cell differentiation via control of a bistable switch that restricts BMP signaling. *Developmental Cell*, *20*, 72–83.

He, L., Sun, J., Gao, Y., Li, B., Wang, Y., Dong, Y., et al. (2021). Kainate receptor modulation by NETO2. *Nature*, *599*, 325–329.

Heckmann, M., & Dudel, J. (1997). Desensitization and resensitization kinetics of glutamate receptor channels from Drosophila larval muscle. *Biophysical Journal*, *72*, 2160–2169.

Hill, C. S. (2016). Transcriptional control by the SMADs. *Cold Spring Harbor Perspectives in Biology*, *8*(10), a022079.

Hogan, B. L. (1996). Bone morphogenetic proteins in development. *Current Opinion in Genetics & Development*, *6*, 432–438.

Huang, H., Liu, S., & Kornberg, T. B. (2019). Glutamate signaling at cytoneme synapses. *Science*, *363*, 948–955.

Jackson, A. C., & Nicoll, R. A. (2011). The expanding social network of ionotropic glutamate receptors: TARPs and other transmembrane auxiliary subunits. *Neuron*, *70*, 178–199.

James, R. E., & Broihier, H. T. (2011). Crimpy inhibits the BMP homolog Gbb in motoneurons to enable proper growth control at the Drosophila neuromuscular junction. *Development*, *138*, 3273–3286.

James, R. E., Hoover, K. M., Bulgari, D., McLaughlin, C. N., Wilson, C. G., Wharton, K. A., et al. (2014). Crimpy enables discrimination of presynaptic and postsynaptic pools of a BMP at the Drosophila neuromuscular junction. *Developmental Cell*, *31*(5), 586–598.

Jan, L. Y., & Jan, Y. N. (1976a). L-glutamate as an excitatory transmitter at the Drosophila larval neuromuscular junction. *The Journal of Physiology*, *262*, 215–236.

Jan, L. Y., & Jan, Y. N. (1976b). Properties of the larval neuromuscular junction in Drosophila melanogaster. *The Journal of Physiology*, *262*, 189–214.

Karr, J., Vagin, V., Chen, K., Ganesan, S., Olenkina, O., Gvozdev, V., et al. (2009). Regulation of glutamate receptor subunit availability by microRNAs. *The Journal of Cell Biology*, *185*, 685–697.

Kashima, R., Roy, S., Ascano, M., Martinez-Cerdeno, V., Ariza-Torres, J., Kim, S., et al. (2016). Augmented noncanonical BMP type II receptor signaling mediates the synaptic abnormality of fragile X syndrome. *Science Signaling*, *9*, ra58.

Kawabata, M., Inoue, H., Hanyu, A., Imamura, T., & Miyazono, K. (1998). Smad proteins exist as monomers in vivo and undergo homo- and hetero-oligomerization upon activation by serine/threonine kinase receptors. *The EMBO Journal*, *17*, 4056–4065.

Keshishian, H., Broadie, K., Chiba, A., & Bate, M. (1996). The drosophila neuromuscular junction: A model system for studying synaptic development and function. *Annual Review of Neuroscience*, *19*, 545–575.

Kim, Y. J., Bao, H., Bonanno, L., Zhang, B., & Serpe, M. (2012). Drosophila Neto is essential for clustering glutamate receptors at the neuromuscular junction. *Genes & Development*, *26*, 974–987.

Kim, Y. J., Igiesuorobo, O., Ramos, C. I., Bao, H., Zhang, B., & Serpe, M. (2015). Prodomain removal enables neto to stabilize glutamate receptors at the Drosophila neuromuscular junction. *PLoS Genetics*, *11*, e1004988.

Kim, N., Kim, S., Nahm, M., Kopke, D., Kim, J., Cho, E., et al. (2019). BMP-dependent synaptic development requires Abi-Abl-Rac signaling of BMP receptor macropinocytosis. *Nature Communications*, *10*, 684.

Kim, M. J., & O'Connor, M. B. (2014). Anterograde activin signaling regulates postsynaptic membrane potential and GluRIIA/B abundance at the Drosophila neuromuscular junction. *PLoS One*, *9*, e107443.

Kim, Y. J., & Serpe, M. (2013). Building a synapse: A complex matter. *Fly*, 7(3), 146–152.

Kim, N., Stiegler, A. L., Cameron, T. O., Hallock, P. T., Gomez, A. M., Huang, J. H., et al. (2008). Lrp4 is a receptor for Agrin and forms a complex with MuSK. *Cell*, *135*, 334–342.

Kiragasi, B., Wondolowski, J., Li, Y., & Dickman, D. K. (2017). A presynaptic glutamate receptor subunit confers robustness to neurotransmission and homeostatic potentiation. *Cell Reports*, *19*, 2694–2706.

Koch, I., Schwarz, H., Beuchle, D., Goellner, B., Langegger, M., & Aberle, H. (2008). Drosophila ankyrin 2 is required for synaptic stability. *Neuron*, *58*, 210–222.

Koh, T. W., Verstreken, P., & Bellen, H. J. (2004). Dap160/intersectin acts as a stabilizing scaffold required for synaptic development and vesicle endocytosis. *Neuron*, *43*, 193–205.

Lee, H. X., Mendes, F. A., Plouhinec, J. L., & De Robertis, E. M. (2009). Enzymatic regulation of pattern: BMP4 binds CUB domains of Tolloids and inhibits proteinase activity. *Genes & Development*, *23*, 2551–2562.

Lee, G., & Schwarz, T. L. (2016). Filamin, a synaptic organizer in Drosophila, determines glutamate receptor composition and membrane growth. *eLife*, *5*, e19991.

Li, J., Ashley, J., Budnik, V., & Bhat, M. A. (2007). Crucial role of Drosophila neurexin in proper active zone apposition to postsynaptic densities, synaptic growth, and synaptic transmission. *Neuron*, *55*, 741–755.

Li, Y., Dharkar, P., Han, T. H., Serpe, M., Lee, C. H., & Mayer, M. L. (2016). Novel functional properties of Drosophila CNS glutamate receptors. *Neuron*, *92*, 1036–1048.

Li, H., Janssens, J., Waegeneer, M. D., Kolluru, S. S., Davie, K., Gardeux, V., et al. (2022). Fly cell Atlas: A single-nucleus transcriptomic atlas of the adult fruit fly. *Science*, *375*, eabk2432.

Li, W., Yao, A., Zhi, H., Kaur, K., Zhu, Y. C., Jia, M., et al. (2016). Angelman syndrome protein Ube3a regulates synaptic growth and endocytosis by inhibiting BMP signaling in Drosophila. *PLoS Genetics*, *12*, e1006062.

Liebl, F. L., & Featherstone, D. E. (2005). Genes involved in Drosophila glutamate receptor expression and localization. *BMC Neuroscience*, *6*, 44.

Liebl, F. L., & Featherstone, D. E. (2008). Identification and investigation of Drosophila postsynaptic density homologs. *Bioinformatics and Biology Insights*, *2*, 375–387.

Liebl, F. L., Werner, K. M., Sheng, Q., Karr, J. E., McCabe, B. D., & Featherstone, D. E. (2006). Genome-wide P-element screen for Drosophila synaptogenesis mutants. *Journal of Neurobiology*, *66*, 332–347.

Ljaschenko, D., Ehmann, N., & Kittel, R. J. (2013). Hebbian plasticity guides maturation of glutamate receptor fields in vivo. *Cell Reports*, *3*, 1407–1413.

Mace, P. D., Cutfield, J. F., & Cutfield, S. M. (2006). High resolution structures of the bone morphogenetic protein type II receptor in two crystal forms: Implications for ligand binding. *Biochemical and Biophysical Research Communications*, *351*, 831–838.

Marie, B., Sweeney, S. T., Poskanzer, K. E., Roos, J., Kelly, R. B., & Davis, G. W. (2004). Dap160/intersectin scaffolds the periactive zone to achieve high-fidelity endocytosis and normal synaptic growth. *Neuron*, *43*, 207–219.

Marmion, R. A., Jevtic, M., Springhorn, A., Pyrowolakis, G., & Yakoby, N. (2013). The Drosophila BMPRII, wishful thinking, is required for eggshell patterning. *Developmental Biology*, *375*, 45–53.
Marom, B., Heining, E., Knaus, P., & Henis, Y. I. (2011). Formation of stable homomeric and transient heteromeric bone morphogenetic protein (BMP) receptor complexes regulates Smad protein signaling. *The Journal of Biological Chemistry*, *286*, 19287–19296.
Marques, G., Haerry, T. E., Crotty, M. L., Xue, M., Zhang, B., & O'Connor, M. B. (2003). Retrograde Gbb signaling through the Bmp type 2 receptor wishful thinking regulates systemic FMRFa expression in Drosophila. *Development*, *130*, 5457–5470.
Marques, G., & Zhang, B. (2006). Retrograde signaling that regulates synaptic development and function at the Drosophila neuromuscular junction. *International Review of Neurobiology*, *75*, 267–285.
Marrus, S. B., & DiAntonio, A. (2004). Preferential localization of glutamate receptors opposite sites of high presynaptic release. *Current Biology*, *14*, 924–931.
Marrus, S. B., Portman, S. L., Allen, M. J., Moffat, K. G., & DiAntonio, A. (2004). Differential localization of glutamate receptor subunits at the Drosophila neuromuscular junction. *The Journal of Neuroscience*, *24*, 1406–1415.
Massague, J. (1990). The transforming growth factor-beta family. *Annual Review of Cell Biology*, *6*, 597–641.
Mayer, M. L. (2021). Structural biology of kainate receptors. *Neuropharmacology*, *190*, 108511.
McCabe, B. D., Hom, S., Aberle, H., Fetter, R. D., Marques, G., Haerry, T. E., et al. (2004). Highwire regulates presynaptic BMP signaling essential for synaptic growth. *Neuron*, *41*, 891–905.
McCabe, B. D., Marques, G., Haghighi, A. P., Fetter, R. D., Crotty, M. L., Haerry, T. E., et al. (2003). The BMP homolog Gbb provides a retrograde signal that regulates synaptic growth at the Drosophila neuromuscular junction. *Neuron*, *39*, 241–254.
McMahan, U. J. (1990). The agrin hypothesis. *Cold Spring Harbor Symposia on Quantitative Biology*, *55*, 407–418.
Menon, K. P., Andrews, S., Murthy, M., Gavis, E. R., & Zinn, K. (2009). The translational repressors Nanos and Pumilio have divergent effects on presynaptic terminal growth and postsynaptic glutamate receptor subunit composition. *The Journal of Neuroscience*, *29*, 5558–5572.
Menon, K. P., Carrillo, R. A., & Zinn, K. (2013). Development and plasticity of the Drosophila larval neuromuscular junction. *Wiley Interdisciplinary Reviews: Developmental Biology*, *2*, 647–670.
Menon, K. P., Sanyal, S., Habara, Y., Sanchez, R., Wharton, R. P., Ramaswami, M., et al. (2004). The translational repressor Pumilio regulates presynaptic morphology and controls postsynaptic accumulation of translation factor eIF-4E. *Neuron*, *44*, 663–676.
Merino, C., Penney, J., Gonzalez, M., Tsurudome, K., Moujahidine, M., O'Connor, M. B., et al. (2009). Nemo kinase interacts with Mad to coordinate synaptic growth at the Drosophila neuromuscular junction. *The Journal of Cell Biology*, *185*, 713–725.
Metwally, E., Zhao, G., Li, W., Wang, Q., & Zhang, Y. Q. (2019). Calcium-activated calpain specifically cleaves glutamate receptor IIA But Not IIB at the <em>Drosophila</em> neuromuscular junction. *The Journal of Neuroscience*, *39*, 2776–2791.
Meyerson, J. R., Chittori, S., Merk, A., Rao, P., Han, T. H., Serpe, M., et al. (2016). Structural basis of kainate subtype glutamate receptor desensitization. *Nature*, *537*, 567–571.
Meyerson, J. R., Kumar, J., Chittori, S., Rao, P., Pierson, J., Bartesaghi, A., et al. (2014). Structural mechanism of glutamate receptor activation and desensitization. *Nature*, *514*, 328–334.

Morimoto, T., Nobechi, M., Komatsu, A., Miyakawa, H., & Nose, A. (2009). Subunit-specific and homeostatic regulation of glutamate receptor localization by CaMKII in Drosophila neuromuscular junctions. *Neuroscience*, *165*, 1284–1292.

Mosca, T. J., Hong, W., Dani, V. S., Favaloro, V., & Luo, L. (2012). Trans-synaptic Teneurin signalling in neuromuscular synapse organization and target choice. *Nature*, *484*, 237–241.

Moustakas, A., & Heldin, C. H. (2005). Non-Smad TGF-beta signals. *Journal of Cell Science*, *118*, 3573–3584.

Newman, Z. L., Hoagland, A., Aghi, K., Worden, K., Levy, S. L., Son, J. H., et al. (2017). Input-specific plasticity and homeostasis at the Drosophila larval neuromuscular junction. *Neuron*, *93*(1388–1404), e1310.

Ng, D., Pitcher, G. M., Szilard, R. K., Sertie, A., Kanisek, M., Clapcote, S. J., et al. (2009). Neto1 is a novel CUB-domain NMDA receptor-interacting protein required for synaptic plasticity and learning. *PLoS Biology*, *7*, e41.

Nguyen, T. H., Han, T. H., Newfeld, S. J., & Serpe, M. (2020). Selective disruption of synaptic BMP signaling by a Smad mutation adjacent to the highly conserved H2 helix *Genetics*, *216*(1), 159–175.

Nicoll, R. A. (2017). A brief history of long-term potentiation. *Neuron*, *93*, 281–290.

O'Connor-Giles, K. M., Ho, L. L., & Ganetzky, B. (2008). Nervous wreck interacts with thickveins and the endocytic machinery to attenuate retrograde BMP signaling during synaptic growth. *Neuron*, *58*, 507–518.

Owald, D., Khorramshahi, O., Gupta, V. K., Banovic, D., Depner, H., Fouquet, W., et al. (2012). Cooperation of Syd-1 with Neurexin synchronizes pre- with postsynaptic assembly. *Nature Neuroscience*, *15*, 1219–1226.

Ozdamar, B., Bose, R., Barrios-Rodiles, M., Wang, H. R., Zhang, Y., & Wrana, J. L. (2005). Regulation of the polarity protein Par6 by TGFbeta receptors controls epithelial cell plasticity. *Science*, *307*, 1603–1609.

Park, H., & Poo, M. M. (2013). Neurotrophin regulation of neural circuit development and function. *Nature Reviews. Neuroscience*, *14*, 7–23.

Parnas, D., Haghighi, A. P., Fetter, R. D., Kim, S. W., & Goodman, C. S. (2001). Regulation of postsynaptic structure and protein localization by the Rho-type guanine nucleotide exchange factor dPix. *Neuron*, *32*, 415–424.

Petersen, S. A., Fetter, R. D., Noordermeer, J. N., Goodman, C. S., & DiAntonio, A. (1997). Genetic analysis of glutamate receptors in Drosophila reveals a retrograde signal regulating presynaptic transmitter release. *Neuron*, *19*, 1237–1248.

Petzoldt, A. G., Lee, Y. H., Khorramshahi, O., Reynolds, E., Plested, A. J., Herzel, H., et al. (2014). Gating characteristics control glutamate receptor distribution and trafficking in vivo. *Current Biology*, *24*, 2059–2065.

Piccioli, Z. D., & Littleton, J. T. (2014). Retrograde BMP signaling modulates rapid activity-dependent synaptic growth via presynaptic LIM kinase regulation of cofilin. *The Journal of Neuroscience*, *34*, 4371–4381.

Qin, G., Schwarz, T., Kittel, R. J., Schmid, A., Rasse, T. M., Kappei, D., et al. (2005). Four different subunits are essential for expressing the synaptic glutamate receptor at neuromuscular junctions of Drosophila. *The Journal of Neuroscience* *25*, 3209–3218.

Ramesh, N., Escher, M. J. F., Mampell, M. M., Bohme, M. A., Gotz, T. W. B., Goel, P., et al. (2021). Antagonistic interactions between two Neuroligins coordinate pre- and postsynaptic assembly. *Current Biology*, *31*(1711–1725), e1715.

Ramos, C. I., Igiesuorobo, O., Wang, Q., & Serpe, M. (2015). Neto-mediated intracellular interactions shape postsynaptic composition at the Drosophila neuromuscular junction. *PLoS Genetics*, *11*, e1005191.

Rasse, T. M., Fouquet, W., Schmid, A., Kittel, R. J., Mertel, S., Sigrist, C. B., et al. (2005). Glutamate receptor dynamics organizing synapse formation in vivo. *Nature Neuroscience*, *8*, 898–905.

Rawson, J. M., Lee, M., Kennedy, E. L., & Selleck, S. B. (2003). Drosophila neuromuscular synapse assembly and function require the TGF-beta type I receptor saxophone and the transcription factor Mad. *Journal of Neurobiology*, *55*, 134–150.

Reichardt, L. F. (2006). Neurotrophin-regulated signalling pathways. *Philosophical Transactions of the Royal Society of London Series B, Biological Sciences*, *361*, 1545–1564.

Reist, N. E., Werle, M. J., & McMahan, U. J. (1992). Agrin released by motor neurons induces the aggregation of acetylcholine receptors at neuromuscular junctions. *Neuron*, *8*, 865–868.

Roche, J. P., Packard, M. C., Moeckel-Cole, S., & Budnik, V. (2002). Regulation of synaptic plasticity and synaptic vesicle dynamics by the PDZ protein Scribble. *The Journal of Neuroscience*, *22*, 6471–6479.

Rodal, A. A., Blunk, A. D., Akbergenova, Y., Jorquera, R. A., Buhl, L. K., & Littleton, J. T. (2011). A presynaptic endosomal trafficking pathway controls synaptic growth signaling. *The Journal of Cell Biology*, *193*, 201–217.

Ruiz-Canada, C., Ashley, J., Moeckel-Cole, S., Drier, E., Yin, J., & Budnik, V. (2004). New synaptic bouton formation is disrupted by misregulation of microtubule stability in aPKC mutants. *Neuron*, *42*, 567–580.

Schmid, A., Hallermann, S., Kittel, R. J., Khorramshahi, O., Frolich, A. M., Quentin, C., et al. (2008). Activity-dependent site-specific changes of glutamate receptor composition in vivo. *Nature Neuroscience*, *11*, 659–666.

Schmid, A., Qin, G., Wichmann, C., Kittel, R. J., Mertel, S., Fouquet, W., et al. (2006). Non-NMDA-type glutamate receptors are essential for maturation but not for initial assembly of synapses at Drosophila neuromuscular junctions. *The Journal of Neuroscience*, *26*, 11267–11277.

Schmierer, B., & Hill, C. S. (2007). TGFbeta-SMAD signal transduction: Molecular specificity and functional flexibility. *Nature Reviews Molecular Cell Biology*, *8*, 970–982.

Shi, W., Chen, Y., Gan, G., Wang, D., Ren, J., Wang, Q., et al. (2013). Brain tumor regulates neuromuscular synapse growth and endocytosis in Drosophila by suppressing mad expression. *The Journal of Neuroscience*, *33*, 12352–12363.

Sigrist, S. J., Reiff, D. F., Thiel, P. R., Steinert, J. R., & Schuster, C. M. (2003). Experience-dependent strengthening of Drosophila neuromuscular junctions. *The Journal of Neuroscience*, *23*, 6546–6556.

Smith, R. B., Machamer, J. B., Kim, N. C., Hays, T. S., & Marques, G. (2012). Relay of retrograde synaptogenic signals through axonal transport of BMP receptors. *Journal of Cell Science*, *125*, 3752–3764.

Sobolevsky, A. I., Rosconi, M. P., & Gouaux, E. (2009). X-ray structure, symmetry and mechanism of an AMPA-subtype glutamate receptor. *Nature*, *462*, 745–756.

Spinner, M. A., Pinter, K., Drerup, C. M., & Herman, T. G. (2020). A conserved role for vezatin proteins in cargo-specific regulation of retrograde axonal transport. *Genetics*, *216*, 431–445.

Stephan, R., Goellner, B., Moreno, E., Frank, C. A., Hugenschmidt, T., Genoud, C., et al. (2015). Hierarchical microtubule organization controls axon caliber and transport and determines synaptic structure and stability. *Developmental Cell*, *33*, 5–21.

Stinchfield, M. J., Takaesu, N. T., Quijano, J. C., Castillo, A. M., Tiusanen, N., Shimmi, O., et al. (2012). Fat facets deubiquitylation of Medea/Smad4 modulates interpretation of a Dpp morphogen gradient. *Development*, *139*, 2721–2729.

Sudhof, T. C. (2012). The presynaptic active zone. *Neuron*, *75*, 11–25.

Sulkowski, M. J., Han, T. H., Ott, C., Wang, Q., Verheyen, E. M., Lippincott-Schwartz, J., et al. (2016). A novel, noncanonical BMP pathway modulates synapse maturation at the drosophila neuromuscular junction. *PLoS Genetics*, *12*, e1005810.

Sulkowski, M., Kim, Y. J., & Serpe, M. (2014). Postsynaptic glutamate receptors regulate local BMP signaling at the Drosophila neuromuscular junction. *Development*, *141*, 436–447.

Sun, K., Westholm, J. O., Tsurudome, K., Hagen, J. W., Lu, Y., Kohwi, M., et al. (2012). Neurophysiological defects and neuronal gene deregulation in Drosophila mir-124 mutants. *PLoS Genetics*, *8*, e1002515.

Sweeney, S. T., & Davis, G. W. (2002). Unrestricted synaptic growth in spinster-a late endosomal protein implicated in TGF-beta-mediated synaptic growth regulation. *Neuron*, *36*, 403–416.

Tang, A. H., Chen, H., Li, T. P., Metzbower, S. R., MacGillavry, H. D., & Blanpied, T. A. (2016). A trans-synaptic nanocolumn aligns neurotransmitter release to receptors. *Nature*, *536*, 210–214.

Thoenen, H., & Barde, Y. A. (1980). Physiology of nerve growth factor. *Physiological Reviews*, *60*, 1284–1335.

Thomas, U., & Sigrist, S. J. (2012). Glutamate receptors in synaptic assembly and plasticity: Case studies on fly NMJs. *Advances in Experimental Medicine and Biology*, *970*, 3–28.

Tomita, S. (2010). Regulation of ionotropic glutamate receptors by their auxiliary subunits. *Physiology (Bethesda)*, *25*, 41–49.

Tomita, S., & Castillo, P. E. (2012). Neto1 and Neto2: Auxiliary subunits that determine key properties of native kainate receptors. *The Journal of Physiology*, *590*, 2217–2223.

Tsang, H. T., Edwards, T. L., Wang, X., Connell, J. W., Davies, R. J., Durrington, H. J., et al. (2009). The hereditary spastic paraplegia proteins NIPA1, spastin and spartin are inhibitors of mammalian BMP signalling. *Human Molecular Genetics*, *18*, 3805–3821.

Urist, M. R. (1965). Bone: Formation by autoinduction. *Science*, *150*, 893–899.

Van Vactor, D., & Sigrist, S. J. (2017). Presynaptic morphogenesis, active zone organization and structural plasticity in Drosophila. *Current Opinion in Neurobiology*, *43*, 119–129.

Vanlandingham, P. A., Fore, T. R., Chastain, L. R., Royer, S. M., Bao, H., Reist, N. E., et al. (2013). Epsin 1 promotes synaptic growth by enhancing BMP signal levels in motoneuron nuclei. *PLoS One*, *8*, e65997.

Wang, Q., Han, T. H., Nguyen, P., Jarnik, M., & Serpe, M. (2018). Tenectin recruits integrin to stabilize bouton architecture and regulate vesicle release at the Drosophila neuromuscular junction. *eLife*, *7*, e35518.

Wang, Y., Lobb-Rabe, M., Ashley, J., Anand, V., & Carrillo, R. A. (2021). Structural and functional synaptic plasticity induced by convergent synapse loss in the Drosophila neuromuscular circuit. *The Journal of Neuroscience*, *41*, 1401–1417.

Wang, R., Mellem, J. E., Jensen, M., Brockie, P. J., Walker, C. S., Hoerndli, F. J., et al. (2012). The SOL-2/Neto auxiliary protein modulates the function of AMPA-subtype ionotropic glutamate receptors. *Neuron*, *75*, 838–850.

Xing, G., Gan, G., Chen, D., Sun, M., Yi, J., Lv, H., et al. (2014). Drosophila neuroligin3 regulates neuromuscular junction development and synaptic differentiation. *The Journal of Biological Chemistry*, *289*, 31867–31877.

Zeng, Y. A., Rahnama, M., Wang, S., Sosu-Sedzorme, W., & Verheyen, E. M. (2007). Drosophila Nemo antagonizes BMP signaling by phosphorylation of Mad and inhibition of its nuclear accumulation. *Development*, *134*, 2061–2071.

Zhai, R. G., & Bellen, H. J. (2004). The architecture of the active zone in the presynaptic nerve terminal. *Physiology (Bethesda)*, *19*, 262–270.

Zhang, Y. E. (2009). Non-Smad pathways in TGF-beta signaling. *Cell Research*, *19*, 128–139.

Zhang, Y. Q., Bailey, A. M., Matthies, H. J., Renden, R. B., Smith, M. A., Speese, S. D., et al. (2001). Drosophila fragile X-related gene regulates the MAP1B homolog Futsch to control synaptic structure and function. *Cell*, *107*, 591–603.

Zhang, Y., Feng, X., We, R., & Derynck, R. (1996). Receptor-associated Mad homologues synergize as effectors of the TGF-beta response. *Nature*, *383*, 168–172.

Zhang, B., Luo, S., Wang, Q., Suzuki, T., Xiong, W. C., & Mei, L. (2008). LRP4 serves as a coreceptor of agrin. *Neuron*, *60*, 285–297.

Zhang, W., St-Gelais, F., Grabner, C. P., Trinidad, J. C., Sumioka, A., Morimoto-Tomita, M., et al. (2009). A transmembrane accessory subunit that modulates kainate-type glutamate receptors. *Neuron*, *61*, 385–396.

Zhao, K., Hong, H., Zhao, L., Huang, S., Gao, Y., Metwally, E., et al. (2020). Postsynaptic cAMP signalling regulates the antagonistic balance of Drosophila glutamate receptor subtypes. *Development*, *147*(24), dev191874.

Zhao, G., Wu, Y., Du, L., Li, W., Xiong, Y., Yao, A., et al. (2015). Drosophila S6 kinase like inhibits neuromuscular junction growth by downregulating the BMP receptor thickveins. *PLoS Genetics*, *11*, e1004984.

Zhu, H., Kavsak, P., Abdollah, S., Wrana, J. L., & Thomsen, G. H. (1999). A SMAD ubiquitin ligase targets the BMP pathway and affects embryonic pattern formation. *Nature*, *400*, 687–693.

CHAPTER SEVEN

# Wnt-Frizzled planar cell polarity signaling in the regulation of cell motility

**Yildiz Koca, Giovanna M. Collu, and Marek Mlodzik***
Department of Cell, Developmental, & Regenerative Biology, Graduate School of Biomedical Sciences, Icahn School of Medicine at Mount Sinai, New York, NY, United States
*Corresponding author: e-mail address: marek.mlodzik@mssm.edu

## Contents



## Abstract

The molecular complexes underlying planar cell polarity (PCP) were first identified in *Drosophila* through analysis of mutant phenotypes in the adult cuticle and the orientation of associated polarized protrusions such as wing hairs and sensory bristles. The same molecules are conserved in vertebrates and are required for the localization of polarized protrusions such as primary or sensory cilia and the orientation of hair follicles. Not only is PCP signaling required to align cellular structures across a tissue, it is also required to coordinate movement during embryonic development and adult homeostasis. PCP signaling allows cells to interpret positional cues within a tissue to move in the appropriate direction and to coordinate this movement with their neighbors. In this review we outline the molecular basis of the core Wnt-Frizzled/PCP pathway, and describe how this signaling orchestrates collective motility in *Drosophila* and vertebrates. Here we cover the paradigms of ommatidial rotation and border cell migration in *Drosophila*, and convergent

*Current Topics in Developmental Biology*, Volume 150
ISSN 0070-2153
https://doi.org/10.1016/bs.ctdb.2022.03.006

extension in vertebrates. The downstream cell biological processes that underlie polarized motility include cytoskeletal reorganization, and adherens junctional and extracellular matrix remodeling. We discuss the contributions of these processes in the respective cell motility contexts. Finally, we address examples of individual cell motility guided by PCP factors during nervous system development and in cancer disease contexts.

## 1. Overview of planar cell polarity (PCP) signaling

Cells in tissues often require directional features for proper cellular and organ function. For example, epithelial cells are uniformly polarized along the apical-basal axis, enabling their vectorial functions like protein secretion into a lumen. Most epithelia are further polarized within the planar axis, which is referred to as planar cell polarity (PCP). PCP provides cells and tissues with positional information allowing them to generate polarized structures oriented with respect to tissue axes, to embed specialized cells (e.g., sensory cells) with a specific orientation or to move in a directed fashion (Adler, 2002; Butler & Wallingford, 2017; Davey & Moens, 2017; Devenport, 2016; Goodrich & Strutt, 2011; Humphries & Mlodzik, 2018; Peng & Axelrod, 2012; Seifert & Mlodzik, 2007).

PCP has been best studied and characterized in *Drosophila*, where it is evident in all adult cuticular structures and the compound eye. In wings, for example, cellular hairs are oriented in the proximo-distal axis and this uniform pattern is disrupted in PCP mutants (Fig. 1A). Similarly, in eyes, the regular arrangement of ommatidia with respect to anterior-posterior (AP) and dorsal-ventral (DV) axes is altered in PCP mutants (Fig. 2) (Adler, 2002; Goodrich & Strutt, 2011; Humphries & Mlodzik, 2018; Peng & Axelrod, 2012; Seifert & Mlodzik, 2007). Based on these phenotypes, a core group of evolutionarily conserved PCP genes have been identified, which is referred to as "the core Frizzled (Fz)/PCP factors." The core Fz/PCP module includes the transmembrane proteins Fz, Flamingo (Fmi, a.k.a. Starry night/Stan), and Van Gogh (Vang; a.k.a. Strabismus/Stbm) and the cytoplasmic factors Dishevelled (Dsh), Prickle (Pk), and Diego (Dgo) (Adler, 2012; Goodrich & Strutt, 2011; Humphries & Mlodzik, 2018; Peng & Axelrod, 2012; Seifert & Mlodzik, 2007; Wu & Mlodzik, 2009). The core Fz/PCP factors interact with each other and localize asymmetrically to generate cellular polarization. This polarity information is then transmitted to downstream effectors to elicit tissue-specific responses (Figs. 1 and 2) (Adler, 2002; Goodrich & Strutt, 2011; Harrison, Shao, Strutt, & Strutt, 2020; Humphries & Mlodzik, 2018; Peng & Axelrod, 2012; Seifert & Mlodzik, 2007; Wu & Mlodzik, 2009).

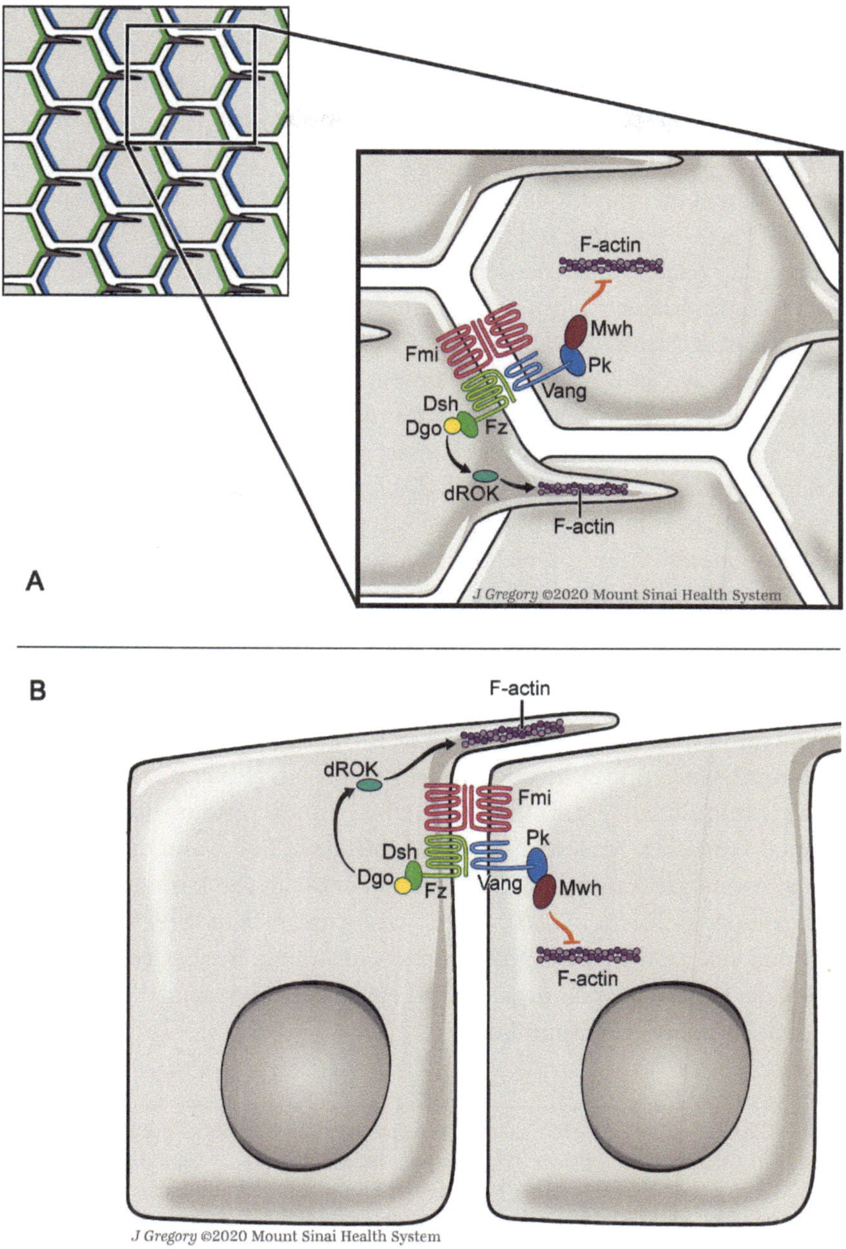


**Fig. 1** Planar cell polarity in the *Drosophila* wing epithelium. Schematic X-Y (top) view in (A) and lateral (Z-section) view in (B) of core PCP complexes is shown. Note that the core PCP complexes are located at/near the adherens junctions in the subapical domain of the cells. In wing epithelia, one of the simplest PCP paradigms, asymmetric distribution *(Continued)*

A second (independent) set of PCP factors identified in *Drosophila* is centered around the proto-cadherins Fat and Dachsous (Ds) which heterophilically interact with each other across cell membranes and have their own set of effectors to generate polarization (Lawrence & Casal, 2018; Matis & Axelrod, 2013; Thomas & Strutt, 2012). Although recent studies suggest that this module also has a role in vertebrate PCP, the mechanism of Fat/Ds signaling and their conservation, and whether they directly or indirectly relate to the core Fz/PCP factors or act upstream or in parallel to them remain largely elusive (Lawrence & Casal, 2018; Matis & Axelrod, 2013; Strutt & Strutt, 2021).

Importantly, and as further outlined below, the core Wnt-Fz/PCP factors regulate a set of cellular responses and read-outs that, like the molecular PCP signaling cassette, appear conserved across animal species (Butler & Wallingford, 2017; Davey & Moens, 2017; Devenport, 2016; Humphries & Mlodzik, 2018; Klein & Mlodzik, 2005; Seifert & Mlodzik, 2007). The two most obvious and common read-outs include (i) polarization of cytoskeletal elements and (ii) regulation of directed cell motility (see also review by Davey & Moens, 2017). While these two read-outs are in many ways linked, they can be separated in certain contexts, as not all PCP associated read-outs cause cell movement or rearrangement of cells relative to each other. While the latter read-out is the focus of this review (see below), the former—affecting cytoskeletal organization—is also linked to the asymmetric, subcellular localization of cilia and centrioles (reviewed in Carvajal-Gonzalez, Mulero-Navarro, & Mlodzik, 2016) and the anchoring of centrosomes required for the orientation of the mitotic spindle (reviewed in Segalen & Bellaiche, 2009). Here, we focus on Wnt/Fz-PCP signaling regulated cell motility in its many aspects including, besides cell migration during development and disease *per se*, cellular rearrangements within a field of cells and neuronal pathfinding.

**Fig. 1—Cont'd** of the core PCP factors starts to emerge at late larval stages and is most obvious in pupal stages (shown in upper left panel schematic, Fz-Dsh-Dgo (blue) and Vang-Pk complexes (green), both stabilized by interactions with Fmi, asymmetrically localize in distal and proximal apical membranes, respectively (Adler, 2012; Goodrich & Strutt, 2011; Humphries & Mlodzik, 2018; Jenny, 2010; Peng & Axelrod, 2012; Seifert & Mlodzik, 2007; Wu & Mlodzik, 2009). The PCP complexes then trigger the polarization of the cytoskeleton through downstream regulators to ensure single spike actin hair formation via the Fz/Dsh/Dgo complex at the distal vertex of each cell (Gault, Olguin, Weber, & Mlodzik, 2012; Strutt & Warrington, 2008; Winter et al., 2001; Yan et al., 2008). Proximal is left. See main text for further details.

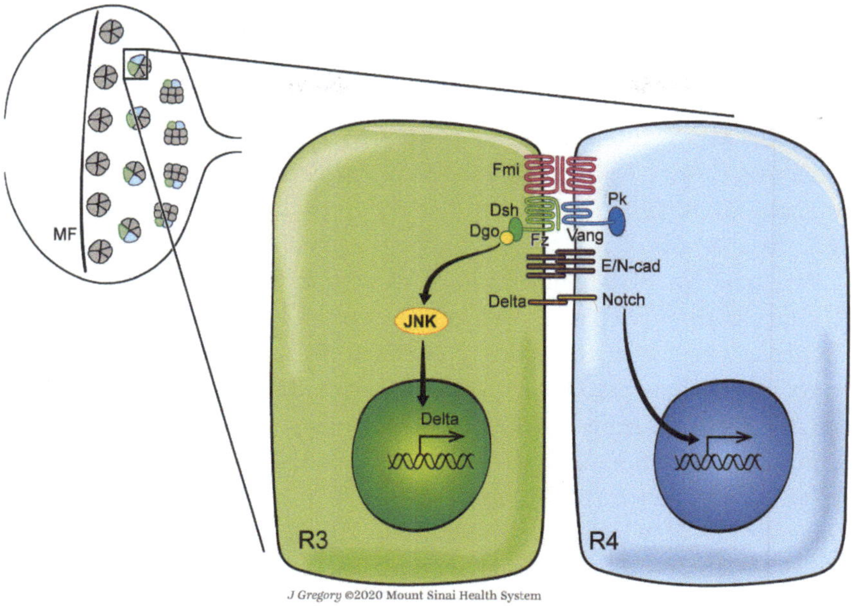


**Fig. 2** Planar cell polarity signaling and ommatidial rotation in the *Drosophila* eye. The *Drosophila* eye develops from an epithelial imaginal disc during larval stages, which is initially composed of identical pluripotent precursor cells ahead of the morphogenetic furrow (MF) (upper left panel). As the MF sweeps across the disc from posterior to anterior, preclusters of differentiating cells start to emerge which will then develop into mature ommatidia (Cagan & Ready, 1989; Roignant & Treisman, 2009; Tomlinson & Ready, 1987). During differentiation and maturation, these clusters rotate 90° in opposite directions in the dorsal and ventral halves of the eye to establish the final mirror-symmetric pattern of ommatidia across the D/V midline ( Jenny, 2010; Mlodzik, 1999). Posterior to the MF, Fz/PCP signaling mainly takes place between the photoreceptors R3 (green) and R4 (blue), inducing R4 fate via Notch signaling activation (Cooper & Bray, 1999; Das, Jenny, Klein, Eaton, & Mlodzik, 2004; Das, Reynolds-Kenneally, & Mlodzik, 2002; del Alamo & Mlodzik, 2006; Fanto & Mlodzik, 1999; Tomlinson & Struhl, 1999; Wolff & Rubin, 1998; Wu, Klein, & Mlodzik, 2004). The core PCP factors become asymmetrically localized in the subapical membrane region of the R3/R4 pair cell boundary, leading to differential downstream nuclear and/or cellular responses in each cell mediated by signaling (JNK, Delta, Notch) and simultaneously adhesion (E- and N-cadherin) molecules which are thought to mediate the cell motility process of ommatidial rotation (Cooper & Bray, 1999; Das et al., 2002, 2004; del Alamo & Mlodzik, 2006; Fanto & Mlodzik, 1999; Mirkovic et al., 2011; Mirkovic & Mlodzik, 2006; Tomlinson & Struhl, 1999; Wolff & Rubin, 1998; Wu et al., 2004). Dorsal is up, anterior is left. See main text for details.

## 1.1 PCP in *Drosophila*

Studies in *Drosophila*, mainly in wings and eyes but also in other tissues like the abdomen, have provided fundamental mechanistic insights into how PCP core factors interact and how PCP is established in individual cells and across tissues (Adler, 2012; Goodrich & Strutt, 2011; Humphries & Mlodzik, 2018; Jenny, 2010; Peng & Axelrod, 2012; Seifert & Mlodzik, 2007; Wu & Mlodzik, 2009). In wings, arguably the best understood PCP system, Fz/PCP signaling leads to asymmetric distribution of core PCP factors from late larval/early pupal stages. Fz-Dsh-Dgo and Vang-Pk complexes become asymmetrically enriched in distal and proximal membranes respectively, as they antagonize each other within the same cell and stabilize each other across cell membranes, to enforce and maintain an initial polarization bias set-up by localized Wnt-expression (Fig. 1) (Adler, 2012; Seifert & Mlodzik, 2007; Wu & Mlodzik, 2009; Wu, Roman, Carvajal-Gonzalez, & Mlodzik, 2013). The well-characterized read-out to this polarized signaling complexes in wing cells is the formation of a single distally pointing actin spike, called trichome, downstream of the Fz-Dsh complex (Fig. 1). In this context, multiple wing hair (Mwh) and Rho-associated kinase (dROK) are two effectors that are employed downstream of the Vang-Pk and Fz-Dsh complexes, respectively, ensuring that a single distal actin hair is generated. Mwh, an anti-Formin, is recruited to the proximal edge of the cells to locally inhibit actin hair formation (Strutt & Warrington, 2008; Yan et al., 2008), whereas Rho and dROK and their associated trafficking acts in the distal vertex to restrict hair formation to a single actin spike (Fig. 1) (Gault et al., 2012; Winter et al., 2001). Thus, polarity is established in the wing through local cytoskeletal rearrangements mediated by the Fz/PCP cassette.

In the *Drosophila* eye, Fz/PCP signaling first instructs R3/R4 photoreceptor cell fate specification, which is followed by directed rotation of ommatidial clusters (Fig. 2) (Jenny, 2010; Mlodzik, 1999). An initial bias in Fz/PCP activity between the R3/R4 precursors specifies the cell that has higher Fz-Dsh/PCP activity as R3 and induces its neighbor to adopt the R4 fate. This specification depends on JNK-mediated transcriptional responses created by Fz-Dsh/PCP signaling that culminate in differential upregulation of *Delta* (*Dl*) and *neuralized* in R3 (Cooper & Bray, 1999; del Alamo & Mlodzik, 2006; Fanto & Mlodzik, 1999; Tomlinson &

Struhl, 1999) which act together to activate Notch signaling in R4 (within the R3/R4 pair), and Notch-signaling specifying the R4 fate (reviewed in Jenny, 2010; Mlodzik, 1999). During this process, the core PCP factors become asymmetrically localized in R3/R4 cells and thus also guide the subsequent directed and coordinated movement of ommatidial clusters via cytoskeletal and junctional rearrangements (Fig. 2) (Jenny, 2010; Seifert & Mlodzik, 2007). Hence, core Fz/PCP signaling triggers distinct cellular responses in wings and eyes by employing tissue-specific downstream effectors, including the regulation of a directed cell movement process in the eye (Jenny, 2010; Klein & Mlodzik, 2005, see dedicated chapter on ommatidial rotation below).

## 1.2 PCP in vertebrates

Data from many vertebrate models reveal that the molecular mechanism(s) and function of core Wnt-Fz/PCP signaling is conserved across species, and that the core PCP factor cassette is involved in polarity establishment in many (if not all) vertebrate tissues (Butler & Wallingford, 2017; Davey & Moens, 2017; Devenport, 2016; Humphries & Mlodzik, 2018; Seifert & Mlodzik, 2007). Similar to PCP establishment in the *Drosophila* wing epithelium (or the *Drosophila* cuticle in general), the mammalian epidermis is planar polarized with hair follicles aligned along the A/P axis and this pattern is disrupted in *Fzd6* mutant mice (Guo, Hawkins, & Nathans, 2004). In this context, Vangl2, Celsr1, and Fzd6 are asymmetrically localized along the A/P axis during hair follicle development (Devenport & Fuchs, 2008), mirroring their polarized localization features from *Drosophila*. In the mammalian inner ear, PCP signaling orients and aligns the sensory hair cells to uniformly polarize the bundles of actin-based stereocilia, a pattern that is randomized in Fzd3/6, Vangl2 and Celsr1 mutants (Curtin et al., 2003; Montcouquiol et al., 2006; Wang et al., 2006). Fzd3 and Vangl2 are enriched at opposite cell membranes, reiterating the asymmetric distribution of the core PCP factors (Montcouquiol et al., 2006). There are many other examples of planar polarized tissues in vertebrates and these prominently include ciliated cells, ranging from multi-ciliated cells of the airway epithelia and kidney tubules (Brzoska et al., 2016; Vladar, Bayly, Sangoram, Scott, & Axelrod, 2012) to cells displaying a polarized localization of the primary cilium, which is critical for their function (Wallingford, 2010).

In addition to polarizing tissues and embedded organ features, PCP has been implicated in multiple morphogenetic processes that require directed cell motility in vertebrates, and associated asymmetric localization of PCP proteins has also been documented in motile cells despite technical challenges (reviewed in Butler & Wallingford, 2017; Davey & Moens, 2017; Devenport, 2016). For example, core Fz/PCP signaling coordinates convergent extension (CE) movements that take place during vertebrate gastrulation and neurulation. Accordingly, PCP mutant embryos show defects in elongation of the body axis and neural tube closure due to the failure and/or misregulation of CE cellular movements and intercalation (Butler & Wallingford, 2017; Davey & Moens, 2017; Devenport, 2016). Similarly, human patient derived mutations in *Vangl* genes associated with neural tube closure defects have been shown to affect core PCP signaling (Humphries, Narang, & Mlodzik, 2020). Overall, in the past 15 years, significant progress has been made in dissecting cellular asymmetries of core PCP factors in the context of CE processes during gastrulation and neurulation and how these might coordinate cellular convergence and extension movements in general (Ciruna, Jenny, Lee, Mlodzik, & Schier, 2006; Nishimura, Honda, & Takeichi, 2012; Roszko, Sepich, Jessen, Chandrasekhar, & Solnica-Krezel, 2015; Williams, Yen, Lu, & Sutherland, 2014; Yin, Kiskowski, Pouille, Farge, & Solnica-Krezel, 2008; reviewed in Butler & Wallingford, 2017; Davey & Moens, 2017; Devenport, 2016). Other PCP-mediated collective motility processes in vertebrates include the migration of facial branchiomotor neurons and growth cone guidance, where perturbing PCP signaling activity causes abnormalities in coordinated cell movements (reviewed in Butler & Wallingford, 2017; Davey & Moens, 2017; Devenport, 2016). Furthermore, mutations in core Fz-PCP factors and associated signaling has been implicated in cancer metastasis (e.g., reviewed in Humphries & Mlodzik, 2018), highlighting the importance of understanding how PCP regulates cell motility in development for addressing related mechanisms in disease contexts. Several of these specific processes will be discussed in this review.

## 2. PCP and cell motility in *Drosophila*

### 2.1 Ommatidial rotation (OR)

During *Drosophila* eye development, Fz/PCP signaling regulates, besides general ommatidial patterning via cell fate specification (Jenny, 2010; Mlodzik, 1999), the directed movement of the ommatidial preclusters.

The eye consists of ~800 regularly arranged ommatidia, with each consisting of 8 photoreceptor neurons arranged into an invariant trapezoidal pattern and 12 accessory (cone, pigment, and bristle) cells. During larval stages, the eye develops from an epithelial imaginal disc, which is initially composed of identical pluripotent precursor cells (Fig. 2). As a wave of cell proliferation and differentiation (called morphogenetic furrow, MF) travels across the disc from posterior to anterior, it leaves in its wake preclusters of differentiating cells that will mature into ommatidia (Cagan & Ready, 1989; Roignant & Treisman, 2009; Tomlinson & Ready, 1987). During differentiation and maturation, these clusters rotate 90° in opposite directions in the dorsal and ventral halves of the eye to establish the mirror-symmetric pattern of adult ommatidia across the D/V midline (Fig. 2) (Jenny, 2010; Mlodzik, 1999). As such, posterior to the MF, Fz/PCP signaling instructs not only R3/R4 cell fate specification, but also coordinates the direction and degree of rotation. The rotation of ommatidial clusters has served as a key model to study PCP regulated cell motility.

Loss-of-function (LOF) mutations in core PCP genes randomize R3/R4 specification and thus chiral ommatidial arrangements or even eliminate chirality, giving rise to R3-R3 or R4-R4 symmetrical ommatidia (Jenny, 2010; Mlodzik, 1999). At the same time, they also cause a randomization of both direction and degree of rotation, suggesting that PCP is critical for both directing and executing rotation and coordinating the process with cell fate induction (Das et al., 2002; Jenny, Darken, Wilson, & Mlodzik, 2003; Rawls & Wolff, 2003; Wolff & Rubin, 1998; Wu et al., 2004). Similarly, interference with PCP signaling and establishment via overexpression of the core components causes phenotypes similar to LOF defects, affecting both cell fate specification and rotation of ommatidia (Das et al., 2002; Fanto & Mlodzik, 1999; Jenny, Reynolds-Kenneally, Das, Burnett, & Mlodzik, 2005; Rawls & Wolff, 2003; Wolff, Guinto, & Rawls, 2007). However, ommatidial rotation (OR) can be genetically uncoupled from cell fate specification, as several genes have been reported to affect OR without interfering with R3/R4 cell fate choice. Genes that affect rotation in a specific manner include *argos* (*aos,* an inhibitory ligand for EGF receptor) (Choi & Benzer, 1994; Gaengel & Mlodzik, 2003; Strutt & Strutt, 2003), *nemo* (*nmo,* a serine-threonine kinase of the MAPK superfamily) (Choi & Benzer, 1994; Fiehler & Wolff, 2008; Mirkovic et al., 2011; Mirkovic & Mlodzik, 2006), *shotgun* (*shg*, DE-cadherin, a cell adhesion molecule) (Mirkovic & Mlodzik, 2006), *zipper* (*zip*, a myosin/motor protein) (Fiehler & Wolff, 2007), *scabrous* (sca, a secreted fibrinogen-related protein)

(Chou & Chien, 2002), and integrins and extracellular matrix (ECM) components (Thuveson et al., 2019). Involvement of these genes in OR suggests that receptor tyrosine kinase (RTK) signaling, cell-cell and cell-matrix interactions, and cytoskeletal dynamics are critical to regulate this cell motility process. These OR-"specific" factors act downstream (or in parallel) to core PCP signaling and translate oriented core PCP factor input to the directional movement of the cell clusters.

Recently, mechanistic insight into how core PCP proteins may regulate OR and cell motility in general is starting to emerge. Nmo, the founding member of the Nlk superfamily of MAP kinases, was the first rotation-specific gene to be identified (Choi & Benzer, 1994). *nmo* null mutants are characterized by severe underrotation of otherwise largely normal ommatidial preclusters (Fiehler & Wolff, 2008; Mirkovic et al., 2011). Although *nmo* is required in all photoreceptors and interommatidial cells for OR to proceed normally, it is enriched at the R4 side of R3/R4 cell border through a physical interaction with the Vang-Pk complex. Nmo phosphorylates β-catenin/E-cadherin complexes whereby it may regulate the dynamics (or adhesive properties) of the E-cad complexes. As *nmo* genetically interacts with *Vang* and *pk*, and β-cat (*arm*)/E-cad (*shg*) in the OR context, its involvement as an effector linking core PCP factors and cell adhesion molecules is functionally supported (Mirkovic et al., 2011). Cadherin-dependent cell adhesion has been implicated in OR and mutations in both *Drosophila* E-cad and N-cad genes have been shown to cause misrotation (Mirkovic & Mlodzik, 2006). Based on their mutant and overexpression phenotypes, they have opposing effects on rotation: DE-cad promotes rotation, whereas DN-cad restricts it. Strikingly, the localization patterns of two cadherins in rotating preclusters are complementary to each other: DE-cad becomes enriched at the membranes between all precluster cells except the R3/R4 border, whereas DN-cad is upregulated in the R3/R4 pair and enriched at their border. E-cad is also detected and required at all adherens junctions in eye imaginal discs and notably at the borders between precluster cells and inter-ommatidial cells (Mirkovic & Mlodzik, 2006). Classically, cadherins may serve a role to hold the cells of the precluster together; nevertheless, normal organization of ommatidia with (under) rotation defects in hypomorphic *shg* mutants indicates that cadherin-based junctional remodeling is critically required for the OR process (Mirkovic & Mlodzik, 2006). In other words, cell adhesion dynamics must be tightly controlled in ommatidial clusters to achieve the correct rate of rotation, and local asymmetries generated by core PCP molecules, via

Nmo or other regulators, may translate into polarized remodeling of adherens junctions, both providing a directional input—with Nmo being enriched in R4 via core PCP factors (Mirkovic et al., 2011)—and also coordinate the OR movement process through its general function in all eye disc cells (Fiehler & Wolff, 2008; Mirkovic et al., 2011; Munoz-Soriano, Ruiz, Perez-Alonso, Mlodzik, & Paricio, 2013).

Cytoskeletal dynamics must also be regulated for proper cell motility during OR. Rotation-specific phenotypes in mutants of *dRhoA* and *dROK* (Strutt, Weber, & Mlodzik, 1997; Winter et al., 2001), which are critical regulators of actin dynamics, support the idea that cytoskeletal rearrangements contribute to the OR process. RhoA genetically interacts with *fz* and *dsh* suggesting that core PCP signaling provides input into cytoskeletal reorganization (Strutt & Strutt, 2003). Similarly, perturbations in the activity of *zip*, encoding *Drosophila* Myosin II heavy chain, also cause ommatidial misrotation (Fiehler & Wolff, 2007). Zip acts downstream of RhoA and dROK in many contexts, and thus a similar mode of signaling may link core PCP factors to Zip regulation during OR (Verdier, Guang Chao, & Settleman, 2006). Importantly, Zip localizes to cell junctions between rotating (ommatidial precluster) and non-rotating (interommatidial) cells, a pattern that is maintained as new cells join the rotating (pre)cluster (Fiehler & Wolff, 2007). In this context, Zip may regulate rotational forces and/or it may affect the remodeling of adherens junctions at cell boundaries, enabling the rotating cluster to slide along the non-rotating cells. These data collectively suggest that Zip (Myosin II) may be involved in spatially and temporally distinct domains of rotation, integrating parallel upstream inputs into cytoskeletal reorganization.

Core Fz/PCP signaling cooperates with other signaling pathways during OR. A mutant allele of *aos*, encoding an inhibitory ligand for *Drosophila* EGF-Receptor (EGFR), was among the first OR specific mutations identified (Choi & Benzer, 1994). In addition, *aos* was isolated in a genetic screen for loci that interact with *fz* (Strutt & Strutt, 2003). Accordingly, *aos* LOF mutants, as well as hypomorphic EGFR alleles, show rotation defects without affecting ommatidial chirality (Brown & Freeman, 2003; Choi & Benzer, 1994; Gaengel & Mlodzik, 2003; Strutt & Strutt, 2003). Although mechanistic details of the involvement of EGFR during OR remain unknown, several lines of data point to an interaction between EGFR signaling and the regulators of cell adhesion and cytoskeleton. Firstly, the expression pattern of Fmi in R3/R4 cells is altered in *aos* mutants suggesting that EGFR signaling affects the establishment of polarity during

R3/R4 specification, at least in part by controlling the adhesive properties of the cluster (Fmi is an atypical cadherin) (Gaengel & Mlodzik, 2003). Rotation defects that arise from reduced EGFR activity are enhanced in *fmi* and *shg* mutants, further linking EGFR signaling to cadherin-based adhesion (Gaengel & Mlodzik, 2003). As junctional dynamics must be regulated throughout rotation, EGFR signaling may feed into cellular adhesion at multiple stages of rotation. Indeed, perturbations of EGFR signaling at larval stages of rotation can lead to misorientation of ommatidia at later pupal stages, potentially by altering the adhesive properties of the rotating clusters (Brown & Freeman, 2003). While the effect of EGFR on rotation is mediated by Ras GTPase activity, expression of various Ras constructs that activate specific subsets of Ras effectors revealed that besides Ras/Raf/MAPK signaling, a Raf/MAPK-independent pathway downstream of Ras is employed during OR: Canoe (Cno/AF-6), a Raf/MAPK-independent Ras effector postulated to link cytoskeleton to cellular junctions acts downstream of EGFR/Ras signaling in the OR setting (Gaengel & Mlodzik, 2003). Consistent with this hypothesis, *cno* mutants show rotation-specific phenotypes and *cno* genetically enhances OR phenotypes associated with the EGFR-ligand regulator Star (Gaengel & Mlodzik, 2003). Taken together, these data suggest that EGFR signaling modulates junctional and cytoskeletal dynamics in parallel to core PCP signaling at multiple stages of ommatidial rotation.

In addition to the above discussed cell-cell adhesion features, cell-ECM interactions are also required for eye morphogenesis (Thuveson et al., 2019). All photoreceptor cells express Myospheroid, the *Drosophila* homolog of β1 Integrin, which localizes to the outside membrane of the photoreceptors within each cluster, forming a cup-like shape inside each cluster, where it interacts with the ECM (Thuveson et al., 2019). Mutations in any of the integrin-ECM associated components lead to rotation defects, with clusters rotating initially too quickly and later asynchronously (Thuveson et al., 2019). Thus, it is worth noting that the basolateral integrin-ECM mediated adhesion acts in the opposite manner to Nmo in regulating OR. Furthermore, an additional gene that could be associated with ECM function(s), *scabrous* (*sca*), which encodes a fibrinogen-related secreted protein (Mlodzik, Baker, & Rubin, 1990), has also been postulated to inhibit rotation, as *sca* mutants are characterized by overrotation of ommatidia (Chou & Chien, 2002), similar to integrin and ECM mutants. Sca is suggested to be transported in vesicles from the furrow to rows 6–8 posterior to it to slow down the OR process (Chou & Chien, 2002). This model is supported by observations that in "furrow stop" mutants ommatidia often overrotate and ectopic expression of Sca in these mutants can rescue this phenotype. Interestingly, sca associated overrotation phenotypes require *nmo* activity, as *sca* and *nmo* double mutants display *nmo* underrotation pattern, whereas underrotation phenotype associated with Sca overexpression is enhanced in *nmo* heterozygosity and suppressed by Nmo overexpression (Chou & Chien, 2002). It has been suggested that Sca may change the properties of the ECM, possibly by creating a "barrier" to the rotating cluster to slow it down. However, as Sca has also been suggested to modulate Notch signaling, or even act as a Notch ligand (Baker, Mlodzik, & Rubin, 1990; Mlodzik et al., 1990; Powell, Wesley, Spencer, & Cagan, 2001), other functional models are possible, either through Notch signaling or the interaction between Notch and EGFR signaling via the Notch-dependent expression of Aos (Koca, Housden, Gault, Bray, & Mlodzik, 2019).

Taken together, ommatidial rotation is a well-studied PCP cell motility process and (at least at the genetic level) it helped significantly to identify and define the role of many genes that are involved in PCP-regulated cell motility in general, covering distinct branches of the cellular machinery ranging from signaling to adhesion and to cytoskeletal regulation. Work on OR also underlines how multiple signaling pathways and local cellular processes are interlinked in achieving precision in a highly regulated cell motility process.

## 2.2 Border cell migration

A second paradigm of collective cell motility in *Drosophila* is border cell migration (BCM) during oogenesis (Fig. 3). The early egg chamber consists of 16 germline cells—an oocyte and 15 nurse cells—that are enveloped by an epithelial layer of (somatic) follicle cells (Wu et al., 2008). At anterior and posterior edges of this follicular epithelium emerges a pair of differentiated cells, referred to as polar cells (Montell, 2003; Montell et al., 2012; Wu et al., 2008). As the egg chamber matures, follicle cells undergo rearrangements during which anterior polar cells signal to their neighbors—so called border cells—and induce their epithelial-to-mesenchymal transition and delamination (Wu et al., 2008). These border cells then migrate posteriorly, toward the anterior border of the oocyte, carrying along the polar cells (Fig. 3) (Montell, 2003; Montell et al., 2012).

Efficient BCM requires the activity of core Fz/PCP signaling. Mutations in *fz, Vang, dsh* and *pk* cause significant delays in migration (Bastock & Strutt, 2007). Knockdown of *fz* and *Vang* specifically in border cells or in

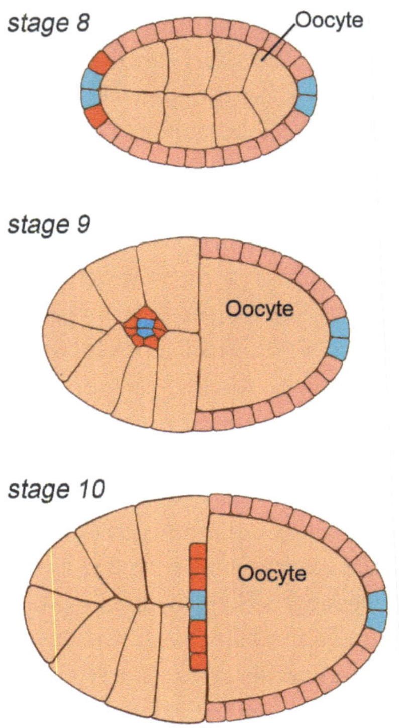


**Fig. 3** Schematic of border cell migration in the *Drosophila* egg chamber. The early egg chamber is composed of 16 germline cells—an oocyte and 15 nurse cells—that are surrounded by an epithelial monolayer of somatic follicle cells (Wu, Tanwar, & Raftery, 2008). As cell divisions in the follicular epithelium continue, a pair of cells at the anterior and posterior ends of the follicular epithelium differentiates into so called "polar cells" (stage 8, blue) (Montell, 2003; Montell, Yoon, & Starz-Gaiano, 2012; Wu et al., 2008). Anterior polar cells start to recruit a cluster from their neighboring epithelial cells, referred to as border cells (stage 8, red), through intercellular signaling events (Wu et al., 2008). While the follicular epithelium undergoes further cellular rearrangements during egg chamber development, the polar-border cell cluster migrates posteriorly, toward the anterior border of the oocyte (stages 9 and 10) (Montell, 2003; Montell et al., 2012). Fz/PCP signaling has been implicated in border cell migration. Anterior is left. See main text for details.

polar cells delays migration, suggesting that PCP is required in both cell types for their normal motility. This was further confirmed by the analysis of genetically mosaic clusters, where only polar or only border cells lacked the core PCP genes *fz* or *Vang*. Importantly, in mosaic clusters with only one polar cell retaining *fz* activity, the polar cell that was Fz+ was always positioned toward the leading edge, suggesting that Fz/PCP activity in a polar cell promotes their migratory behavior although the inverse pattern was not observed for Vang mutants (Bastock & Strutt, 2007). In addition, Fz and Vang localize to adherens junctions between polar and border cells, consistent with the idea that they mediate the "communication" between polar and border cells (Bastock & Strutt, 2007).

Very similar to ommatidial rotation, the core Fz/PCP cassette seems to cooperate with RTK signaling to promote BCM and the migration process entails junctional and cytoskeletal remodeling in the cells of the cluster. EGFR and PVR signaling pathways act redundantly to guide the process, as blocking the activity of both receptors, EGFR and PVR, severely inhibits migration (Duchek & Rorth, 2001; Duchek, Somogyi, Jekely, Beccari, & Rorth, 2001). In fact, individual border cells show differential RTK activity due to a gradient of the respective ligands emanating from the oocyte. This generates distinct cellular responses between the front and the back of the cluster, enabling the border cell cluster to move forward (Bianco et al., 2007). In particular, Rac activity was shown to be asymmetric in border cell clusters with highest levels in cells at the front, a pattern that was lost upon elimination of RTK signaling (Cai et al., 2014; Wang, He, Wu, Hahn, & Montell, 2010). This pattern was accompanied by a difference in the protrusive behavior of the front and the rear cells such that the protrusion number and speed was higher in the front, whereas loss of Rac activity or RTK signaling abolished this asymmetric feature. Border cells communicate with each other to generate and maintain this polarized behavior and DE-cad is an essential element of this communication (Cai et al., 2014; Wang et al., 2010). Cells of the border cell cluster are held together by increased levels of E-cad on the inner cell membranes within the cluster (Niewiadomska, Godt, & Tepass, 1999), which resembles the increased E-cad levels within the photoreceptor cluster during the OR process (Mirkovic & Mlodzik, 2006). Knockdown of E-cad in border cells, or in nurse cells, causes migration to ectopic positions, although the border cells retain their cluster formation and some motility, suggesting that E-cad levels sustain directed motility (Niewiadomska et al., 1999), which is again very similar to E-cad requirements during OR (Mirkovic & Mlodzik, 2006). A detailed analysis of E-cad engagement in border cells revealed that E-cad tension levels also show an RTK signaling- and Rac-dependent gradient in the border cells, with higher tension in the front of the cluster and this gradient is required for polarized Rac activity and protrusive behavior in border cells (Cai et al., 2014), suggesting that E-cad acts in a positive feedback loop with Rac downstream of RTK signaling to reinforce asymmetric cell behavior in border cells during BCM (Cai et al., 2014).

How Fz/PCP cooperates with RTK signaling in border cell migration remains unclear. Notably, elimination of PCP signaling delays BCM but does not block it, unlike RTK signaling (Bastock & Strutt, 2007). Whether there are Wnt gradients in the oocyte at the stage of BCM has not been reported. However, considering the involvement of Rac and E-cad downstream of PCP signaling in *Drosophila* OR, the involvement of Fz/PCP signaling may provide additional input into these molecules to make BCM more efficient. In line with this notion, Fz/PCP signaling was shown to regulate the protrusive activity of border cells during migration, and removal of *fz, Vang,* or *dsh* activity in border cells leads to a loss of RhoA-mediated protrusions (Bastock & Strutt, 2007). Furthermore, knockdown of *fz* altered distribution of RhoA in border cells and suppressed "no-protrusion phenotypes" associated with RhoA overactivation. This suggests that Fz/PCP signaling affects the actin cytoskeleton by positively regulating RhoA activity (Bastock & Strutt, 2007). On the other hand, Rac-mediated asymmetric protrusive activity in border cells was lost upon inhibition of JNK signaling (Wang et al., 2010), which is often activated downstream of Fz/PCP signaling in the *Drosophila* eye (Boutros, Paricio, Strutt, & Mlodzik, 1998; Strutt et al., 1997), consistent with a cross-talk between Fz/PCP and RTK signaling pathways in the BCM context.

It is worth noting here that in both these cell migration contexts in *Drosophila*, ommatidial rotation and border cell migration, the core Fz/PCP pathway and RTK signaling cooperate and largely act on the same downstream effectors, both regarding the cytoskeletal regulation as well as cell-adhesion mediated by cadherins, in particular, E-cad.

# 3. PCP regulated cell motility processes in vertebrates

## 3.1 Convergent extension cellular movements and the core PCP factors

Convergent Extension (abbreviated here as CE) is a key morphogenetic phenomenon that shapes the body plan during embryogenesis (Keller et al., 2000; Tada & Heisenberg, 2012; Yin, Ciruna, & Solnica-Krezel, 2009). CE was the first morphogenetic process in vertebrates linked to Wnt/Fz-PCP signaling. Cellular movements during CE are best characterized during vertebrate gastrulation, where mesendodermal cells move dorsally toward the midline of the gastrulating embryo and intercalate between their neighbors along the A/P-axis (Fig. 4A) (Keller et al., 2000; Tada & Heisenberg, 2012; Yin et al., 2009). These cells extend bipolar protrusions at medial and lateral ends that make stable contacts with neighboring cells

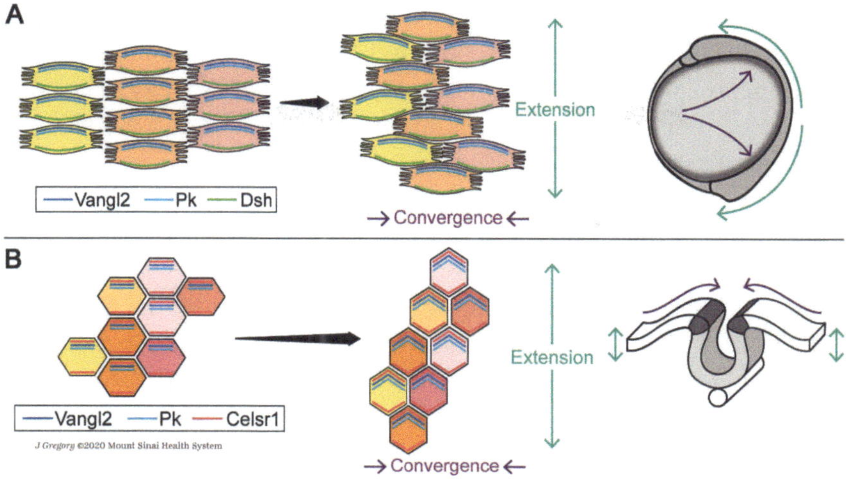


**Fig. 4** Schematic of planar cell polarity core factor localization during CE movements of gastrulation and neurulation. (A) During vertebrate gastrulation, mesendodermal cells move dorsally toward the midline of the embryo and intercalate between their neighbors along the A/P-axis (Keller et al., 2000; Tada & Heisenberg, 2012; Yin et al., 2009), narrowing the body along the medio-lateral (M/L) axis and elongate it along the anterior-posterior (A/P) axis. Bipolar protrusions emanating from mesendodermal cells at medial and lateral edges facilitate this process by allowing for stable contacts with the neighboring cells and generating traction (Keller et al., 2000; Tada & Heisenberg, 2012; Yin et al., 2009). The core PCP factors localize asymmetrically in the converging cells, as shown here schematically for Vangl2/Pk (note that there are multiple Pk factors) and Dsh (Dvl in mammals), and mediate their protrusive behavior (Roszko et al., 2015; Yin et al., 2008). (B) During neurulation, the neural folds that emerge from apical thickening of the neuroepithelium unite dorsally and close the neural tube via CE rearrangements toward the midline (Keller et al., 2000; Keller, Shih, & Sater, 1992; Nikolopoulou, Galea, Rolo, Greene, & Copp, 2017; Tada & Heisenberg, 2012). In this context, cell-cell intercalations were reported to be driven by polarized actomyosin contractions within the neuroepithelium, regulated by asymmetric core PCP factor localization. Vangl2, Pk and Celsr1/Fmi bridges the cells in the A/P axis, signaling in the apical domain (Butler & Wallingford, 2018; Ciruna et al., 2006; McGreevy, Vijayraghavan, Davidson, & Hildebrand, 2015; Nishimura et al., 2012; Ossipova, Kim, & Sokol, 2015). Anterior is up in all images. See main text for details.

and generate traction, which allows them to elongate and intercalate (Keller et al., 2000; Tada & Heisenberg, 2012; Yin et al., 2009). These coordinated cellular movements and rearrangements narrow the body along the medio-lateral axis and elongate it along the anterior-posterior (A/P) axis. Similarly, during neurulation, the neuroepithelium thickens apico-basally via CE rearrangements of neural plate cells, with apical wedging elevating the neural folds which then fuse dorsally and continue CE rearrangements toward the midline to close the neural tube (Fig. 4B) (Keller et al., 1992, 2000; Nikolopoulou et al., 2017; Tada & Heisenberg, 2012). Notably, cells of the neuroepithelium show monopolar protrusive activity with lamellipodia orienting medially, which might suggest that there are differences in the mechanisms regulating CE during gastrulation and neurulation.

Around the time the Wnt-Fz/PCP pathway was defined as a signaling pathway in *Drosophila* (Boutros et al., 1998; Strutt et al., 1997), several groups reported that an equivalent non-canonical Wnt-dependent Fz/PCP signaling system is critical for CE movements in vertebrates. The first implications of non-canonical Wnt-signaling in gastrulation movements came from studies in Xenopus frogs where perturbations in Wnt5a, Fz8 and Dvl2 disrupted gastrulation processes and axis elongation without causing phenotypes associated with disrupting Wnt/β-catenin (canonical) signaling, and subsequent studies confirmed this during CE in neurulation (Deardorff, Tan, Conrad, & Klein, 1998; Moon et al., 1993; Sokol, 1996). In Xenopus, interference with Wnt11 function caused CE defects marked by shortening of the A/P axis with normal cell specification; and Wnt11-dependent inhibition of elongation in animal pole explants was rescued by expression of a truncated form of Dsh impaired in canonical Wnt-signaling capability (Tada & Smith, 2000). Importantly, a non-canonical Wnt-PCP signaling requirement in CE was confirmed genetically, when zebrafish screens identified *wnt11/silberblick* mutants, which caused a shortening of axial mesoderm and central nervous system (CNS) and their broadening due to the impairment of convergence and cell intercalations, which was again rescued by expression of a Dsh construct that lacked canonical signaling activity (Heisenberg et al., 2000). It was later shown that in Xenopus embryos, targeted (over)expression of mutant Xdsh that disrupted Wnt-PCP signaling specifically caused a lack of elongation of the A/P-axis and neural epithelium, and a failure in neural tube closure (Wallingford & Harland, 2002). Consistently, targeted overexpression of Wnt5a and XFz-8 phenocopied Xdsh overexpression phenotypes, confirming that a Wnt-Fz/PCP-pathway, related to the Fz/PCP signaling casette in *Drosophila*, is at work regulating CE in vertebrates (Wallingford & Harland, 2001, 2002).

Genetic screens in zebrafish and subsequent functional studies on core PCP gene homologs in Xenopus and zebrafish confirmed and refined the involvement of a conserved Wnt-Fz/PCP signaling cassette in vertebrate CE movements assembling the vertebrate PCP factors into largely the same molecular pathway as in *Drosophila* (e.g., reviewed in Montero & Heisenberg, 2004; Roszko, Sawada, & Solnica-Krezel, 2009; Tada, Concha, & Heisenberg, 2002; Veeman, Axelrod, & Moon, 2003; Wallingford, Fraser, & Harland, 2002). For example, both gain-of-function and LOF of Vangl2/Stbm, Fz7, and Pk in Xenopus and zebrafish severely perturbed CE processes often marked by shortening of the notochord and neural plate, without affecting cell fate specification (Carreira-Barbosa et al., 2003; Ciruna et al., 2006; Darken et al., 2002; Djiane, Riou, Umbhauer, Boucaut, & Shi, 2000; Goto & Keller, 2002; Takeuchi et al., 2003; Veeman, Slusarski, Kaykas, Louie, & Moon, 2003; Yin et al., 2008). In zebrafish, shortening of the A/P-axis in *trilobite* (*tri*: the zebrafish Vangl2 gene) mutants was shown to be caused by the inability of cells to elongate in the mediolateral axis, which resulted in the impairment of dorsal cell migration and subsequent CE-type intercalation (Jessen et al., 2002; Roszko et al., 2015). In this context, Tri/Vangl2 is required in both the elongating cells and the neighboring tissue, suggesting a cell autonomous requirement with a non-cell autonomous effect, similar to its function in *Drosophila* (Jessen et al., 2002). In Xenopus, Stbm/Vangl-associated neural tube closure defects was also reported to be due to lack of neural CE and cell intercalation movements (Park & Moon, 2002). Here, engrafted single cells mutant for Stbm/Vangl function were able to intercalate between their wild type neighbors, supporting a non-cell autonomous requirement during CE. Besides the highly conserved core PCP genes, several additional regulators of Wnt-Fz/PCP-signaling have also been identified in vertebrates based on their reported CE phenotypes (for other CE focused reviews see (Butler & Wallingford, 2017; Huebner & Wallingford, 2018; Keller et al., 2000; Montero & Heisenberg, 2004; Roszko et al., 2009; Tada & Heisenberg, 2012; Veeman, Axelrod, & Moon, 2003; Wallingford, 2012). Taken together, these studies are consistent with Wnt-Fz/PCP signaling acting critically in the context of CE processes, both during gastrulation and neurulation, to provide cells with directional information that guides their migration and intercalation behavior. However, it should be noted that interactions among the PCP genes are more complicated than in *Drosophila*, as there are several gene family members for each *Drosophila* gene, and additional vertebrate specific PCP regulators exist.

## 3.2 Downstream effectors of PCP in CE movements

As in *Drosophila* eye PCP establishment (Boutros et al., 1998; Strutt et al., 1997), the JNK pathway is employed downstream of Fz/PCP signaling during CE processes. Hyperactivation or depletion of JNK interferes with gastrulation movements. For example, Wnt5a can activate JNK in cultured cells and CE phenotype induced by Wnt5a overexpression in Xenopus can be rescued by expression of a dominant negative JNK

(Yamanaka et al., 2002). Similarly, Wnt11/Fz7 signaling can activate JNK during Xenopus gastrulation through Rac GTPase activity (Habas, Dawid, & He, 2003; Habas, Kato, & He, 2001). Although Vangl/Stbm and Pk can also activate JNK in Xenopus and synergize with Dsh to elevate JNK activity, the JNK activation read-out here might be more complicated (Habas et al., 2001, 2003). However, it remains unclear how JNK activation regulates CE processes. Although nuclear responses are likely involved via JNK activation, its upstream activators of the Rho family of GTPases (acting as effectors of Fz-Dsh signaling (Boutros et al., 1998; Strutt et al., 1997)) appear as global regulators of CE movements and, in addition to their nuclear effects via JNK, they regulate multiple other cellular processes (Habas et al., 2001, 2003) as discussed below.

Coordination of multiple signaling inputs and cellular processes is essential during CE regulation, including protrusive activity, actomyosin contractility, cell-cell adhesion, and cell-matrix interactions (Butler & Wallingford, 2017; Huebner & Wallingford, 2018; Keller et al., 2000; Montero & Heisenberg, 2004; Roszko et al., 2009; Skoglund & Keller, 2010; Tada & Heisenberg, 2012; Veeman, Axelrod, & Moon, 2003; Wallingford, 2012). Besides the conserved requirement of Rho-family GTPases and JNK activation downstream of the Wnt-Fz/PCP pathway, FGF signaling has also been linked to CE defects (Nutt, Dingwell, Holt, & Amaya, 2001). Thus, a scenario of Fz/PCP and RTK signaling cooperation, similar to what has been observed in *Drosophila* ommatidial rotation and border cell migration, is likely in play during CE regulation.

### *3.2.1 Cytoskeletal rearrangements*

As in *Drosophila* OR and border cell migration, Wnt-Fz/PCP signaling mediates cytoskeletal rearrangements during CE processes. A key aspect of CE regulation is polarized protrusive cellular activity, which has often been linked to PCP signaling (Butler & Wallingford, 2017; Huebner & Wallingford, 2018; Keller et al., 2000; Montero & Heisenberg, 2004; Roszko et al., 2009; Tada & Heisenberg, 2012; Veeman, Axelrod, & Moon, 2003; Wallingford, 2012). It was initially discovered that expression of Dsh isoforms that interfere with Wnt-Fz/PCP signaling causes alterations in the protrusive behavior of cells (Wallingford et al., 2000). Lamellipodia that extend from mediolateral ends of cells during CE fail to do so upon expression of PCP-interfering Dsh isoforms. Such loss of polarity was accompanied by a decrease in length to width ratio of the respective cells and a loss of alignment in the mediolateral axis during CE (Wallingford et al., 2000). Many studies later confirmed protrusive activity as an important effector read-out of PCP signaling that enables cells to elongate, migrate, and intercalate while being tightly linked to multiple other local cellular processes that need to be coordinated during CE movements, as discussed below.

An elegant study using the chick neural tube linked Fz/PCP signaling to polarized actomyosin activity during CE (Nishimura et al., 2012). During neural plate bending, active myosin and F-actin were enriched at mediolateral edges of neural-plate cells and time-lapse images revealed that contractions of these cells occur in mediolateral directions, which helps them intercalate and elongate (Nishimura et al., 2012). This cellular behavior was dependent on Rho kinase (ROCK) activation and lost upon ROCK inhibition. Importantly, Celsr1/Fmi was enriched at mediolateral edges of these cells along with PDZ-RhoGEF, and upon knockdown of Celsr1/Fmi or expression of a dominant negative Dsh, the polarized pattern of PDZ-RhoGEF and actomyosin cables was lost. PDZ-RhoGEF physically interacts with Dsh via the adapter protein Daam1 (Habas et al., 2001, 2003), suggesting that the Rho regulator may be recruited to mediolateral edges through Dsh (Nishimura et al., 2012). Consistently, during Xenopus neural tube closure, PCP signaling was reported to be required for polarized actomyosin contractility (Butler & Wallingford, 2018). In this context, Vangl2 and Pk2 localization displayed a significant bias to anterior cell junctions, while being dynamically enriched in shrinking junctions in temporal correlation with actomyosin contractions (Butler & Wallingford, 2018; Ossipova et al., 2015). Taken together, Fz/PCP signaling provides input into the regulators of cytoskeletal dynamics, namely actin machinery and actomyosin network, to confer polarized protrusive and contractile activity required to drive convergence and extension (Fig. 4).

### 3.2.2 Regulation of junctional adhesion

Intercellular interactions through cell adhesion molecules are crucial for CE movements and they appear to act in the context of Fz/PCP signaling. Classical cadherins in vertebrates, C-cad, N-cad and E-cad, are all highly expressed in mesoderm cells during gastrula stages. Overexpression of wild type or dominant-negative C-cad was shown to perturb gastrulation movements in Xenopus. Remarkably, the adhesive activity of C-cad decreased during activin-induced elongation of animal cap explants, whereas co-treatment with C-cad activating antibodies inhibited this elongation (Zhong, Brieher, & Gumbiner, 1999). Accordingly, *cdh1* zygotic mutant zebrafish embryos exhibit mild defects in convergence and extension with wider somites at the segmentation stage (Shimizu et al., 2005). N-cad LOF in zebrafish impairs the extension of posterior axial mesodermal cells, thereby shortening the A/P axis (Harrington, Hong, Fasanmi, & Brewster, 2007). N-cad was further shown to be involved in neural CE processes in zebrafish: in N-cad-depleted embryos, lateral/dorsal neural keel cells failed to form stable protrusions, elongate and radially intercalate (Hong & Brewster, 2006). In addition to classical cadherins, paraxial protocadherin was shown to have role in vertebrate gastrulation. Overexpression of XPAPC or its depletion by morpholinos in Keller explants inhibited constriction but not elongation in the involuting marginal zone (IMZ) (Unterseher et al., 2004). Although XPAPC morpholinos do not severely affect the general morphology of Xenopus embryos, a closer examination with mesodermal markers revealed that the axial and paraxial tissue was broadened due to its lack of constriction. Consistently, XPAPC depletion randomized the polarization axis of mesodermal cells in explants and slowed down their dorsal migration. Nevertheless, PAPC could have a signaling activity in this context, as JNK activation was decreased upon XPAPC depletion and this was rescued by expression of a constitutively active RhoA, which was also deactivated upon PAPC depletion (Unterseher et al., 2004). Cell adhesion molecules have often been linked to the Fz/PCP pathway. Wnt5a/Fz7/Ror2 signaling, morpholinos of which phenocopy the PAPC like CE defects, was shown to regulate PAPC expression in a JNK-dependent manner (Schambony & Wedlich, 2007; Unterseher et al., 2004). Wnt11/Fz7 signaling, on the other hand, stabilizes PAPC on the membrane by blocking its clathrin- and dynamin-1-mediated internalization (Kraft, Berger, Wallkamm, Steinbeisser, & Wedlich, 2012). Strikingly, Fz7 binds C-cad and PAPC on the membrane, where PAPC abundance inhibits C-cad clustering, pointing to an interplay between different adhesion molecules (Kraft et al., 2012). Consistently, during Xenopus CE movements, C-cad clustering, which is essential for successful CE movements, was further shown to be dependent on the Fz/PCP pathway (Huebner et al., 2021).

During gastrulation in zebrafish, there are many modes of cell movements that contribute to convergence and extension of tissue and elongation of the A/P axis in each of the germ layers (see Williams & Solnica-Krezel, 2020, for a comprehensive review). The genetic and imaging tools available in zebrafish have allowed for a greater understanding of the localization and contribution of PCP components during these cell rearrangement processes. During gastrulation, cells in the endoderm undergo CE movements, similarly to the mesodermal cells as described above. Recent high resolution imaging studies of endodermal CE have shown that components of the Wnt-Fz/PCP pathway (including *gpc4/knypek,* a heparan-sulfate proteoglycan that influences non-canonical Wnt signaling), are required for polarization of cell shape and behavior (Balaraju, Hu, Rodriguez, Murry, & Lin, 2021). GFP-Vangl2 is localized to the anterior of these cells, which intercalate in a mediolateral direction. In addition to the polarized protrusions discussed above, junctional remodeling is also required for directed migration. The level of adhesive contacts during CE must be tightly balanced to maintain tissue integrity and yet allow cells to change position within the tissue to alter tissue dimensions. *gpc4*-mutant embryos show elevated levels of Cdh2 (N-cadherin) at the membrane, which through increased adhesive contacts counteracts the remodeling of adhesive junctions as cells try to exchange neighbors during intercalation. Gpc4 was shown to regulate the Rab5c-dependent endocytosis of Cdh2 to control the level of Cdh2 at the membrane, and thus neighbor adhesion.

The link between Wnt-Fz/PCP and Rab5c-dependent cadherin trafficking is strengthened by experiments investigating the directed migration of prechordal plate progenitors toward the animal pole following cellular internalization (Ulrich et al., 2005) (see also review Williams & Solnica-Krezel, 2020, for comprehensive outline of this process). Wnt11/Slb is required to coordinate prechordal plate progenitor movement toward the animal pole. In *wnt11/slb* mutant mesendodermal cells, cohesion between cells is reduced and this is accompanied by changes in E-cad membrane localization. Through genetic interactions and biophysical methods, it was determined that Wnt11 acts via Rab-5c- and dynamin1-dependent endocytosis of E-cad to regulate mesendodermal cell adhesion.

At earlier stages of development, during blastoderm spreading, termed "doming," Wnt-Fz/PCP signaling is also required to spatially regulate adhesion to increase cell movement (Petridou, Grigolon, Salbreux, Hannezo, & Heisenberg, 2019). During doming, the cells in the blastoderm begin to spread over the yolk, and this cohesive movement of cells represents a decrease in tissue viscosity. Prior to doming, cells are tightly packed together and resemble a solid, whereas at the onset of doming cell-cell adhesion is reduced and cells are able to flow around the yolk, akin to the movement of molecules in a fluid. This is termed tissue fluidization. E-cad contacts are destabilized, allowing for increased neighbor exchange and cell movements, and these processes are reduced in *wnt11/slb* mutants (Petridou et al., 2019). Interestingly, a similar process of tissue fluidization is evident during posterior elongation, in which mesodermal precursors migrate and progressively differentiate into mature mesodermal cells that are then incorporated into the presomitic mesoderm (Mongera et al., 2018). This process is also dependent upon N-cad regulation, and it will be interesting to see whether this regulation is also dependent upon the Wnt/PCP-regulated endocytosis.

A link between convergent-extension movements and cadherin trafficking is also conserved in *Drosophila* polarized cell rearrangements in the early embryo. A/P axis elongation occurs during a process known as germband extension (reviewed in Pare & Zallen, 2020). This process involves polarized cellular intercalations that are impacted by multiple signals, with receptors of the Toll superfamily being the predominant drivers of this process during *Drosophila* germband extension. Interestingly, the core PCP factors are not required in the *Drosophila* embryo for such intercalations; it is likely that the fast speed of the cell intercalation process and the abundance of maternal product of some of the core PCP factors contributes to their lack of involvement. Imaging and biophysical methods of a related process, tracheal tube morphogenesis during later *Drosophila* embryogenesis, determined that at junctions along the A/P borders of cells, PCP components are enriched and, through recruitment of RhoGEF2, there is a corresponding decrease in E-cadherin levels, which would allow for polarized junctional remodeling (Warrington, Strutt, & Strutt, 2013). Here also E-cadherin levels are controlled by polarized regulation of clathrin- and dynamin-dependent endocytosis (Levayer & Lecuit, 2012). This relationship extends to intercalations in the pupal wing (Classen, Anderson, Marois, & Eaton, 2005; Warrington et al., 2013). During pupal wing development, cell packing increases through junctional remodeling to produce a hexagonal array of cells (depicted in Fig. 1). PCP proteins are required for this process and regulate Rab11- and dynamin-dependent E-cad recycling (Classen et al., 2005). Fmi was also shown to localize E-cad-containing exocyst vesicles, further supporting a role for PCP complexes in regulating E-cad junctional dynamics to allow for neighbor exchange and tissue remodeling (Classen et al., 2005).

Together these data highlight both the importance and complexity of differential regulation of adhesion molecules during CE processes and the requirement of Wnt-Fz/PCP signaling in regulating complex adhesive behavior. PCP-dependent fine-tuning of adhesive behavior is likely to share common regulators between *Drosophila* and vertebrates. For example, a vertebrate homolog of *Drosophila* Nmo, Nlk1, was shown to affect Wnt11 signaling and CE movements during Xenopus gastrulation (Thorpe & Moon, 2004). Moreover, Nlk1 and PAPC genetically and physically interact, and reciprocally stabilize each other during Wnt11 signaling and Xenopus CE processes (Kumar, Ciprianidis, Theiss, Steinbeisser, & Kaufmann, 2017). Despite the existence of Nmo, as a conserved downstream Wnt-Fz/PCP effector, the regulation of adhesive behavior by the core PCP factors is likely to be more complex in vertebrates than in *Drosophila*.

### 3.2.3 ECM remodeling and signaling

Wnt-Fz/PCP signaling has also been linked to the remodeling of ECM during CE processes in vertebrates, possibly a similar manner to *Drosophila* ommatidial rotation (Thuveson et al., 2019). Cell-matrix interactions through integrin and its ligand fibronectin (FN) are critical for CE gastrulation processes. Xenopus embryos that lack FN fibrils showed a shorter and broader A/P axis (Marsden & DeSimone, 2003). For example, neural and mesodermal cells of Keller explants injected with FN morpholinos or anti-integrin function-blocking antibodies failed to polarize their protrusions, elongate, and intercalate, whereas this behavior was rescued when the explants were cultured on fibronectin (Davidson, Marsden, Keller, & Desimone, 2006). Fibronectin organization has been linked to PCP signaling. Essentially, the assembly of fibronectin fibrils starts at the onset of gastrulation and is observed only along the surface of the mesoderm. This restricted or "polarized" pattern is critical for proper CE movements. Overexpression of Vangl2/Stbm, Pk or Fz in Xenopus embryos disrupts this localized assembly of fibronectin fibrils along the surface of the mesoderm (Goto, Davidson, Asashima, & Keller, 2005). PCP signaling appears to have a dual role in this context: While it provides the cues to polarize the ECM, it is further required for ECM-dependent polarization of protrusions, as evident by experiments when culturing Xenopus notochord explants on fibronectin was not sufficient to rescue the protrusion, elongation, and intercalation defects caused by the PCP factor overexpression (Goto et al., 2005). Similarly to these results, fibrillar FN matrix failed to form on the surface of the dorsal marginal zone (DMZ) upon dominant negative Wnt11 expression in animal caps and this effect was rescued by expression of a Dsh isoform that only activates Wnt/PCP signaling, or by activation of Rho and Rac GTPases (Dzamba, Jakab, Marsden, Schwartz, & DeSimone, 2009). As these rescue experiments required integrin activity, these data implicated Wnt-Fz/PCP signaling in the (re)organization of ECM through integrin signaling. However, core PCP factors might affect fibronectin organization through various mechanisms. For example, loss of Vangl2 and Pk1a function in zebrafish causes a reduction in fibronectin levels as a result of increased matrix metalloproteinase activity during gastrulation, and here Vangl2 was shown to promote MMP endocytosis *in vitro* by antagonizing focal adhesion kinase (FAK) activity (Williams et al., 2012). Conversely, *glypican4* and *fz7a/7b* zebrafish mutants have increased fibronectin assembly albeit normal matrix metalloproteinase activity (Dohn, Mundell, Sawyer, Dunlap, & Jessen, 2013), and surprisingly this effect was due to high accumulation of N-cad on cell membranes, as it was rescued by N-cad knockdown.

Taken together, these studies all emphasize the significance of Wnt-Fz/PCP signaling in regulating cellular behavior at multiple layers during mesendodermal and neural CE processes, ranging from polarized cytoskeletal protrusions to cell-cell adhesion and to the communication between cells and the ECM (Butler & Wallingford, 2017; Huebner & Wallingford, 2018; Keller et al., 2000; Montero & Heisenberg, 2004; Roszko et al., 2009; Skoglund & Keller, 2010; Tada et al., 2002; Tada & Heisenberg, 2012; Veeman, Axelrod, & Moon, 2003; Wallingford, 2012). A handful of pathways have been shown to be critical in CE processes that likely co-operate with Wnt-Fz/PCP signaling. For example, similarly to ommatidial rotation and border cell migration in *Drosophila*, RTK signaling pathways appear important contributors to CE regulation. A negative regulator of FGF signaling in Xenopus, XSprouty or XSpry, causes the shortening of the A/P-axis by impairing the CE processes when overexpressed (Nutt et al., 2001). Unlike dominant negative FGFR-expressing embryos, which show severe developmental defects, mesoderm induction was normal in XSpry overexpressing embryos. Xspry overexpression halted FGF-dependent Ca2+ efflux but did not affect MAPK activity in oocytes, suggesting that a Ras/MAPK-independent pathway downstream of FGF signaling may modulate CE movements, reminiscent of the Ras/MAPK-independent involvement of EGFR signaling during ommatidial rotation (Gaengel & Mlodzik, 2003). Moreover, as XSpry was shown to inhibit Fz/PCP signaling, by decreasing Dsh recruitment to the membrane in embryos, crosstalk between the FGF and PCP signaling pathways is likely (Wang et al., 2008). As XSpry also genetically and physically interacts with PAPC during Xenopus gastrulation (Wang et al., 2008), it may be feeding into multiple effector pathways regulating CE processes. In summary, taking data from *Drosophila* PCP processes and vertebrate gastrulation together, Wnt-Fz/PCP and RTK signaling are likely to co-operate in many such contexts, orchestrating the local cellular readouts necessary for cell motility and intercalation throughout animal development. Although there are significant differences between mammalian embryogenesis, which occurs *in utero*, and the examples outlined above, Wnt-Fz/PCP signaling is still required for directed cell movements during such processes (Williams et al., 2014) including neural tube closure (reviewed in Wang, Marco, Capra, & Kibar, 2019) and cardiac outflow tract morphogenesis (Sinha, Wang, Evans, Wynshaw-Boris, & Wang, 2012).

## 3.3 Facial branchiomotor neuron migration

An interesting migratory process is tangential migration of facial branchiomotor neurons (FBMNs) in the vertebrate hindbrain, which has been shown to require Wnt-Fz/PCP signaling (Fig. 5). FBMNs are a group of cranial branchiomotor neurons that are born in rhombomere 4 (r4) and the cell bodies undergo a posterior migration to r6 and r7, where they form the facial motor nucleus whilst their axons remain in r4 and then exit to innervate those muscles derived from the second branchial arch (Chandrasekhar, 2004).

Forward genetic screens in zebrafish revealed that core PCP components are required for FBMN migration. LOF alleles of *tri/Vangl2*,

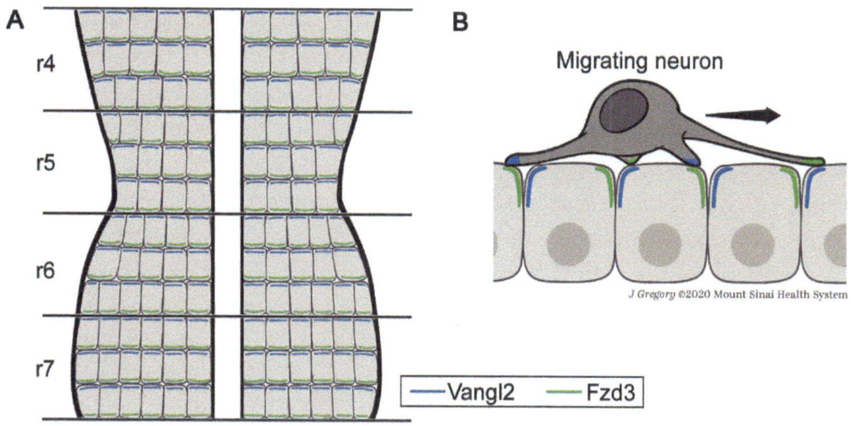


**Fig. 5** Model of PCP-mediated FBMN migration in the vertebrate hindbrain. The vertebrate hindbrain is segmented into developmental units called rhombomeres during embryonic development. FBMNs are born in rhombomere 4 (r4) and tangentially migrate to the more posterior r7 (Chandrasekhar, 2004). In zebrafish, rhombomeres have been shown to be planar polarized with Fzd3 and Vangl2 being asymmetrically enriched in anterior and posterior (sub)apical membranes, respectively (A) (Davey, Mathewson, & Moens, 2016). Migrating FBMNs enrich filopodial protrusions over the neuroepithelium in the direction of migration. Vangl2 becomes transiently enriched at the tips of filopodia in FBMNs preceding retraction, suggesting that transient PCP-mediated signaling events between FBMNs and the polarized neuroepithelium may promote FBMN migration (B) (Davey et al., 2016). Anterior is up. See text for details.

*Pk1a/Pk1b*, *Fz3a*, *Celsr2* and *scribble1* in zebrafish cause an impairment in caudal migration of FBMNs (Bingham, Higashijima, Okamoto, & Chandrasekhar, 2002; Carreira-Barbosa et al., 2003; Jessen et al., 2002; Rohrschneider, Elsen, & Prince, 2007; Wada et al., 2005; Wada, Tanaka, Nakayama, Iwasaki, & Okamoto, 2006). Chimeric analyses showed that the transmembrane core PCP factors Vangl2, Fzd3a and Celsr2 have a cell autonomous role in the migrating neurons as well as a non-autonomous role in the cellular environment of the migration process (Davey et al., 2016; Jessen et al., 2002; Wada et al., 2006), whereas the cytoplasmic Pk1b has a mostly cell autonomous involvement (Rohrschneider et al., 2007). Consistently, FBMNs that express a truncated form of Dvl that is dominant negative for the non-canonical pathway largely fail to migrate in a wild-type environment and expression of an equivalent Dvl construct in the migratory environment blocks the migration of wild type FBMNs (Davey et al., 2016). These data suggest that the communication between FBMNs and the migratory environment through core Fz/PCP signaling drives the migration of FBMNs. Nevertheless, in this context, interference with the established PCP Wnt-ligands, *silberblick*/Wnt11, *pipetail*/Wnt5a and the glypican *knypek*, does not impair FBMN migration in zebrafish (although they are critical for CE movements, see above) (Bingham et al., 2002; Jessen et al., 2002). These data suggest that there might be different extracellular upstream regulators of PCP-dependent FBMN migration in zebrafish, as compared to CE processes, or that there is functional redundancy among the zebrafish Wnts in this context. Nonetheless, the core Fz/PCP-dependence in FBMN migration is conserved in vertebrates, and in mice Wnt involvement has been suggested (Glasco et al., 2016; Qu et al., 2010; Vivancos et al., 2009). Further research in vertebrates will be required to better understand the exact nature of the Fz/PCP signaling involvement during this process.

During migration, FBMNs generate filopodial protrusions which have often been associated with neuronal cell migration and this protrusive activity appears to be regulated by Fz/PCP signaling (Fig. 5). As FBMNs migrate posteriorly through r5 and r6, they enrich their filopodia in the direction of migration (Davey et al., 2016). Strikingly, FBMNs fail to polarize their protrusive activity in *tri/vangl2* mutants. Chimeric analyses showed that Vangl2 and Fzd3a have opposing cell autonomous and non-autonomous functions in regulating protrusive activity. In FBMNs, Vangl2 destabilizes filopodia whereas Fzd3a stabilizes them. Conversely, in the migratory environment, Vangl2 acts to stabilize FBMN protrusions while Fzd3a has a destabilizing role (Davey et al., 2016). These findings are reminiscent of the antagonism between Fz and Vang in regulating actin polymerization during hair formation in the *Drosophila* wing epithelium (Adler, 2012; Klein & Mlodzik, 2005). Strikingly, Vangl2 is transiently enriched at the tips of filopodia in FBMNs preceding retraction, and the migratory environment or the "substrate" on which the neurons migrate display polarized localization of Vangl2 and Fzd3a to opposite sides of each cell, along the A/P axis, suggesting that transient interactions between FBMNs and the "substrate" neuroepithelium through these two core PCP factors (and associated signaling) may drive the migration of FBMNs (Fig. 5) (Davey et al., 2016). It has been proposed that cell-autonomous functions of Fzd3a and Vangl2 are enabled in filopodia, as they contact polarized Vangl2 and Fzd3a domains of neuroepithelial cells respectively to promote migration. In this context, the cell-autonomous activities of the Fz/PCP factors are likely to involve not only cytoskeletal remodeling (as seen by the formation or retraction of filopodia), but also cellular adhesion (Davey et al., 2016). Additional studies will be required to elucidate potential other interactors of PCP signaling including cell adhesion factors to promote FBMN migration.

## 3.4 Wnt/Fz-PCP signaling regulated axonal pathfinding

During nervous system development and patterning, axonal pathfinding can be considered a specialized cell movement process, through which the growth cone of the axon moves along and across tissues and guideposts to reach its target location. While the cell body stays "local," the growth cone "migrates" through tissues, guided by activating and inhibitory cues. Wnt signals have been identified as a set of such conserved growth cone guiding cues (reviewed in Dickson, 2005; Zou, 2006). In this context, Wnt/Fz-PCP signaling has been established as a major player, with Wnts working as guidance cues and the core Fz/PCP module mediating the cellular responses to these cues to direct the growth cone (reviewed in Goodrich, 2008; Zou, 2020). In particular, graded expression of Wnt molecules along the A/P axis of the developing central nervous system guides axons along this axis (reviewed in Hollis 2nd & Zou, 2012) and this feature of Wnt molecules is evolutionarily conserved. Wnt family members have been initially demonstrated to affect axonal pathfinding along the A/P axis in *Drosophila* (Yoshikawa, McKinnon, Kokel, & Thomas, 2003), *C. elegans* (Hilliard & Bargmann, 2006; Pan et al., 2006), and mouse (Lyuksyutova et al., 2003) and subsequently shown to do so in other vertebrates as well (reviewed in Zou, 2020). Moreover, injury to the vertebrate spinal cord has been shown to cause reactivation of expression of several Wnt members that are thought to regulate the associated *de novo* growth cone guidance and migration (reviewed in Hollis 2nd & Zou, 2012).

The impact of Wnts on axonal pathfinding in several parts of the nervous system has been firmly linked to the core Fz/PCP module both in *Drosophila* (Gombos et al., 2015; Mrkusich, Flanagan, & Whitington, 2011; Shimizu, Sato, & Tabata, 2011; Yuan et al., 2016) and in vertebrates (Fenstermaker et al., 2010; Hua, Smallwood, & Nathans, 2013; Onishi et al., 2013; Shafer, Onishi, Lo, Colakoglu, & Zou, 2011). The mechanistic understanding of how Wnts regulate growth cone extension and guidance via the Wnt/Fz-PCP pathway is, however, less well developed. Studies in mouse suggested that within the growth cone, there are separated domains of either Vangl2 or Fzd3 signaling units, and the protrusive behavior depends on complex interactions between these factors and the Dvl isoforms, which control Fzd3 internalization and the activation of the JNK signaling (Fig. 6A) (reviewed in Onishi, Hollis, & Zou, 2014; Zou, 2020). During growth cone migration, the core PCP factors are thought to be cooperating mainly with N-cad based adhesion, as compared to major E-cad involvement in many of the processes discussed above. As growth cone based "migration" shares the same principles with other migratory processes, regarding the involvement of adhesion junctions and polarity proteins, both with respect to PCP as well as apical-basal polarity factors, the growth cone has been coined a "half adherens junction" (reviewed in Zou, 2020).

While the mechanistic aspects of core PCP regulated growth cone guidance are very difficult to study *in vivo* in the mouse central nervous system, a model of action has recently received strong support through an elegant set of experiments with cochlear mouse neurons, linking Vangl2 complex function in axonal growth cones to N-cad dynamics and actin flow (Dos-Santos Carvalho et al., 2020). Here, the authors demonstrate that Vangl2 activity restricts neuronal outgrowth by controlling N-cad dynamics and increasing retrograde actin flow in spikes of the growth cone. Vangl2 is thought to positively regulate N-cad diffusion and mobility at the membrane, likely via affecting the mechanical coupling between N-cad and actin filaments. This type of regulatory interactions between the core PCP factors and cell adhesion machinery is again similar to the cadherin based regulations in the context of ommatidial rotation, border cell migration, and convergence extension movements (see chapters above). Interestingly, the effects of Vangl2 (or PCP complexes in general) on N-cad activity might be reciprocal

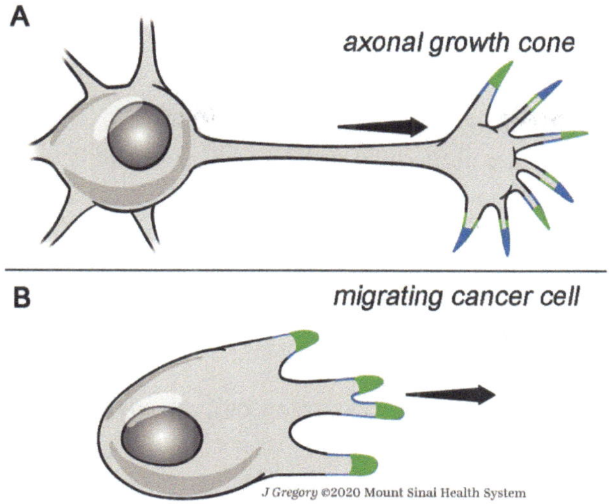


**Fig. 6** Schematics of axonal growth cone migration (A) and cancer cell migration (B) as mediated by core PCP factors. (A) Model for the growth cone migration of commissural axons as guided by Wnt signaling via the core Fz/PCP factors. Vangl2 complexes (green) and Fzd3-Dvl complexes (blue) are subcellularly asymmetrically localized in the filopodial tips of the growth cone protrusions, mediating a regulated Fzd3/Dvl endocytosis and associated effects on actin protrusions. Note that in the absence of Wnts these core PCP complex localization domains become randomized and the formation/retraction of protrusions is also randomized, causing stalling of the axonal extension (Dos-Santos Carvalho et al., 2020). (B) Schematic of migrating cancer cell with localized core PCP complex domains that also lead to organized Fzd/Dvl endocytosis and protrusion formation. It is not yet clearly defined how Wnt signals influence cancer cell migration regarding paracrine and autocrine signaling in general. See main text for details and references.

as N-cad substrates also affect Vangl2 localization and distribution, and by extension also the core PCP complexes in general (Dos-Santos Carvalho et al., 2020).

Although not linked to neuronal and axonal cell motility directly, it is worth noting in this context that the core PCP complexes have also been linked to synapse formation and stability. For example, in the developing glutamatergic neurons of mice, Vangl2 and Celsr3/Fmi have been shown to have opposing functions, with Vangl2 inhibiting synapse formation (Thakar et al., 2017). Here the Celsr3/Fzd3 complexes sit on the bouton side of the synapse, while Celsr3/Vangl2 complexes are located on the spine. As the synapse formation heavily relies on regulated cell adhesive behavior, it again highlights a strong regulatory link between the core PCP complexes and cell adhesion regulation in general and in this specialized context.

The molecular scenario employed by the core PCP module during growth cone migration and guidance is likely complicated with (at least) the receptor tyrosine kinase (RTK) family member Ryk/Derailed also playing an important role as a Wnt receptor (Liu et al., 2005; Yoshikawa et al., 2003). Whether and how it influences the core PCP module biology remain unclear, and even if Ryk/Derailed would act in parallel to the core PCP module factors, it is likely to function through some of the downstream effectors of the Fz and Vang complexes. For additional insight and references, please see a comprehensive recent review of Wnt-PCP signaling regulation of axonal growth cone guidance and migration (Zou, 2020).

## 3.5 Wnt/PCP signaling and cancer cell dissemination and migration

The molecular core PCP factor cassette has a broad applicability during many developmental processes, including cell migration (see above). It is thus not a surprise that defective PCP signaling plays a causative role in many disorders ranging from ciliopathies to neural tube closure and organ defects (Butler & Wallingford, 2017; Simons & Mlodzik, 2008). Recent analyses also highlight a critical role for Wnt-Fz/PCP signaling in cancer and cancer cell dissemination or migration. While the canonical Wnt/β-catenin signaling pathway has long been causatively linked to cancer initiation and progression, the function of Wnt/PCP signaling in cancer has been underappreciated. However, recent work has revealed a strong correlation between core PCP component upregulation, which also includes several Wnt ligands, and unfavorable prognosis in several different cancer types (Daulat & Borg, 2017). Nonetheless, the situation is more complex as PCP factors can also act as tumor suppressors (Daulat & Borg, 2017). It is thus apparent that the contribution of Wnt/PCP signaling to cancer progression can differ significantly depending on the type of cancer and, importantly, the stage of the disease.

Along with the function of Wnt/PCP signaling in axonal growth guidance and developmental cell migration in general, an important role for PCP in cancer is linked to cell dissemination (for example, Daulat & Borg, 2017). In particular, PCP features have been analyzed in detail in mouse models of breast cancer, where fibroblast-derived exosomes promote autocrine Wnt11/PCP signaling and cause invasive cellular behavior (Luga et al., 2012). Importantly, migrating breast cancer cells displayed mutually exclusive localization of the core PCP complexes, with Vangl and Fzd anchored complexes being separated into adjacent domains (see schematic in Fig. 6B), which is somewhat reminiscent of the developmental asymmetric localization of the PCP factors during growth cone guidance (Fig. 6A). Here again, the Fzd-Dvl complex induces actin-based protrusions and the Vangl-Pk complex antagonizes this process (Luga et al., 2012). Along these lines, a pathway termed "lateral signaling" has been defined in breast cancer cells, which builds upon the principles of the spatially separated, but locally antagonistic functions of the core PCP complexes, with again Fzd-Dvl complexes promoting protrusive activity (Zhang et al., 2016). Furthermore, Pk1 can form a complex with Arhgap21/23 to downregulate RhoA activity, providing a mechanistic insight into how the Vangl/Pk complexes antagonize protrusive activities (Zhang et al., 2016). Taken together with the general functions of the molecular PCP cassette, this builds a model whereby the antagonistic behavior of the two PCP complexes establishes a spatially regulated activation of RhoA, modulating actomyosin activity and focal adhesion to promote efficient cell migration (Zhang et al., 2016). While the interaction of Pk1 with Arhgap21/23 was detected by a mass spectrometry analysis in breast cancer cells (and not yet pursued further), it is very likely that a similar link could be detected in other PCP regulated cell migration and growth cone guidance scenarios (see above). The protrusion promoting effects of the Fzd-Dvl complexes, with focus on Fzd6, are further discussed elsewhere (Corda & Sala, 2017). It is highly probable that multiple signaling pathways, both downstream of PCP and RTK signaling for example, converge in cancer cells to impact actin dynamics and promote cancer cell dissemination.

## 4. Concluding remarks

In this chapter, we have outlined the molecular details of the Wnt-Fz/PCP complexes and how they lead to the polarized activity of downstream effectors in processes that are direct cell migration or related morphological process. The outcome of core Wnt/Fz-PCP activity on motility depends upon the cell type, but in each case covered here, PCP signaling allows a cell to move in a directed manner relative to its position within a whole tissue. In general, PCP signaling can affect (i) the cytoskeleton through actomyosin regulation, which is required for force generation; (ii) junctional remodeling through cadherin trafficking, which is required for detaching from old neighbors and attaching to new ones; and (iii) modification of the ECM substrate upon which the cell is migrating. Each of these processes is vital for a cell to undergo directed movement. Core PCP signaling does not function in isolation, and in many instances cells integrate PCP signaling with input from other pathways, particularly RTKs, to ensure motility is appropriately regulated within the respective cellular environment.

PCP information is propagated across a tissue and can be used to coordinate the orientation of all cells within a field; this coordination is required for the CE movements that occur in each of the germ layers during gastrulation. In addition to responding to global cues such as A/P or D/V axes, PCP signaling can also allow groups of cells to respond to more localized cues, such as ommatidial clusters during rotation and border cells during migration to the oocyte. Even on an individual cell level, particularly in the nervous system, PCP complexes can direct the growth and movement of individual cells. Much of the work uncovering the molecular basis of PCP and its impact on motility has come from model organisms such as *Drosophila*, zebrafish, and *Xenopus*, however these processes are highly conserved and are at play during most if not all stages of mammalian development.

The impact of dysregulated PCP signaling can be seen in the many congenital syndromes that accompany mutations in core PCP genes, and many such disease links are now being discovered and studied. Similarly, a contribution of core Wnt/Fz-PCP signaling to cancer cell dissemination and migration has been established, but many questions remain unanswered in the disease contexts. Despite a growing knowledge base, much remains to be discovered and further research into specific functions and facets of PCP signaling will improve our understanding of both development and disease. As discussed in this chapter, the core PCP pathway has been mainly associated with the asymmetric localization of the core components with relevance in contexts of both cell polarity and cell migration during development and disease. There is still much to learn about the temporal organization of core complexes and their links to downstream effectors. With the advent of innovative live-imaging tools and biophysical methods to complement genetic and molecular studies, we look forward to further advances in the understanding of PCP-regulated motility at a molecular and cellular level. There are many exciting discoveries still to come in the core Wnt/PCP field and its impact on cell motility.

## Acknowledgments

We are grateful to all Mlodzik lab members for their helpful discussions and support. We wish to thank Jill Gregory at the Academic Medical Illustration Department at the Icahn School of Medicine at Mount Sinai for excellent assistance with figure preparation. Related research in the Mlodzik laboratory is supported by National Institutes of Health grants from NIGMS and NEI.

## References


Adler, P. N. (2002). Planar signaling and morphogenesis in Drosophila. *Developmental Cell, 2*, 525–535.

Adler, P. N. (2012). The frizzled/stan pathway and planar cell polarity in the Drosophila wing. *Current Topics in Developmental Biology, 101*, 1–31.

Baker, N. E., Mlodzik, M., & Rubin, G. M. (1990). Spacing differentiation in the developing Drosophila eye: A fibrinogen-related lateral inhibitor encoded by scabrous. *Science, 250*, 1370–1377.

Balaraju, A. K., Hu, B., Rodriguez, J. J., Murry, M., & Lin, F. (2021). Glypican 4 regulates planar cell polarity of endoderm cells by controlling the localization of Cadherin 2. *Development, 148*.

Bastock, R., & Strutt, D. (2007). The planar polarity pathway promotes coordinated cell migration during Drosophila oogenesis. *Development, 134*, 3055–3064.

Bianco, A., Poukkula, M., Cliffe, A., Mathieu, J., Luque, C. M., Fulga, T. A., et al. (2007). Two distinct modes of guidance signalling during collective migration of border cells. *Nature, 448*, 362–365.

Bingham, S., Higashijima, S., Okamoto, H., & Chandrasekhar, A. (2002). The Zebrafish trilobite gene is essential for tangential migration of branchiomotor neurons. *Developmental Biology, 242*, 149–160.

Boutros, M., Paricio, N., Strutt, D. I., & Mlodzik, M. (1998). Dishevelled activates JNK and discriminates between JNK pathways in planar polarity and wingless signaling. *Cell, 94*, 109–118.

Brown, K. E., & Freeman, M. (2003). Egfr signalling defines a protective function for ommatidial orientation in the Drosophila eye. *Development, 130*, 5401–5412.

Brzoska, H. L., d'Esposito, A. M., Kolatsi-Joannou, M., Patel, V., Igarashi, P., Lei, Y., et al. (2016). Planar cell polarity genes Celsr1 and Vangl2 are necessary for kidney growth, differentiation, and rostrocaudal patterning. *Kidney International, 90*, 1274–1284.

Butler, M. T., & Wallingford, J. B. (2017). Planar cell polarity in development and disease. *Nature Reviews. Molecular Cell Biology, 18*, 375–388.

Butler, M. T., & Wallingford, J. B. (2018). Spatial and temporal analysis of PCP protein dynamics during neural tube closure. *eLife, 7*.

Cagan, R. L., & Ready, D. F. (1989). The emergence of order in the Drosophila pupal retina. *Developmental Biology, 136*, 346–362.

Cai, D., Chen, S. C., Prasad, M., He, L., Wang, X., Choesmel-Cadamuro, V., et al. (2014). Mechanical feedback through E-cadherin promotes direction sensing during collective cell migration. *Cell, 157*, 1146–1159.

Carreira-Barbosa, F., Concha, M. L., Takeuchi, M., Ueno, N., Wilson, S. W., & Tada, M. (2003). Prickle 1 regulates cell movements during gastrulation and neuronal migration in zebrafish. *Development, 130*, 4037–4046.

Carvajal-Gonzalez, J. M., Mulero-Navarro, S., & Mlodzik, M. (2016). Centriole positioning in epithelial cells and its intimate relationship with planar cell polarity. *BioEssays, 38*, 1234–1245.

Chandrasekhar, A. (2004). Turning heads: Development of vertebrate branchiomotor neurons. *Developmental Dynamics, 229*, 143–161.

Choi, K. W., & Benzer, S. (1994). Rotation of photoreceptor clusters in the developing Drosophila eye requires the nemo gene. *Cell, 78*, 125–136.

Chou, Y. H., & Chien, C. T. (2002). Scabrous controls ommatidial rotation in the Drosophila compound eye. *Developmental Cell, 3*, 839–850.

Ciruna, B., Jenny, A., Lee, D., Mlodzik, M., & Schier, A. F. (2006). Planar cell polarity signalling couples cell division and morphogenesis during neurulation. *Nature, 439*, 220–224.

Classen, A. K., Anderson, K. I., Marois, E., & Eaton, S. (2005). Hexagonal packing of Drosophila wing epithelial cells by the planar cell polarity pathway. *Developmental Cell, 9*, 805–817.

Cooper, M. T., & Bray, S. J. (1999). Frizzled regulation of Notch signalling polarizes cell fate in the Drosophila eye. *Nature*, *397*, 526–530.
Corda, G., & Sala, A. (2017). Non-canonical WNT/PCP signalling in cancer: Fzd6 takes centre stage. *Oncogensis*, *6*, e364.
Curtin, J. A., Quint, E., Tsipouri, V., Arkell, R. M., Cattanach, B., Copp, A. J., et al. (2003). Mutation of Celsr1 disrupts planar polarity of inner ear hair cells and causes severe neural tube defects in the mouse. *Current Biology*, *13*, 1129–1133.
Darken, R. S., Scola, A. M., Rakeman, A. S., Das, G., Mlodzik, M., & Wilson, P. A. (2002). The planar polarity gene strabismus regulates convergent extension movements in Xenopus. *The EMBO Journal*, *21*, 976–985.
Das, G., Jenny, A., Klein, T. J., Eaton, S., & Mlodzik, M. (2004). Diego interacts with Prickle and Strabismus/Van Gogh to localize planar cell polarity complexes. *Development*, *131*, 4467–4476.
Das, G., Reynolds-Kenneally, J., & Mlodzik, M. (2002). The atypical cadherin Flamingo links Frizzled and Notch signaling in planar polarity establishment in the Drosophila eye. *Developmental Cell*, *2*, 655–666.
Daulat, A. M., & Borg, J. P. (2017). Wnt/planar cell polarity signaling: New opportunities for cancer treatment. *Trends in Cancer*, *3*, 113–125.
Davey, C. F., Mathewson, A. W., & Moens, C. B. (2016). PCP signaling between migrating neurons and their planar-polarized neuroepithelial environment controls filopodial dynamics and directional migration. *PLoS Genetics*, *12*, e1005934.
Davey, C. F., & Moens, C. B. (2017). Planar cell polarity in moving cells: Think globally, act locally. *Development*, *144*, 187–200.
Davidson, L. A., Marsden, M., Keller, R., & Desimone, D. W. (2006). Integrin alpha5beta1 and fibronectin regulate polarized cell protrusions required for Xenopus convergence and extension. *Current Biology*, *16*, 833–844.
Deardorff, M. A., Tan, C., Conrad, L. J., & Klein, P. S. (1998). Frizzled-8 is expressed in the Spemann organizer and plays a role in early morphogenesis. *Development*, *125*, 2687–2700.
del Alamo, D., & Mlodzik, M. (2006). Frizzled/PCP-dependent asymmetric neuralized expression determines R3/R4 fates in the Drosophila eye. *Developmental Cell*, *11*, 887–894.
Devenport, D. (2016). Tissue morphodynamics: Translating planar polarity cues into polarized cell behaviors. *Seminars in Cell & Developmental Biology*, *55*, 99–110.
Devenport, D., & Fuchs, E. (2008). Planar polarization in embryonic epidermis orchestrates global asymmetric morphogenesis of hair follicles. *Nature Cell Biology*, *10*, 1257–1268.
Dickson, B. J. (2005). Wnts send axons up and down the spinal cord. *Nature Neuroscience*, *8*, 1130–1132.
Djiane, A., Riou, J., Umbhauer, M., Boucaut, J., & Shi, D. (2000). Role of frizzled 7 in the regulation of convergent extension movements during gastrulation in Xenopus laevis. *Development*, *127*, 3091–3100.
Dohn, M. R., Mundell, N. A., Sawyer, L. M., Dunlap, J. A., & Jessen, J. R. (2013). Planar cell polarity proteins differentially regulate extracellular matrix organization and assembly during zebrafish gastrulation. *Developmental Biology*, *383*, 39–51.
Dos-Santos Carvalho, S., Moreau, M. M., Hien, Y. E., Garcia, M., Aubailly, N., Henderson, D. J., et al. (2020). Vangl2 acts at the interface between actin and N-cadherin to modulate mammalian neuronal outgrowth. *eLife*, *9*.
Duchek, P., & Rorth, P. (2001). Guidance of cell migration by EGF receptor signaling during Drosophila oogenesis. *Science*, *291*, 131–133.
Duchek, P., Somogyi, K., Jekely, G., Beccari, S., & Rorth, P. (2001). Guidance of cell migration by the Drosophila PDGF/VEGF receptor. *Cell*, *107*, 17–26.

Dzamba, B. J., Jakab, K. R., Marsden, M., Schwartz, M. A., & DeSimone, D. W. (2009). Cadherin adhesion, tissue tension, and noncanonical Wnt signaling regulate fibronectin matrix organization. *Developmental Cell*, *16*, 421–432.

Fanto, M., & Mlodzik, M. (1999). Asymmetric Notch activation specifies photoreceptors R3 and R4 and planar polarity in the Drosophila eye. *Nature*, *397*, 523–526.

Fenstermaker, A. G., Prasad, A. A., Bechara, A., Adolfs, Y., Tissir, F., Goffinet, A., et al. (2010). Wnt/planar cell polarity signaling controls the anterior-posterior organization of monoaminergic axons in the brainstem. *The Journal of Neuroscience*, *30*, 16053–16064.

Fiehler, R. W., & Wolff, T. (2007). Drosophila Myosin II, Zipper, is essential for ommatidial rotation. *Developmental Biology*, *310*, 348–362.

Fiehler, R. W., & Wolff, T. (2008). Nemo is required in a subset of photoreceptors to regulate the speed of ommatidial rotation. *Developmental Biology*, *313*, 533–544.

Gaengel, K., & Mlodzik, M. (2003). Egfr signaling regulates ommatidial rotation and cell motility in the Drosophila eye via MAPK/Pnt signaling and the Ras effector Canoe/AF6. *Development*, *130*, 5413–5423.

Gault, W. J., Olguin, P., Weber, U., & Mlodzik, M. (2012). Drosophila CK1-gamma, gilgamesh, controls PCP-mediated morphogenesis through regulation of vesicle trafficking. *The Journal of Cell Biology*, *196*, 605–621.

Glasco, D. M., Pike, W., Qu, Y., Reustle, L., Misra, K., Di Bonito, M., et al. (2016). The atypical cadherin Celsr1 functions non-cell autonomously to block rostral migration of facial branchiomotor neurons in mice. *Developmental Biology*, *417*, 40–49.

Gombos, R., Migh, E., Antal, O., Mukherjee, A., Jenny, A., & Mihaly, J. (2015). The formin DAAM functions as molecular effector of the planar cell polarity pathway during axonal development in Drosophila. *The Journal of Neuroscience*, *35*, 10154–10167.

Goodrich, L. V. (2008). The plane facts of PCP in the CNS. *Neuron*, *60*, 9–16.

Goodrich, L. V., & Strutt, D. (2011). Principles of planar polarity in animal development. *Development*, *138*, 1877–1892.

Goto, T., Davidson, L., Asashima, M., & Keller, R. (2005). Planar cell polarity genes regulate polarized extracellular matrix deposition during frog gastrulation. *Current Biology*, *15*, 787–793.

Goto, T., & Keller, R. (2002). The planar cell polarity gene strabismus regulates convergence and extension and neural fold closure in Xenopus. *Developmental Biology*, *247*, 165–181.

Guo, N., Hawkins, C., & Nathans, J. (2004). Frizzled6 controls hair patterning in mice. *Proceedings of the National Academy of Sciences of the United States of America*, *101*, 9277–9281.

Habas, R., Dawid, I. B., & He, X. (2003). Coactivation of Rac and Rho by Wnt/Frizzled signaling is required for vertebrate gastrulation. *Genes & Development*, *17*, 295–309.

Habas, R., Kato, Y., & He, X. (2001). Wnt/Frizzled activation of Rho regulates vertebrate gastrulation and requires a novel Formin homology protein Daam1. *Cell*, *107*, 843–854.

Harrington, M. J., Hong, E., Fasanmi, O., & Brewster, R. (2007). Cadherin-mediated adhesion regulates posterior body formation. *BMC Developmental Biology*, *7*, 130.

Harrison, C., Shao, H., Strutt, H., & Strutt, D. (2020). Molecular mechanisms mediating asymmetric subcellular localisation of the core planar polarity pathway proteins. *Biochemical Society Transactions*, *48*, 1297–1308.

Heisenberg, C. P., Tada, M., Rauch, G. J., Saude, L., Concha, M. L., Geisler, R., et al. (2000). Silberblick/Wnt11 mediates convergent extension movements during zebrafish gastrulation. *Nature*, *405*, 76–81.

Hilliard, M. A., & Bargmann, C. I. (2006). Wnt signals and frizzled activity orient anterior-posterior axon outgrowth in C. elegans. *Developmental Cell*, *10*, 379–390.

Hollis, E. R., 2nd, & Zou, Y. (2012). Expression of the Wnt signaling system in central nervous system axon guidance and regeneration. *Frontiers in Molecular Neuroscience*, *5*, 5.

Hong, E., & Brewster, R. (2006). N-cadherin is required for the polarized cell behaviors that drive neurulation in the zebrafish. *Development*, *133*, 3895–3905.

Hua, Z. L., Smallwood, P. M., & Nathans, J. (2013). Frizzled3 controls axonal development in distinct populations of cranial and spinal motor neurons. *eLife*, *2*, e01482.

Huebner, R. J., Malmi-Kakkada, A. N., Sarikaya, S., Weng, S., Thirumalai, D., & Wallingford, J. B. (2021). Mechanical heterogeneity along single cell-cell junctions is driven by lateral clustering of cadherins during vertebrate axis elongation. *eLife*, *10*.

Huebner, R. J., & Wallingford, J. B. (2018). Coming to consensus: A unifying model emerges for convergent extension. *Developmental Cell*, *46*, 389–396.

Humphries, A. C., & Mlodzik, M. (2018). From instruction to output: Wnt/PCP signaling in development and cancer. *Current Opinion in Cell Biology*, *51*, 110–116.

Humphries, A. C., Narang, S., & Mlodzik, M. (2020). Mutations associated with human neural tube defects display disrupted planar cell polarity in Drosophila. *eLife*, *9*, e535322.

Jenny, A. (2010). Planar cell polarity signaling in the Drosophila eye. *Current Topics in Developmental Biology*, *93*, 189–227.

Jenny, A., Darken, R. S., Wilson, P. A., & Mlodzik, M. (2003). Prickle and Strabismus form a functional complex to generate a correct axis during planar cell polarity signaling. *The EMBO Journal*, *22*, 4409–4420.

Jenny, A., Reynolds-Kenneally, J., Das, G., Burnett, M., & Mlodzik, M. (2005). Diego and Prickle regulate Frizzled planar cell polarity signalling by competing for Dishevelled binding. *Nature Cell Biology*, *7*, 691–697.

Jessen, J. R., Topczewski, J., Bingham, S., Sepich, D. S., Marlow, F., Chandrasekhar, A., et al. (2002). Zebrafish trilobite identifies new roles for Strabismus in gastrulation and neuronal movements. *Nature Cell Biology*, *4*, 610–615.

Keller, R., Davidson, L., Edlund, A., Elul, T., Ezin, M., Shook, D., et al. (2000). Mechanisms of convergence and extension by cell intercalation. *Philosophical Transactions of the Royal Society of London. Series B, Biological Sciences*, *355*, 897–922.

Keller, R., Shih, J., & Sater, A. (1992). The cellular basis of the convergence and extension of the Xenopus neural plate. *Developmental Dynamics*, *193*, 199–217.

Klein, T. J., & Mlodzik, M. (2005). Planar cell polarization: An emerging model points in the right direction. *Annual Review of Cell and Developmental Biology*, *21*, 155–176.

Koca, Y., Housden, B. E., Gault, W. J., Bray, S. J., & Mlodzik, M. (2019). Notch signaling coordinates ommatidial rotation in the Drosophila eye via transcriptional regulation of the EGF-receptor ligand Argos. *Scientific Reports*, *9*, 18628.

Kraft, B., Berger, C. D., Wallkamm, V., Steinbeisser, H., & Wedlich, D. (2012). Wnt-11 and Fz7 reduce cell adhesion in convergent extension by sequestration of PAPC and C-cadherin. *The Journal of Cell Biology*, *198*, 695–709.

Kumar, R., Ciprianidis, A., Theiss, S., Steinbeisser, H., & Kaufmann, L. T. (2017). Nemo-like kinase 1 (Nlk1) and paraxial protocadherin (PAPC) cooperatively control Xenopus gastrulation through regulation of Wnt/planar cell polarity (PCP) signaling. *Differentiation*, *93*, 27–38.

Lawrence, P. A., & Casal, J. (2018). Planar cell polarity: Two genetic systems use one mechanism to read gradients. *Development*, *145*.

Levayer, R., & Lecuit, T. (2012). Biomechanical regulation of contractility: Spatial control and dynamics. *Trends in Cell Biology*, *22*, 61–81.

Liu, Y., Shi, J., Lu, C. C., Wang, Z. B., Lyuksyutova, A. I., Song, X. J., et al. (2005). Ryk-mediated Wnt repulsion regulates posterior-directed growth of corticospinal tract. *Nature Neuroscience*, *8*, 1151–1159.

Luga, V., Zhang, L., Viloria-Petit, A. M., Ogunjimi, A. A., Inanlou, M. R., Chiu, E., et al. (2012). Exosomes mediate stromal mobilization of autocrine Wnt-PCP signaling in breast cancer cell migration. *Cell*, *151*, 1542–1556.

Lyuksyutova, A. I., Lu, C. C., Milanesio, N., King, L. A., Guo, N., Wang, Y., et al. (2003). Anterior-posterior guidance of commissural axons by Wnt-frizzled signaling. *Science, 302*, 1984–1988.

Marsden, M., & DeSimone, D. W. (2003). Integrin-ECM interactions regulate cadherin-dependent cell adhesion and are required for convergent extension in Xenopus. *Current Biology, 13*, 1182–1191.

Matis, M., & Axelrod, J. D. (2013). Regulation of PCP by the fat signaling pathway. *Genes & Development, 27*, 2207–2220.

McGreevy, E. M., Vijayraghavan, D., Davidson, L. A., & Hildebrand, J. D. (2015). Shroom3 functions downstream of planar cell polarity to regulate myosin II distribution and cellular organization during neural tube closure. *Biology Open, 4*, 186–196.

Mirkovic, I., Gault, W. J., Rahnama, M., Jenny, A., Gaengel, K., Bessette, D., et al. (2011). Nemo kinase phosphorylates beta-catenin to promote ommatidial rotation and connects core PCP factors to E-cadherin-beta-catenin. *Nature Structural & Molecular Biology, 18*, 665–672.

Mirkovic, I., & Mlodzik, M. (2006). Cooperative activities of drosophila DE-cadherin and DN-cadherin regulate the cell motility process of ommatidial rotation. *Development, 133*, 3283–3293.

Mlodzik, M. (1999). Planar polarity in the Drosophila eye: A multifaceted view of signaling specificity and cross-talk. *The EMBO Journal, 18*, 6873–6879.

Mlodzik, M., Baker, N. E., & Rubin, G. M. (1990). Isolation and expression of scabrous, a gene regulating neurogenesis in Drosophila. *Genes & Development, 4*, 1848–1861.

Mongera, A., Rowghanian, P., Gustafson, H. J., Shelton, E., Kealhofer, D. A., Carn, E. K., et al. (2018). A fluid-to-solid jamming transition underlies vertebrate body axis elongation. *Nature, 561*, 401–405.

Montcouquiol, M., Sans, N., Huss, D., Kach, J., Dickman, J. D., Forge, A., et al. (2006). Asymmetric localization of Vangl2 and Fz3 indicate novel mechanisms for planar cell polarity in mammals. *The Journal of Neuroscience, 26*, 5265–5275.

Montell, D. J. (2003). Border-cell migration: The race is on. *Nature Reviews. Molecular Cell Biology, 4*, 13–24.

Montell, D. J., Yoon, W. H., & Starz-Gaiano, M. (2012). Group choreography: Mechanisms orchestrating the collective movement of border cells. *Nature Reviews. Molecular Cell Biology, 13*, 631–645.

Montero, J. A., & Heisenberg, C. P. (2004). Gastrulation dynamics: Cells move into focus. *Trends in Cell Biology, 14*, 620–627.

Moon, R. T., Campbell, R. M., Christian, J. L., McGrew, L. L., Shih, J., & Fraser, S. (1993). Xwnt-5A: A maternal Wnt that affects morphogenetic movements after overexpression in embryos of Xenopus laevis. *Development, 119*, 97–111.

Mrkusich, E. M., Flanagan, D. J., & Whitington, P. M. (2011). The core planar cell polarity gene prickle interacts with flamingo to promote sensory axon advance in the Drosophila embryo. *Developmental Biology, 358*, 224–230.

Munoz-Soriano, V., Ruiz, C., Perez-Alonso, M., Mlodzik, M., & Paricio, N. (2013). Nemo regulates cell dynamics and represses the expression of miple, a midkine/pleiotrophin cytokine, during ommatidial rotation. *Developmental Biology, 377*, 113–125.

Niewiadomska, P., Godt, D., & Tepass, U. (1999). DE-Cadherin is required for intercellular motility during Drosophila oogenesis. *The Journal of Cell Biology, 144*, 533–547.

Nikolopoulou, E., Galea, G. L., Rolo, A., Greene, N. D., & Copp, A. J. (2017). Neural tube closure: Cellular, molecular and biomechanical mechanisms. *Development, 144*, 552–566.

Nishimura, T., Honda, H., & Takeichi, M. (2012). Planar cell polarity links axes of spatial dynamics in neural-tube closure. *Cell, 149*, 1084–1097.

Nutt, S. L., Dingwell, K. S., Holt, C. E., & Amaya, E. (2001). Xenopus Sprouty2 inhibits FGF-mediated gastrulation movements but does not affect mesoderm induction and patterning. *Genes & Development, 15*, 1152–1166.

Onishi, K., Hollis, E., & Zou, Y. (2014). Axon guidance and injury-lessons from Wnts and Wnt signaling. *Current Opinion in Neurobiology, 27*, 232–240.

Onishi, K., Shafer, B., Lo, C., Tissir, F., Goffinet, A. M., & Zou, Y. (2013). Antagonistic functions of Dishevelleds regulate Frizzled3 endocytosis via filopodia tips in Wnt-mediated growth cone guidance. *The Journal of Neuroscience, 33*, 19071–19085.

Ossipova, O., Kim, K., & Sokol, S. Y. (2015). Planar polarization of Vangl2 in the vertebrate neural plate is controlled by Wnt and Myosin II signaling. *Biology Open, 4*, 722–730.

Pan, C. L., Howell, J. E., Clark, S. G., Hilliard, M., Cordes, S., Bargmann, C. I., et al. (2006). Multiple Wnts and frizzled receptors regulate anteriorly directed cell and growth cone migrations in Caenorhabditis elegans. *Developmental Cell, 10*, 367–377.

Pare, A. C., & Zallen, J. A. (2020). Cellular, molecular, and biophysical control of epithelial cell intercalation. *Current Topics in Developmental Biology, 136*, 167–193.

Park, M., & Moon, R. T. (2002). The planar cell-polarity gene stbm regulates cell behaviour and cell fate in vertebrate embryos. *Nature Cell Biology, 4*, 20–25.

Peng, Y., & Axelrod, J. D. (2012). Asymmetric protein localization in planar cell polarity: Mechanisms, puzzles, and challenges. *Current Topics in Developmental Biology, 101*, 33–53.

Petridou, N. I., Grigolon, S., Salbreux, G., Hannezo, E., & Heisenberg, C. P. (2019). Fluidization-mediated tissue spreading by mitotic cell rounding and non-canonical Wnt signalling. *Nature Cell Biology, 21*, 169–178.

Powell, P. A., Wesley, C., Spencer, S., & Cagan, R. L. (2001). Scabrous complexes with Notch to mediate boundary formation. *Nature, 409*, 626–630.

Qu, Y., Glasco, D. M., Zhou, L., Sawant, A., Ravni, A., Fritzsch, B., et al. (2010). Atypical cadherins Celsr1-3 differentially regulate migration of facial branchiomotor neurons in mice. *The Journal of Neuroscience, 30*, 9392–9401.

Rawls, A. S., & Wolff, T. (2003). Strabismus requires Flamingo and Prickle function to regulate tissue polarity in the Drosophila eye. *Development, 130*, 1877–1887.

Rohrschneider, M. R., Elsen, G. E., & Prince, V. E. (2007). Zebrafish Hoxb1a regulates multiple downstream genes including prickle1b. *Developmental Biology, 309*, 358–372.

Roignant, J. Y., & Treisman, J. E. (2009). Pattern formation in the Drosophila eye disc. *The International Journal of Developmental Biology, 53*, 795–804.

Roszko, I., Sawada, A., & Solnica-Krezel, L. (2009). Regulation of convergence and extension movements during vertebrate gastrulation by the Wnt/PCP pathway. *Seminars in Cell & Developmental Biology, 20*, 986–997.

Roszko, I., Sepich, D. S., Jessen, J. R., Chandrasekhar, A., & Solnica-Krezel, L. (2015). A dynamic intracellular distribution of Vangl2 accompanies cell polarization during zebrafish gastrulation. *Development, 142*, 2508–2520.

Schambony, A., & Wedlich, D. (2007). Wnt-5A/Ror2 regulate expression of XPAPC through an alternative noncanonical signaling pathway. *Developmental Cell, 12*, 779–792.

Segalen, M., & Bellaiche, Y. (2009). Cell division orientation and planar cell polarity pathways. *Seminars in Cell & Developmental Biology, 20*, 972–977.

Seifert, J. R., & Mlodzik, M. (2007). Frizzled/PCP signalling: A conserved mechanism regulating cell polarity and directed motility. *Nature Reviews, 8*, 126–138.

Shafer, B., Onishi, K., Lo, C., Colakoglu, G., & Zou, Y. (2011). Vangl2 promotes Wnt/planar cell polarity-like signaling by antagonizing Dvl1-mediated feedback inhibition in growth cone guidance. *Developmental Cell, 20*, 177–191.

Shimizu, K., Sato, M., & Tabata, T. (2011). The Wnt5/planar cell polarity pathway regulates axonal development of the Drosophila mushroom body neuron. *The Journal of Neuroscience, 31*, 4944–4954.

Shimizu, T., Yabe, T., Muraoka, O., Yonemura, S., Aramaki, S., Hatta, K., et al. (2005). E-cadherin is required for gastrulation cell movements in zebrafish. *Mechanisms of Development, 122*, 747–763.
Simons, M., & Mlodzik, M. (2008). Planar cell polarity signaling: From fly development to human disease. *Annual Review of Genetics, 42*, 517–540.
Sinha, T., Wang, B., Evans, S., Wynshaw-Boris, A., & Wang, J. (2012). Disheveled mediated planar cell polarity signaling is required in the second heart field lineage for outflow tract morphogenesis. *Developmental Biology, 370*, 135–144.
Skoglund, P., & Keller, R. (2010). Integration of planar cell polarity and ECM signaling in elongation of the vertebrate body plan. *Current Opinion in Cell Biology, 22*, 589–596.
Sokol, S. Y. (1996). Analysis of Dishevelled signalling pathways during Xenopus development. *Current Biology, 6*, 1456–1467.
Strutt, H., & Strutt, D. (2003). EGF signaling and ommatidial rotation in the Drosophila eye. *Current Biology, 13*, 1451–1457.
Strutt, H., & Strutt, D. (2021). How do the Fat-Dachsous and core planar polarity pathways act together and independently to coordinate polarized cell behaviours? *Open Biology, 11*, 200356.
Strutt, D., & Warrington, S. J. (2008). Planar polarity genes in the Drosophila wing regulate the localisation of the FH3-domain protein Multiple Wing Hairs to control the site of hair production. *Development, 135*, 3103–3111.
Strutt, D. I., Weber, U., & Mlodzik, M. (1997). The role of RhoA in tissue polarity and Frizzled signalling. *Nature, 387*, 292–295.
Tada, M., Concha, M. L., & Heisenberg, C. P. (2002). Non-canonical Wnt signalling and regulation of gastrulation movements. *Seminars in Cell & Developmental Biology, 13*, 251–260.
Tada, M., & Heisenberg, C. P. (2012). Convergent extension: Using collective cell migration and cell intercalation to shape embryos. *Development, 139*, 3897–3904.
Tada, M., & Smith, J. C. (2000). Xwnt11 is a target of Xenopus Brachyury: Regulation of gastrulation movements via Dishevelled, but not through the canonical Wnt pathway. *Development, 127*, 2227–2238.
Takeuchi, M., Nakabayashi, J., Sakaguchi, T., Yamamoto, T. S., Takahashi, H., Takeda, H., et al. (2003). The prickle-related gene in vertebrates is essential for gastrulation cell movements. *Current Biology, 13*, 674–679.
Thakar, S., Wang, L., Yu, T., Ye, M., Onishi, K., Scott, J., et al. (2017). Evidence for opposing roles of Celsr3 and Vangl2 in glutamatergic synapse formation. *Proceedings of the National Academy of Sciences of the United States of America, 114*, E610–E618.
Thomas, C., & Strutt, D. (2012). The roles of the cadherins Fat and Dachsous in planar polarity specification in Drosophila. *Developmental Dynamics, 241*, 27–39.
Thorpe, C. J., & Moon, R. T. (2004). nemo-like kinase is an essential co-activator of Wnt signaling during early zebrafish development. *Development, 131*, 2899–2909.
Thuveson, M., Gaengel, K., Collu, G. M., Chin, M. L., Singh, J., & Mlodzik, M. (2019). Integrins are required for synchronous ommatidial rotation in the Drosophila eye linking planar cell polarity signalling to the extracellular matrix. *Open Biology, 9*, 190148.
Tomlinson, A., & Ready, D. F. (1987). Neuronal differentiation in Drosophila ommatidium. *Developmental Biology, 120*, 366–376.
Tomlinson, A., & Struhl, G. (1999). Decoding vectorial information from a gradient: Sequential roles of the receptors Frizzled and Notch in establishing planar polarity in the Drosophila eye. *Development, 126*, 5725–5738.
Ulrich, F., Krieg, M., Schotz, E. M., Link, V., Castanon, I., Schnabel, V., et al. (2005). Wnt11 functions in gastrulation by controlling cell cohesion through Rab5c and E-cadherin. *Developmental Cell, 9*, 555–564.

Unterseher, F., Hefele, J. A., Giehl, K., De Robertis, E. M., Wedlich, D., & Schambony, A. (2004). Paraxial protocadherin coordinates cell polarity during convergent extension via Rho A and JNK. *The EMBO Journal*, *23*, 3259–3269.

Veeman, M. T., Axelrod, J. D., & Moon, R. T. (2003). A second canon. Functions and mechanisms of beta-catenin-independent Wnt signaling. *Developmental Cell*, *5*, 367–377.

Veeman, M. T., Slusarski, D. C., Kaykas, A., Louie, S. H., & Moon, R. T. (2003). Zebrafish Prickle, a modulator of noncanonical Wnt/Fz signaling, regulates gastrulation movements. *Current Biology*, *13*, 680–685.

Verdier, V., Guang Chao, C., & Settleman, J. (2006). Rho-kinase regulates tissue morphogenesis via non-muscle myosin and LIM-kinase during Drosophila development. *BMC Developmental Biology*, *6*, 38.

Vivancos, V., Chen, P., Spassky, N., Qian, D., Dabdoub, A., Kelley, M., et al. (2009). Wnt activity guides facial branchiomotor neuron migration, and involves the PCP pathway and JNK and ROCK kinases. *Neural Development*, *4*, 7.

Vladar, E. K., Bayly, R. D., Sangoram, A. M., Scott, M. P., & Axelrod, J. D. (2012). Microtubules enable the planar cell polarity of airway cilia. *Current Biology*, *22*, 2203–2212.

Wada, H., Iwasaki, M., Sato, T., Masai, I., Nishiwaki, Y., Tanaka, H., et al. (2005). Dual roles of zygotic and maternal Scribble1 in neural migration and convergent extension movements in zebrafish embryos. *Development*, *132*, 2273–2285.

Wada, H., Tanaka, H., Nakayama, S., Iwasaki, M., & Okamoto, H. (2006). Frizzled3a and Celsr2 function in the neuroepithelium to regulate migration of facial motor neurons in the developing zebrafish hindbrain. *Development*, *133*, 4749–4759.

Wallingford, J. B. (2010). Planar cell polarity signaling, cilia and polarized ciliary beating. *Current Opinion in Cell Biology*, *22*, 597–604.

Wallingford, J. B. (2012). Planar cell polarity and the developmental control of cell behavior in vertebrate embryos. *Annual Review of Cell and Developmental Biology*, *28*, 627–653.

Wallingford, J. B., Fraser, S. E., & Harland, R. M. (2002). Convergent extension: The molecular control of polarized cell movement during embryonic development. *Developmental Cell*, *2*, 695–706.

Wallingford, J. B., & Harland, R. M. (2001). Xenopus Dishevelled signaling regulates both neural and mesodermal convergent extension: Parallel forces elongating the body axis. *Development*, *128*, 2581–2592.

Wallingford, J. B., & Harland, R. M. (2002). Neural tube closure requires Dishevelled-dependent convergent extension of the midline. *Development*, *129*, 5815–5825.

Wallingford, J. B., Rowning, B. A., Vogeli, K. M., Rothbacher, U., Fraser, S. E., & Harland, R. M. (2000). Dishevelled controls cell polarity during Xenopus gastrulation. *Nature*, *405*, 81–85.

Wang, J., Hamblet, N. S., Mark, S., Dickinson, M. E., Brinkman, B. C., Segil, N., et al. (2006). Dishevelled genes mediate a conserved mammalian PCP pathway to regulate convergent extension during neurulation. *Development*, *133*, 1767–1778.

Wang, X., He, L., Wu, Y. I., Hahn, K. M., & Montell, D. J. (2010). Light-mediated activation reveals a key role for Rac in collective guidance of cell movement in vivo. *Nature Cell Biology*, *12*, 591–597.

Wang, Y., Janicki, P., Koster, I., Berger, C. D., Wenzl, C., Grosshans, J., et al. (2008). Xenopus paraxial protocadherin regulates morphogenesis by antagonizing Sprouty. *Genes & Development*, *22*, 878–883.

Wang, M., Marco, P., Capra, V., & Kibar, Z. (2019). Update on the role of the non-canonical Wnt/planar cell polarity pathway in neural tube defects. *Cell*, *8*.

Warrington, S. J., Strutt, H., & Strutt, D. (2013). The Frizzled-dependent planar polarity pathway locally promotes E-cadherin turnover via recruitment of RhoGEF2. *Development*, *140*, 1045–1054.

Williams, B. B., Cantrell, V. A., Mundell, N. A., Bennett, A. C., Quick, R. E., & Jessen, J. R. (2012). VANGL2 regulates membrane trafficking of MMP14 to control cell polarity and migration. *Journal of Cell Science, 125*, 2141–2147.

Williams, M. L. K., & Solnica-Krezel, L. (2020). Cellular and molecular mechanisms of convergence and extension in zebrafish. *Current Topics in Developmental Biology, 136*, 377–407.

Williams, M., Yen, W., Lu, X., & Sutherland, A. (2014). Distinct apical and basolateral mechanisms drive planar cell polarity-dependent convergent extension of the mouse neural plate. *Developmental Cell, 29*, 34–46.

Winter, C. G., Wang, B., Ballew, A., Royou, A., Karess, R., Axelrod, J. D., et al. (2001). Drosophila Rho-associated kinase (Drok) links Frizzled-mediated planar cell polarity signaling to the actin cytoskeleton. *Cell, 105*, 81–91.

Wolff, T., Guinto, J. B., & Rawls, A. S. (2007). Screen for genetic modifiers of stbm reveals that photoreceptor fate and rotation can be genetically uncoupled in the Drosophila eye. *PLoS One, 2*, e453.

Wolff, T., & Rubin, G. M. (1998). Strabismus, a novel gene that regulates tissue polarity and cell fate decisions in Drosophila. *Development, 125*, 1149–1159.

Wu, J., Klein, T. J., & Mlodzik, M. (2004). Subcellular localization of frizzled receptors, mediated by their cytoplasmic tails, regulates signaling pathway specificity. *PLoS Biology, 2*, E158.

Wu, J., & Mlodzik, M. (2009). A quest for the mechanism regulating global planar cell polarity of tissues. *Trends in Cell Biology, 19*, 295–305.

Wu, J., Roman, A. C., Carvajal-Gonzalez, J. M., & Mlodzik, M. (2013). Wg and Wnt4 provide long-range directional input to planar cell polarity orientation in Drosophila. *Nature Cell Biology, 15*, 1045–1055.

Wu, X., Tanwar, P. S., & Raftery, L. A. (2008). Drosophila follicle cells: Morphogenesis in an eggshell. *Seminars in Cell & Developmental Biology, 19*, 271–282.

Yamanaka, H., Moriguchi, T., Masuyama, N., Kusakabe, M., Hanafusa, H., Takada, R., et al. (2002). JNK functions in the non-canonical Wnt pathway to regulate convergent extension movements in vertebrates. *EMBO Reports, 3*, 69–75.

Yan, J., Huen, D., Morely, T., Johnson, G., Gubb, D., Roote, J., et al. (2008). The multiple-wing-hairs gene encodes a novel GBD-FH3 domain-containing protein that functions both prior to and after wing hair initiation. *Genetics, 180*, 219–228.

Yin, C., Ciruna, B., & Solnica-Krezel, L. (2009). Convergence and extension movements during vertebrate gastrulation. *Current Topics in Developmental Biology, 89*, 163–192.

Yin, C., Kiskowski, M., Pouille, P. A., Farge, E., & Solnica-Krezel, L. (2008). Cooperation of polarized cell intercalations drives convergence and extension of presomitic mesoderm during zebrafish gastrulation. *The Journal of Cell Biology, 180*, 221–232.

Yoshikawa, S., McKinnon, R. D., Kokel, M., & Thomas, J. B. (2003). Wnt-mediated axon guidance via the Drosophila Derailed receptor. *Nature, 422*, 583–588.

Yuan, L., Hu, S., Okray, Z., Ren, X., De Geest, N., Claeys, A., et al. (2016). The Drosophila neurogenin Tap functionally interacts with the Wnt-PCP pathway to regulate neuronal extension and guidance. *Development, 143*, 2760–2766.

Zhang, L., Luga, V., Armitage, S. K., Musiol, M., Won, A., Yip, C. M., et al. (2016). A lateral signalling pathway coordinates shape volatility during cell migration. *Nature Communications, 7*, 11714.

Zhong, Y., Brieher, W. M., & Gumbiner, B. M. (1999). Analysis of C-cadherin regulation during tissue morphogenesis with an activating antibody. *The Journal of Cell Biology, 144*, 351–359.

Zou, Y. (2006). Navigating the anterior-posterior axis with Wnts. *Neuron, 49*, 787–789.

Zou, Y. (2020). Breaking symmetry—Cell polarity signaling pathways in growth cone guidance and synapse formation. *Current Opinion in Neurobiology, 63*, 77–86.

CHAPTER EIGHT

# Talking to your neighbors across scales: Long-distance Notch signaling during patterning

**Zena Hadjivasiliou[a,b,c,*] and Ginger Hunter[d,*]**
[a]Department of Physics and Astronomy, University College, London, United Kingdom
[b]Institute for the Physics of Living Systems, University College, London, United Kingdom
[c]Mathematical and Physical Biology Laboratory, The Francis Crick Institute, London, United Kingdom
[d]Department of Biology, Clarkson University, Potsdam, NY, United States
*Corresponding authors: e-mail address: zena.hadjivasiliou@ucl.ac.uk; ghunter@clarkson.edu

## Contents



## Abstract

Tissue patterning is a critical part of animal development. Here we review the role that length- and timescales play in shaping patterns during development, focusing on the mechanisms by which Notch-mediated lateral inhibition signaling generates periodic tissue patterns. Because Notch ligands and receptors are membrane bound, the signaling that underlies lateral inhibition depends on direct cell-cell contacts. Nevertheless, there are many biological examples where effective Notch signaling occurs over distances larger than adjacent cells. Here, we summarize the theoretical and experimental evidence for mechanisms that modify the scale of Notch-mediated lateral inhibition.

*Current Topics in Developmental Biology*, Volume 150
ISSN 0070-2153
https://doi.org/10.1016/bs.ctdb.2022.04.002

We focus on how cell protrusions, in addition to other cell behaviors like proliferation and neighbor exchange, allow for Notch signaling to both extend lateral inhibition beyond nearest neighbors and impact the timescale of patterning. Using recent examples, we examine how dynamic cell behaviors like the formation of protrusions affect the timing of Notch-mediated lateral inhibition as well as the density of the final tissue pattern. We suggest that mechanisms that affect the length and timescale of Notch signaling may have key implications for the evolution of patterns. This review highlights the role of cell behaviors in controlling the temporal and spatial dynamics of pattern formation across scales.

## 1. Introduction

The transformation of a single cell into an elaborate and reproducible body plan during development is achieved through a plethora of mechanisms that allow cells to dynamically organize themselves in time and space. This transformation requires cells in growing tissues to continuously transmit and gather information about timing and position, both locally and at the scale of the tissue or the entire organism. Although cells can directly sense only their local environment, mechanisms have evolved that allow cells to overcome this constraint and transcend the length and time scales imposed by their individual positions and life cycles.

Classically, one dominant theory of developmental patterning mechanisms is morphogen gradients, where molecules that are transcribed in localized regions can spread to form graded concentration profiles in a target tissue (Stapornwongkul & Vincent, 2021; Wolpert, 1969) (Fig. 1A). As a result, the morphogen concentration at any location becomes a readout of position that cells can use to turn on appropriate fates, generating spatial patterns at the scale of the tissue. However, more recently it is clear that tissue-scale gradients of gene activation in many cases are generated through signals mediated via local cell-cell contact alone (Bischoff et al., 2013; Fancher & Mugler, 2020; Hall et al., 2021; Kornberg, 2017; Zhang & Scholpp, 2019). Examples of long-range signaling via direct contact are the long range gradients of Vg1 and activin seen in early Xenopus embryos, which are formed through a signaling relay between adjacent cells that has the overall effect of a morphogen being transported over longer distances (Reilly & Melton, 1996). Alternatively, information transfer can occur rapidly between cells through force generation and mechanotransduction. When cells exert stress on their direct neighbors through cell junctions, these local stresses can lead to rapid mechanical waves that travel across cells and

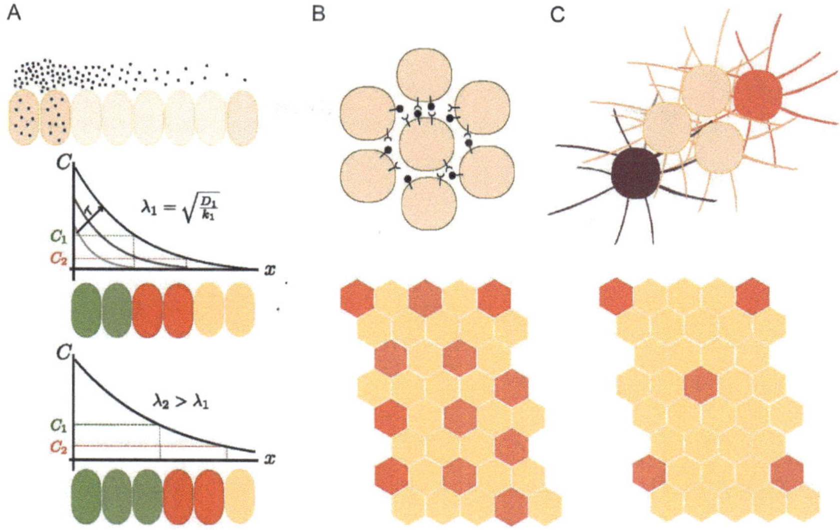


**Fig. 1** Length and timescales of signaling and patterning. (A) Morphogen molecules diffuse away from a localized source to form a graded concentration profile in a target tissue. Morphogen profiles can often be described by exponential curves whose decay length, $\lambda$, defines the range of the gradient. The value of the decay length is given by $\lambda = \sqrt{\frac{D}{k}}$ where $D$ is the morphogen effective diffusion coefficient and $k$ is the effective degradation rate. The morphogen profile builds up and reaches steady state at a timescale $\tau = 1/k$. Cells take on different fates according to the morphogen concentration in space. Changes in the diffusion coefficient or degradation rate impact the morphogen decay length and the downstream pattern. (B) Local signaling interactions between cells can lead to symmetry breaking and periodic pattern formation at the level of the tissue. Without any mechanisms to expand signaling interactions beyond immediate neighbors, emerging spatial patterns are dense. (C) Cells can extend long cellular protrusions that mediate signaling through contact. A dense network of signaling protrusions emerges that leads to sparser tissue level patterns.

direct symmetry breaking and morphogenesis at the tissue level (Duque & Gorfinkiel, 2016; Serra-Picamal et al., 2012; Vasquez, Tworoger, & Martin, 2014).

Another example of tissue level patterning mediated by local cell-cell interactions is that of lateral inhibition. Lateral inhibition is a conserved juxtacrine signaling mechanism that, during the development of most organisms, drives the formation of diverse fine-grained patterns such as spots and striped boundaries in tissues that are, initially, nearly homogenous (Cohen, Georgiou, Stevenson, Miodownik, & Baum, 2010; Collier, Monk, Maini, & Lewis, 1996; Hamada et al., 2014). During this type of patterning, each cell within the initial group has the potential to acquire either a signal receiving or signal sending fate. Small differences in the initial group of cells are amplified through feedback loops until one cell adopts a signal sending state. Since cells that have adopted the signal sending state inhibit the adoption of the same state in their contacting neighbors (Meinhardt & Gierer, 2000; Simpson, 1990), their neighbors adopt the signal receiving state, leading to periodic patterns at the tissue level.

In such cases, the effective length and time scales over which cells can interact are not immediately obvious. At first glance, only the signaling dynamics of cells that are in direct contact with one another should be expected to be coupled. However, cells can extend their zone of influence, for example through signaling filopodia or cellular rearrangements (Kornberg, 2017; Maroto, Dale, Dequéant, Petit, & Pourquié, 2005; Uriu, Morishita, & Iwasa, 2010). Furthermore, the spatiotemporal signaling dynamics of cells far from one another may become coupled even for cells that have never been in direct contact, for instance when relay mechanisms or trigger waves are in place so that the signaling states of cells at a distance come in or out of phase (Serra-Picamal et al., 2012).

Here, in exploring the power of this type of process to pattern tissues, we first introduce the role that length- and timescales play in patterning events in general. We will then focus on the Notch pathway, briefly reviewing examples of lateral inhibition and the processes that modify the range of Notch signaling. Next, we will summarize significant mathematical models of lateral inhibition and how they address the problem of scale. We next present specific examples where the spatiotemporal dynamics of patterning are affected by cellular protrusions. Finally, we discuss the potential of contact-mediated patterning for diversification in the course of evolution.

## 2. Patterning in space and time

### 2.1 Length scales

#### *2.1.1 Length scale of positional information*

The range of signaling at the molecular and cellular level ultimately specifies the length scale of spatial patterns generated at the tissue level (Fig. 1). This is most easily understood when we consider patterning along a single axis, e.g., along the anterior-posterior (AP) axis of a developing embryo. The positional information model of patterning posits that, in such cases, a cell's response to a morphogen depends on the concentration of morphogen and the cell state (which may define a threshold for response to the morphogen) (Wolpert, 1969). In this model, morphogen is typically produced at a pattern boundary and is transported away from the source in order to generate a gradient of morphogen across a region of the tissue (Fig. 1A).

In the case of the establishment of the AP body plan during the initial stages of embryogenesis in *Drosophila melanogaster*, patterning occurs by the iterative use of morphogen gradients to drive the formation of successively smaller stripes of gene expression that confer segment identities to cells along the AP axis (Jaeger et al., 2004). Probably the best studied of these is the early Bicoid gradient (Ali-Murthy & Kornberg, 2016; Clark & Akam, 2016; Driever & Nüsslein-Volhard, 1988; Spirov et al., 2009). The transcriptional response of the nuclei in each segment depends on the morphogen's decay length—which in principle depends on protein diffusivity and degradation, and the spatial profile of mRNA which may itself form a gradient that is diffusion independent, e.g., through dispersion along cytoskeletal cables (Ali-Murthy & Kornberg, 2016; Spirov et al., 2009). Changes that impact events at the molecular level, such as mRNA production and transport, molecular diffusion and turnover, will be reflected at the tissue level through changes in the morphogen decay length and the spatial organization of boundaries (Drocco, Grimm, Tank, & Wieschaus, 2011) (Fig. 1A).

The Bicoid gradient that occurs in the embryonic syncytial blastoderm of Drosophila is essentially an intracellular event. In principle, gradients in multicellular tissues rely on similar processes, although, in the case of tissues, morphogens are secreted or presented externally. A well-studied example of this is the gradient of the TGFβ homolog Dpp in the Drosophila wing imaginal disc (Ben-Zvi, Pyrowolakis, Barkai, & Shilo, 2011; Kicheva et al., 2007; Stapornwongkul, de Gennes, Cocconi, Salbreux, & Vincent, 2020; Wartlick et al., 2011; Zhu, Qiu, Chen, Nie, & Lander, 2020). Here, the concentration of Dpp in the imaginal disc specifies the position of the veins in the adult wing (Bosch, Ziukaite, Alexandre, Basler, & Vincent, 2017; Campbell & Tomlinson, 1999). The concentration and decay length of Dpp increases over time to scale with the overall growth of the tissue, ensuring that the underlying pattern remains proportionate to the organ size (Wartlick et al., 2011). In principle, the ability of the Dpp gradient to scale with the tissue size may depend on a number of processes including extracellular ligand dispersion, the binding and unbinding of the ligand to receptors in the cell membrane, receptor-ligand internalization, recycling and active transport via cells (Hatori, Wood, Oliveira Barbosa, & Kornberg, 2021; Huang, Liu, & Kornberg, 2019; Zhu et al., 2020). Scaling, in this case, is achieved by tuning the contributions that these processes play over time (Romanova-Michaelides et al., 2022).

### 2.1.2 Length scale of periodic patterns

The generation of periodic patterns, like those seen on the skin of animals from butterflies to birds, represent another type of challenge to developing organisms. Examples include the spacing between hair or feather follicles, pigmentation stripes, or sensory bristles (Cohen et al., 2010; Glover et al., 2017; Shyer et al., 2017; Yamaguchi, Yoshimoto, & Kondo, 2007). These patterns are characterized by the frequency at which motifs repeat themselves in space. In such cases, the spatial frequency itself depends on the effective length scale of the signaling molecules that pattern the tissue. In principle, periodic patterns can be achieved in an initially homogeneous tissue through reaction-diffusion processes, like those first described by Turing (Turing, 1952). This model proposes that an activator locally stimulates its own production together with that of an inhibitor. Differences in the relative rates of diffusion of the activator and inhibitor can lead to the emergence of stable, periodic patterns, where the wavelength of the spatial pattern depends on the diffusion coefficients of the activator and inhibitor, and the kinetic functions that specify interactions between the activator and inhibitor (Meinhardt & Gierer, 2000; Ouyang, Li, Li, & Swinney, 1995).

However, there are other ways to organize periodic tissue patterns. Periodic patterns can also arise from contact-mediated interactions without the need for prepatterns, diffusible molecules, or information at the global level (Kondo & Miura, 2010). One can show that local interactions between cells in a noisy environment are sufficient to lead to symmetry breaking and patterning in an initially homogeneous tissue (Collier et al., 1996) (Fig. 1B). In this case, the frequency of the developing pattern depends on the range over which cells can send and receive signals (Fig. 1B, C). When signaling only occurs between direct neighbors, patterning is expected to be denser whereas sparser patterns emerge when cells expand their sphere of influence through movement or signaling at protrusions (Fig. 1C). Contact dependent signaling at a distance can occur via the formation of cellular protrusions, including filopodia, cytonemes, and tunneling nanotubes. Cells in developing tissues and in culture have been observed to extend long, thin, processes that allow them to signal to distant cells (González-Méndez, Gradilla, & Guerrero, 2019). These protrusions are observed to participate in several signaling paradigms (e.g., TGFß, Wnt, Shh) (Bischoff et al., 2013; Hall et al., 2021; Hsiung, Ramirez-Weber, David Iwaki, & Kornberg, 2005; Inaba, Buszczak, & Yamashita, 2015; Mattes et al., 2018). The primary evidence that they participate in the dispersion of local signals includes (1) the localization of signaling molecules along the length of the protrusion; (2) that interfering with protrusion length leads to the disruption of signaling gradients; and (3) that downstream effectors are activated in the vicinity of protrusion contacts. Indeed, the models of Turing-like patterns does not specify how the activator and inhibitor are transported in space, and cellular protrusions likely represents just one of the mechanisms by which morphogen movement occurs.

## 2.2 Time scales

Developmental patterning takes place during a finite window of time: patterning is initiated when certain signaling pathways become activated and ends when cells have received the appropriate signals and have become committed to their fate. The entire process depends on a range of events, each characterized by its own timescale. For example, sub-cellular timescales comprise events like metabolite turnover, transcription, or intracellular trafficking, while cellular timescales comprise events such as cell division and migration. Returning to the example of the Bicoid gradient in Drosophila embryos, multiple events associated to different timescales occur for the Bicoid transcription factor to form a gradient: this includes the translation of protein from the maternally deposited mRNA (minutes; Petkova, Little, Liu, & Gregor, 2014), diffusion or active transport of protein and mRNA away from the anterior pole (Ali-Murthy & Kornberg, 2016; Durrieu et al., 2018; Spirov et al., 2009), the binding of transcription factors to DNA targets (seconds), protein turnover lifetime ($t_{1/2}$–30 min; Durrieu et al., 2018), and nuclear division cycles (minutes; Foe & Alberts, 1983). Together, these events ensure a robust and reproducible morphogen gradient along the AP axis of the developing embryo that is initiated at fertilization and maintained until cellularization ~3 h later.

For multicellular patterns, similar considerations for transcription and translation can be made, but the timescale of trafficking, extracellular dispersion and degradation, as well as receptor dynamics must also be accounted for. Other cell behaviors in epithelia also need to be considered, for example the cell cycle for patterning tissues that are simultaneously growing; the timescale for the formation of cellular structures, like filopodia, which can deliver membrane-bound ligands and receptors; the timescale of cell movements—including neighbor exchanges, and individual or collective migrations. Together these determine the spatiotemporal dynamics of pattern formation.

The development of cutting-edge live-imaging techniques have allowed researchers to quantify many of these processes. Some examples include, FRAP and FCS assays to quantify the diffusivity and degradation rate of morphogens (Kicheva et al., 2007; Zhou et al., 2012) or nanobody assays that can quantify rates of molecular internalization and recycling (Buser, Schleicher, Prescianotto-Baschong, & Spiess, 2018; Stapornwongkul et al., 2020). However, several questions remain. How do the multiple events occurring at the molecular, cellular and tissue level together dictate the time window in which patterning can happen? And can some processes be neglected when defining the timescale of patterning? Theoretical modeling can help shed light on these questions. For example, it can be shown that in a system where several processes control morphogen transport—such as extracellular diffusion, internalization, recycling and degradation of molecules—different effective timescales emerge that each depend on all these phenomena. Therefore, the timescale at which the morphogen profile is expected to reach steady state is not determined by a single process but is a nontrivial function of the rates at which molecules are trafficked and degraded (Aguilar-Hidalgo, Hadjivasiliou, Romanova-Michaelides, González-Gaitán, & Jülicher, 2019). At the same time, there are regimes in the parameter space that suggest certain events may dominate morphogen gradient formation and others can be neglected. A rigorous theoretical framework together with appropriate quantitative assays can help define which processes specify the timescale of gradient formation and patterning (Romanova-Michaelides et al., 2022). In the following section we approach this problem in the context of Notch signaling, then discuss how quantitative modeling has helped to integrate the cell and molecular complexities that drive to Notch-mediated tissue patterning.

# 3. Notch mediated patterning across scales

## 3.1 Notch signaling overview

Notch signaling is one of the best-studied examples of lateral inhibition, and is employed throughout development and across evolution to specify distinct signaling identities in neighboring cells. Examples of processes that depend on Notch signaling include, among many more, the selection of small sensory bristles in the Drosophila notum (Corson, Couturier, Rouault, Mazouni, & Schweisguth, 2017), the differentiation of cells into neurons in the zebrafish spinal cord (Hadjivasiliou et al., 2019), blood vessel formation (Zakirov et al., 2021), hair cell patterning in the inner ear (Lanford et al., 1999) and the synchronization of oscillations between neighboring cells during vertebrate segmentation (Ozbudak & Lewis, 2008). Notch signaling requires that signal sending and receiving cells are in contact with one another (Fig. 2A). Notch mediated patterning can require the coordination

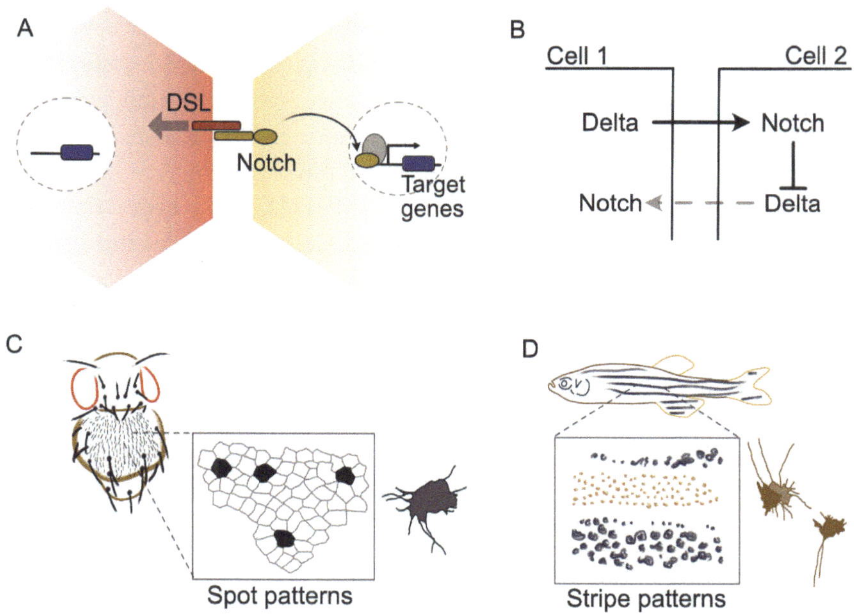


**Fig. 2** Notch signaling during pattern formation. (A) A simplified schematic of Notch signaling. Pink cell to the left expresses high levels of DSL ligand, upon endocytosis (gray arrow) of DSL bound to Notch receptor in trans, Notch is cleaved and the intracellular domain translocates (black arrow) to the nucleus to regulate the expression of target genes with co-transcription factors (gray oval). Thus the yellow cell to the right becomes Notch activated. (B) The overall feedback loop of Notch signaling during lateral inhibition. Notch activation in Cell 2 leads to the repression of further Delta expression in Cell 2, which decreases the activation of Notch in Cell 1. Without Notch-mediated repression of Delta expression, Cell 1 maintains higher levels of Delta ligand. (C) The small sensory hairs on the dorsal thorax of *Drosophila melanogaster* is a model system for long-range lateral inhibition. Notch signaling mediates the spacing of Delta-expressing bristle precursor cells (black cells) among Notch activated epithelial cells (white cells). Cells which are more than one cell diameter away from each other may contact each other via actin-rich protrusions, shown to the right. (C) The pigment stripes of Zebrafish are generated by long-range lateral inhibition. Xanthophores (yellow) and melanophores (dark gray) extend protrusions towards each other (shown to the right) that support Notch signaling.

of many processes including feedback dynamics that underlie the signaling pathway, interactions between the Notch ligand and its receptor within and between cells, cell motility, and cellular protrusions. As such, Notch signaling offers an ideal framework to discuss how events at the molecular, cellular and tissue level together specify the length and timescale of patterning, and how signaling at the single cell level can effectively incorporate spatial and temporal information of a length and time scale that transcends those of individual cells.

Excellent reviews of the mechanisms of Notch activation have been written (e.g., Binshtok & Sprinzak, 2018; Bocci, Onuchic, & Jolly, 2020; Bray, 2016; Kopan & Ilagan, 2009), therefore we will only briefly introduce Notch signaling here. Notch is a type I transmembrane receptor whose canonical signaling pathway is activated upon binding with the type I transmembrane ligands Delta, Serrate, or Jagged (DSL) in a contacting cell. Cleavage of Notch, and thus its overall activation state, depends on the presence of (1) mechanical pulling forces orthogonal to the surface of the receptor expressing cell, provided by endocytosis of DSL by the DSL expressing cell (Gordon et al., 2015; Langridge & Struhl, 2017) and (2) proteases which cleave Notch first towards the C-terminal end of the extracellular domain and next in the transmembrane domain. The Notch intracellular domain (NICD) is released from the membrane via cleavage by the gamma-secretase complex in the transmembrane domain (Struhl & Greenwald, 1999). Once translocated to the nucleus, NICD is free to associate with other transcription factors (e.g., DNA-binding protein CBF-1/Suppressor of Hairless/LAG1; CSL) to collectively modulate the transcription of target genes. The downstream activity of NICD, and its co-factors, on gene expression is highly context dependent (Bray, 2016).

The presence of Notch protein on the cell surface is tightly controlled through the activity of the endosomal regulators (Johnson, Zitserman, & Roegiers, 2016). Movement of Notch from internal pools to the cell surface is dependent on signaling contexts, but the half-life for many signaling receptors at the cell surface occurs on the order of hours (Hervé, Derangeon, Bahbouhi, Mesnil, & Sarrouilhe, 2007). For Notch molecules actively engaging in signaling with a DSL ligand in trans, endocytosis (~minutes), enzymatic cleavages (~seconds) and translocation to the nucleus (~minutes) occur rapidly relative to other steps in the signaling pathway (Ubezio et al., 2016). Once cleaved, the stability of the NICD fragment is in part regulated by post-translational modifications (e.g., phosphorylation of the PEST domain; NICD half-life ranges from minutes to hours) (Fryer, White, &

Jones, 2004). Interactions with NICD also leads to increased stability of CSL-DNA interactions, which may contribute to stabilizing the transcriptional response to Notch signaling (Falo-Sanjuan, Lammers, Garcia, & Bray, 2019). Negative feedback on the pathway occurs through the repression of pro-neural genes and DSL. This feedback loop occurs at longer timescales than other molecular events, in part because of the need to turnover existing Delta (Fig. 2B). All these steps in Notch signaling, as well as the cell behaviors that support the ability of cells to engage in cell-cell contact—motility (minutes–hours) protrusion formation (minutes), cell division (hours)—feed into the overall signaling dynamics and timescale of patterning at the tissue level.

In addition to activation in trans, the Notch signaling pathway is also subject to regulation in cis (i.e., ligand and receptor interactions in the same cell). *Cis*-inhibition is the sequestering of unactivated Notch receptor by higher concentrations of DSL ligand in the same cell. The details of the dynamics and regulation of this sequestration is unknown, but both experimental evidence and mathematical modeling results (del Álamo, Rouault, & Schweisguth, 2011; Palmer, Jia, & Deng, 2014; Sprinzak, Lakhanpal, LeBon, Garcia-Ojalvo, & Elowitz, 2011; Sprinzak et al., 2010) indicate that cis-inhibition contributes to efficient lateral inhibition, and the generation of sharp boundaries during certain pattern formation processes. It is important to note that molecular and cellular noise contributes to every step in the signaling pathway, and therefore stochastic noise in molecular processes need to be averaged over space and/or time in order to generate a precise patterning result. However, these heterogeneities may also help the progression of reproducible patterns in vivo. The ability of experimental approaches to measure the collective effect of different molecular and cellular timescales, as well as the noise averaging associated with them to achieve a coherent signal, is currently limited.

## 3.2 The interdependence of space and time in Notch signaling

A key aspect of Notch signaling is the negative feedback which exists downstream of receptor activation. One of the indirect targets of NICD transcriptional regulation is DSL itself, such that expression of ligand is repressed by activation of Notch (Fig. 2B). It follows that neighboring cells inhibit one another from producing the DSL ligands. As a result, patterns of cells of alternating fates emerge in an initially homogeneous tissue (Fig. 1B). The emerging pattern is expected to be dense with about 2–3 cell diameters between subsequent DSL cells, although this also depends on the dimensionality of the tissue (Cohen et al., 2010). The spatial patterns observed in many experimental systems, however, are more sparse suggesting that the effective range over which Notch signaling takes place exceeds that of direct neighbors alone (Fig. 2C, D). What is more, the duration of the process of lateral inhibition relies on the transcriptional feedback that dictate the signaling pathway. For example, it is estimated that a single round of lateral inhibition in mouse cells would take up to 6 h, suggesting that the process of patterning an entire tissue may take several days (Zakirov et al., 2021). Nonetheless, patterning often takes effect in a substantially smaller time window. These observations suggest that additional mechanisms may be in place that expand the length and timescales over which Notch mediated patterning occurs. In the following sections we discuss the role of processes that impact the effective length scale and temporal dynamics of Notch signaling.

## 3.3 Mechanisms for the spatiotemporal control of Notch signaling

### *3.3.1 Protrusions*

Evidence for Notch signaling through cellular protrusions include several examples from Drosophila and Zebrafish. Epithelial cells that are precursors to the small sensory bristles in the Drosophila pupal notum extend long, actin-rich, protrusions from their apical (De Joussineau et al., 2003; Renaud & Simpson, 2001) or basal surface (Cohen et al., 2010) (Fig. 2C). The former only have been suggested to appear under mechanical stress. Since Delta protein localizes to these protrusions, it was proposed that they might help increase the range of signaling for any given sensory bristle precursor cell. Indeed, the spacing between bristle precursor cells is coupled to protrusion length and mutations which disrupt protrusion length increase bristle density (Cohen et al., 2010; Georgiou & Baum, 2010; Hunter et al., 2019). The protrusions are present on all epithelial cells in the notum, in addition to the bristle precursor cells, which should further increase the range of lateral inhibition.

In the notum, protrusions also play a role in regulating the time window over which patterning occurs. This is achieved by coupling Notch signaling to the cell cycle that ultimately determines when cells become committed to their fates. Once cells reach a threshold of Notch activation, G2-exit is triggered; epithelial cells which have divided no longer participate in lateral inhibition (Hunter et al., 2016). Notch activation in cells that receive Delta signals through protrusions alone increases more slowly compared to cells with larger contacts, leading to a longer time window over which patterning takes place. The formation of this pattern features a refinement process, whereby cells in the tissue adjust the pattern prior to the pattern being fixed. Adjustments include switching cell fates and apoptosis (Cohen et al., 2010; Koto, Kuranaga, & Miura, 2011), which help to ensure regular spacing but occur by yet unknown mechanisms. Therefore, allowing cells distant from the signal sending bristle precursor cell to have an increased window of time for plasticity may be a mechanism that promotes pattern refinement.

During development of the zebrafish pigment stripes, xanthophores (yellow pigment cells) express Delta ligands while melanophores (dark pigment cells) express Notch receptors, suggesting that Notch signaling plays a role in patterning the zebrafish skin (Hamada et al., 2014). Ectopic expression of Delta or Notch changes the boundaries and thickness of the stripes, supporting this hypothesis. Hamada et al. (2014) show that melanophores extend long protrusions towards xanthophores that mediate Notch signaling and contribute to the organization of the pigment stripes (Fig. 2D). Eom, Bain, Patterson, Grout, and Parichy (2015) demonstrated how Notch signaling via protrusions extending from xanthophore cells promotes the generation of stripes of alternating color that are several cell diameters wide. These long ($\sim$60 μm, or 5–6 cell diameters) rapid protrusions, or airinemes, are supported by both the actin and microtubule cytoskeleton, such that the genetic or pharmacological disruption of the cytoskeleton leads to the failure to segregate melanophores from interstripe regions. Delta ligand is observed to be carried in vesicles in airinemes and are released at the tip of the protrusion. In addition, disruption of xanthopore airinemes leads to a decreased Notch response in contacting melanophore cells. Finally, expression of constitutively active Notch in melanophores resulted in stripes that were broader than in wild type, supporting a role for Notch in promoting pigment sorting through melanophore migration. Although the role of airineme length and dynamics in the generation of the final pigment pattern is not yet known, these results demonstrate a role for long-range Notch signaling in the development of a Turing-like pattern.

Dynamic protrusions also specify the spatiotemporal dynamics of Notch signaling and subsequent neuronal differentiation in the zebrafish spinal cord (Hadjivasiliou et al., 2019; Moore & Alexandre, 2020), and the selection of tip cells in branching angiogenesis (Page et al., 2019). In these model systems, protrusion length and dynamics impact both the spatial and temporal dynamics of Notch signaling and pattern formation. We discuss these examples in more detail in the next section.

In other instances, it is less clear if the activity of protrusions is important for signaling length or time scales. Notch or Delta carrying protrusions have been shown to be essential for the development of the air sac primordium and stem cell niches in Drosophila (Huang & Kornberg, 2015; Yatsenko & Shcherbata, 2021). In the first example, the activation of Notch signaling in the air sac primordium is due to Delta on cytonemes, expressed in nearby myoblasts during wing disc development. In the second, the activation of Notch signaling in terminal filament cells during oogenesis is due to Delta presented on cellular projections from primordial germ cells. In both cases, the role of these protrusions is not a function of their length. In these examples, protrusion mediated signaling appears to ensure the targeted delivery of morphogens between cell types or tissues, comparable to the connectivity and function of neurons. When these extensions are shorter, for example, signaling simply fails.

The dynamic behavior of signaling protrusions can be measured using markers of cell shape visualized over time (González-Méndez, Seijo-Barandiarán, & Guerrero, 2017; Hunter et al., 2019). These studies show that protrusions have behaviors beyond simply extending and retracting, including pausing or trapezoid behavior, or even collective behaviors. Protrusions have lifetimes on the order of minutes to hours, and the signaling proteins on the protrusions often also appear motile. However, the tools needed to systematically and specifically manipulate protrusion mediated signaling have been lacking. Most evidence relies on cell-wide genetic manipulation of cytoskeleton regulators that also effect other essential processes in the cell that may contribute to signaling. For example, decreased Cdc42 activity leads to the decreased formation and maximum length of some protrusions, but Cdc42 is also can also play a role in endocytosis, which is essential for Notch signaling. Recently, promising optogenetic tools have been developed based on engineered myosin motors that allow the specific manipulation of morphogens along protrusions (Zhang et al., 2021). Tools such as these will help experimentalists address remaining questions about Notch signaling via protrusions. Direct evidence for Notch signaling via protrusions is still needed in most systems, comparable to the recent observations of Shh signaling via cytonemes in cell culture (Hall et al., 2021).

### *3.3.2 Cell division*

For tissues undergoing proliferation during patterning, cell divisions can in principle affect both the length- and timescales of Notch signaling. First, proliferation can increase the distance between two signal sending cells.

Unless there is a mechanism to increase the range of Notch signaling and allow the two cells to continue contacting each other, this can lead to local Notch signal minima. In the case of repeating patterns, local Notch minima leads to "filling in" of the pattern. This mechanism may be a feature of pattern refinement. During bristle patterning, the onset of cell division signals the end of lateral inhibition between bristle precursors and epithelial cell neighbors; in order to create the appropriate pattern, Notch activated cells divide first, followed by the bristle precursor. Disruption of this relative timing leads to inappropriate filling-in of the pattern and errors in bristle placement (Cohen et al., 2010; Hunter et al., 2016; Nègre, Ghysen, & Martinez, 2003).

Second, changes in cell morphology and tissue organization associated with cell division may have an impact on the lengthscale of Notch signaling. Proliferation can modify connectivity through the formation of new contacts between daughter cells, and so increase Notch/Delta cell interactions within tissues. However, cell rounding associated with mitosis inhibits the ability of cells to extend protrusions. In the patterning notum epithelium, for example, basal protrusions that mediate Notch signaling are retracted as cells enter mitosis (Rosa, Vlassaks, Pichaud, & Baum, 2015). Changes in cell morphology can alter the contact area between neighboring cells, which can have a dramatic effect on the ability of cells to engage in Notch signaling (Shaya et al., 2017). For example, Shaya et al. have shown that smaller cell-cell contacts result in weaker Notch activation dynamics, dominated by the diffusion of ligand and receptor in and out of the contact area. In contrast, larger contacts maintain a stronger Notch activation. Therefore, impact of cell behaviors like rounding during mitosis on Notch signaling will be context dependent.

Finally, one feature of development is that the window of time during which cell fate decisions occur is often coordinated with extrinsic clocks like the cell cycle (Ayeni et al., 2016; Hunter et al., 2016; Slowik & Bermingham-McDonogh, 2016). Not only can the cell cycle help define the window of time during which Notch signaling may occur, it can also be used iteratively to pattern different cell fates in a lineage. Sensory bristle precursor cells in the notum undergo four rounds of cell division, and Notch signaling is critical in between each division in order to specify the supporting cell types of the adult sensory bristle (Guo, Jan, & Jan, 1996). Mistiming the divisions of the bristle precursor lineage leads to the transformation of daughter cells into the wrong cell fates (Ayeni et al., 2016).

### 3.3.3 Cell migration and rearrangements

If a signaling cell migrates in space the number of neighbor cells it contacts increases and the effective range over which lateral inhibition can act grows. For juxtacrine signaling-based patterning systems, this can play an essential role and impact both the spatial and temporal dynamics of resulting patterns. In the case of the mammalian intestinal crypts, Notch-mediated lateral inhibition is required for the generation of secretory and absorptive epithelial cells (Tóth, Ben-Moshe, Gavish, Barkai, & Itzkovitz, 2017). As cells acquire their fates, they collectively migrate towards the lumenal tip of the villi. Tóth et al. shows that a clear delineation between the region in which lateral inhibition can occur and the region in which cell migration can occur is essential for maintaining the correct numbers of each cell type. This separation of behaviors acts to restrict the range of Notch signaling.

In contrast to the villi model, Notch signaling during cell mixing is essential for development of the AP axis in vertebrates (Lawton et al., 2013). The increase in range of Notch signaling provided by movements of cells in the posterior pre-somitic mesoderm allows for the coordination of genetic oscillators (Uriu et al., 2010). Notch signaling plays a key role during vertebrate somatogenesis by synchronizing the oscillations of gene expression in neighboring cells (Liao & Oates, 2017). During somatogenesis, cell movement and division leads to dynamic rearrangements of relative cell position in the pre-somitic mesoderm, the part of the tissue where cells have not yet differentiated into somites. These cellular rearrangements imply that the range of Notch signaling increases as cells exchange neighbors and the number of neighbors they interact with expands. Theoretical work suggests that cell motility in this context promotes synchronization of the oscillations by making oscillations more robust to external perturbations and expanding the parameter space where cell synchronization can be achieved (Uriu et al., 2010).

In the context of epithelia where there is no individual cell migration, neighbor exchanges may instead play a role in increasing the signaling range of a given signal sending cell. For example, at the level of the apical junction, notum epithelial cells do not move large distances even though they may undergo T1 transitions and exchange neighbors (Curran et al., 2017). However, basal to these junctions, the bulk of the cell bodies do exhibit some shuffling (Renaud & Simpson, 2001) but it is unclear if this random motion is required for increasing the scale of sensory bristle patterning.

### 3.3.4 cis-Interactions

During lateral inhibition patterning events, cis-inhibition allows initial heterogeneities in an individual cell's surface levels of Notch or DSL ligand to

## BOX 1 Mathematical descriptions of Notch signaling.

A general model of Notch mediated lateral inhibition can be defined by a set of coupled differential equations that describe the dynamical process of gene activation and inhibition as follows:

$$\frac{dR_i}{dt} = \beta_R \frac{\langle D \rangle_i^m}{q^m + \langle D \rangle_i^m} - \gamma_R R_i \quad (1) \qquad \frac{dD_i}{dt} = \beta_D \frac{p^l}{p^l + R_i^l} - \gamma_D D_i \quad (2)$$

Here $R_i$ and $D_i$ are the expression of Notch signaling and the Delta ligand in cell $i$, and $\langle D \rangle_i$ is the total amount of Delta received by cell $i$. The parameters $\beta_R$ and $\beta_D$ capture the baseline production rates for Notch signaling and Delta respectively and their ratio determines if heterogeneous spatial patterns are possible. The parameters $\gamma_R$ and $\gamma_D$ are the degradation rates of the Notch signal and Delta and specify the timescale at which the system reaches steady state. Finally, $q$, $m$ and $p$, $l$ capture the strength of the feedback between Notch signaling and Delta production.

A more detailed model describes the concentration of free ligand and receptors on individuals cells together with the levels of intracellular Notch singing (Eq. 3-4). Here, $N_i$, $D_i$, and $R_i$ are the concentration of the free Notch receptors, free Delta and activated Notch in cell $i$. The parameters $k_t$ and $k_c$ capture the strength of receptor-ligand binding between and within cells respectively [ref].

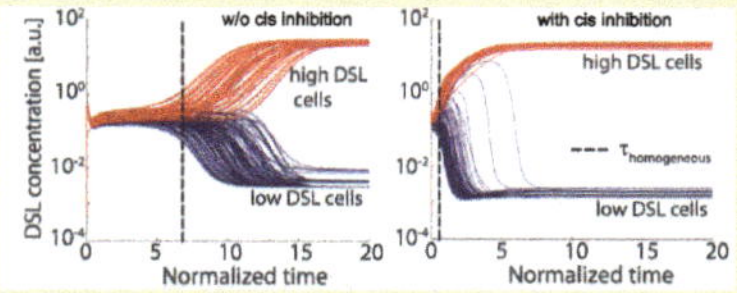


*Dynamics of cell signalling state in an initially homogeneous environment. The timescale of patterning ($\tau_{homogeneous}$) is substantially lower with cis inhibition. Figure adapted from Sprinzak et al (2012).*

$$\frac{dN_i}{dt} = \beta_N - \gamma_N N_i - k_t N_i \langle D \rangle_i - k_c N_i D_i \quad (3)$$

$$\frac{dD_i}{dt} = \beta_D \frac{p^l}{p^l + R_i^l} - \gamma_D D_i - k_t D_i \langle N \rangle_i - k_c N_i D_i \quad (4)$$

$$\frac{dR_i}{dt} = \beta_R \frac{\left(\frac{k_t}{\gamma_s} N_i \langle D \rangle_i\right)^m}{q^m + \left(\frac{k_t}{\gamma_s} N_i \langle D \rangle_i\right)^m} - \gamma_R R_i \quad (5)$$

Eq. (3-5) explicitly consider events that happen at the molecular level, like binding at the cell membrane between and within cells, and associated timescales. Computational analysis of this model combined with experiments showed that cis binding without receptor activation speeds up patterning dynamics, promotes the formation of boundaries and expands the palette of possible patterns Notch signaling can sustain.

Signaling at protrusions can be implemented by expanding the range over which cells can come in contact with one another according to the range and directionality of protrusions [refs]. In addition, different assumptions about the efficiency of binding at protrusion versus cell body contacts can be made by assigning weights to the incoming signals (Eq. 6).

$$\langle D \rangle_i = w_b \left( \sum_{j \in \text{cell body contact}} D_j \right)_i + w_p \left( \sum_{j \in \text{protrusion contact}} D_j \right)_i \quad (6)$$

be quickly amplified without the need to complete several feedback loops (Box 1). In particular, cis-inhibition is thought to circumvent the slowest steps of the signaling pathway, which include the downregulation and degradation of DSL in the Notch activated cell. Experimental evidence in cell culture supports the findings of mathematical models demonstrating that cells with feedback from cis-inhibition spend *less* time in the pre-committed, bipotential state (Sprinzak et al., 2011, 2010), than cells without cis-inhibition. At larger length-scales, the less time is spent in pre-committed cell states, the faster overall patterning can occur. Furthermore, cis-inhibition promotes multistability of patterns and robustness via error minimization (Formosa-Jordan & Ibañes, 2014; Sprinzak et al., 2011). Error minimization in the overall pattern is a consequence of increasing the speed of lateral inhibition: in the absence of cis-inhibition, the longer delays associated with the Notch signaling response in pre-committed cells with similar levels of Notch/DSL (in trans) can lead to scenarios where adjacent cells both initially adopt the same fate and fail to respond to each other before committing (Barad, Rosin, Hornstein, & Barkai, 2010). This may be especially problematic when cell fate commitment is linked to an extrinsic clock, like the cell cycle. Theoretical modeling also suggests an expansion in the phase space of possible patterns as well as parameters that lead to stable solutions is achieved when cis inhibition is in place (Formosa-Jordan & Ibañes, 2014).

### 3.3.5 Other regulators

Although the core Notch signaling pathway requires cell-cell contact, experimental evidence shows that Notch signaling can be modified, both directly and indirectly, by secreted regulators that directly bind to Notch. In principle, these regulators could help define the length or time-scales over which Notch signaling occurs. For example, there is evidence that *scabrous* acts as a direct regulator of Notch signaling, contributing to the length scale of lateral inhibition patterns. *Scabrous* encodes a secreted protein that has been shown to bind directly to Notch receptor and regulate its downstream signaling (Corson et al., 2017; Gavish et al., 2016; Mlodzik, Baker, & Rubin, 1990; Powell, Wesley, Spencer, & Cagan, 2001; Renaud & Simpson, 2001), however evidence suggests that *scabrous* may also be distributed via cellular protrusions (Lacoste et al., 2022). In the notum, *scabrous* mutants display decreased bristle spacing relative to wildtype flies (Renaud & Simpson, 2001) suggesting that the normal function of this protein is to promote long-range lateral inhibition, perhaps by increasing the sensitivity of distant cells to the weak DSL signals occurring in smaller, protrusion mediated, contacts. In the developing fly wing, ectopic expression of *scabrous* phenocopies wing defects in animals heterozygous for Notch, as well as decreasing the expression of Notch target genes (e.g., *Espl-m8*) (Lee, Yu, & Baker, 2000). These results are consistent with Scabrous as an antagonist of Notch activity. Together these studies indicate that the Scabrous can act as an antagonist or activator of Notch signaling, which may be context dependent.

Cells can release extracellular vesicles as a means of dispersing membrane-associated morphogens (McGough & Vincent, 2016). Evidence from cell culture experiments show that Notch ligand Delta-like 4 (Dll-4) can be packaged into exosomes and released into extracellular space (Sharghi-Namini, Tan,

Ong, Ge, & Asada, 2014; Sheldon et al., 2010). When a vesicle containing Dll-4 comes into contact with a Notch expressing cell, Notch target genes are expressed, indicating that the signaling pathway has been activated. Furthermore, in the developing vulva of *C. elegans*, several of Notch ligands and modifiers have been shown to be secreted, without the requirement of vesicles (Chen & Greenwald, 2004; Komatsu et al., 2008). Given evidence that a mechanical pulling force is required for the activation of Notch in trans (Langridge & Struhl, 2017), how effective signaling can occur via secreted Notch ligand is an open question. The distribution of these modifiers in extracellular space will determine the effect that they have on the range of lateral inhibition.

Other morphogens that prepattern tissues may also play an indirect role in the length and timescale of the Notch response. Continuing with the fly bristle pattern example, both Wg and Dpp signaling are required for the positioning of the invariant, large sensory bristle (Sato, Kojima, Michiue, & Saigo, 1999). In this example, Notch signaling occurs in the location specified by Wg and Dpp positional information. While prepatterning followed by lateral inhibition is a recurring theme in developmental patterning, in principle simultaneously combining a prepattern signal with a reaction-diffusion system can alter the length scale of the latter (Green & Sharpe, 2015). As will be discussed below, concurrent VEGF and Notch signaling during angiogenesis can speed up lateral inhibition, through the incorporation of additional feed forward loops.

## 4. Modeling long-range Notch signaling

The emergence of complex patterns during development has classically been attributed to molecules that disperse in tissues over large distances. However, recently theoretical modeling indicates that tissues can self-organize into diverse and complex patterns through contact mediated signaling alone, without the need to invoke diffusible factors (Binshtok & Sprinzak, 2018). Therefore, signaling pathways that require juxtacrine interactions such as Notch can in principle lead to a real diversity of spatial patterns of varying density, from spots to stripes and labyrinths (Formosa-Jordan & Ibañes, 2014; Hadjivasiliou, Hunter, & Baum, 2016). Theoretical work offers a framework that, combined with experiments, can improve our understanding of how the length and timescales involved in Notch signaling impact the spatiotemporal dynamics of patterning at the tissue level.

A mathematical description that captures the signaling dynamics of Notch was first introduced by Collier et al. (1996). This model describes Notch signaling by a pair of differential equations that capture the dynamics of activated Notch and Delta in individual cells (Box 1). In this context, the timescale of signaling interactions, and ultimately patterning at the tissue level, depend on key parameters like the degradation rate of the ligand and signal, and the strength of the negative feedback between Notch signaling and Delta production. A more detailed theoretical framework of the signaling dynamics was later put forward (Sprinzak et al., 2011, 2010). Here, a set of three differential equations describe the concentration of the Notch receptor, ligand and signaling levels over time (Box 1). This more explicit approach allows processes like binding and unbinding of ligands within and between cells to be directly considered. In this way, theoretical predictions about how the signaling and patterning dynamics depend on rates of molecular interactions can be obtained. Different approaches to model Notch signaling have been recently reviewed (Binshtok & Sprinzak, 2018). Below we focus on how cellular protrusions can be incorporated when modeling Notch, and the insights that such models can provide.

## 4.1 Modeling protrusion signaling

The role of protrusion mediated Notch signaling can be modeled by implementing an extended radius of influence for individual cells, so that the effective number of neighbors a cell is in contact with expands according to the protrusion length and polarization (Box 1). The strength of signaling at different contact types (cell body to cell body, protrusion to cell body, and protrusion to protrusion) can assume different weights to reflect variations in ligand or receptor concentrations at protrusions vs cell body, or variations in signaling efficiency. Furthermore, interactions between the Notch receptor and ligand can be modeled to incorporate inefficient receptor activation between cells (e.g., when cleavage of the receptor upon binding is not successful) leading to the effective sequestration of receptors (Hadjivasiliou et al., 2016). Exploring this framework suggests that tuning the interactions between Notch and its ligands at different contact types, as well as varying protrusion length and dynamics can substantially expand the phase space of patterns possible through Notch mediated lateral signaling alone to a range of patterns akin to those seen in diffusion-based systems. The spatial density of resulting patterns depends on protrusion length and the relative efficiency of activating Notch signaling at different cell contacts

(Hadjivasiliou et al., 2016; Vasilopoulos & Painter, 2016). Finally, the timescale of protrusion dynamics relative to transcriptional feedback, signal degradation and the duration of cell cycle together determine the scale of the spatial patterns and a time window over which cell fate determination becomes locked.

An alternative approach has been to capture protrusion mediated Notch signaling as structural noise (Cohen, Baum, & Miodownik, 2011). This has been done in the context of a cellular automaton model whereby the probability of becoming a signaling cell depends on the number of active neighbors, where direct and more distant neighbors are considered. In this context, the number of neighbors required to inactivate a signaling cell reflects a threshold in the incoming signal required for inactivation. Depending on the signaling range and inhibitory thresholds a range of spatial patterns become possible like those found in models that describe protrusion length and dynamics explicitly. In this study, the effects of spatial and temporal noise were explored and it was shown that intermediate levels of noise lead to optimized patterns. This result highlights that signaling noise, which is a feature of signaling systems and here introduced through signaling protrusions, can aid patterning by enabling cells to reverse their signaling state during pattern refinement. Importantly, the levels of noise must be finely tuned to the spatiotemporal dynamics of patterning and unsuitably high levels of noise lead to disordered spatial patterns (Cohen et al., 2011).

In other instances, protrusions are transient and their appearance correlates with the levels of Delta in a cell so that only cells that express high enough levels of the Delta ligand extend protrusions (Hadjivasiliou et al., 2019; Page et al., 2019). Here, protrusions can be modeled as dynamic processes that extend and retract as a response to Notch signaling. Cells that extend the long protrusions are able to upregulate Notch signaling in cells within their reach and so inhibit them from accumulating higher levels of Delta. In this context, theoretical modeling has shown that the speed of protrusion extension and restriction impacts the spatiotemporal dynamics of cell fate determination. Slow or shorter protrusions result in denser patterns that develop in a short period of time and vice versa (Hadjivasiliou et al., 2019).

Incorporating protrusions in models of Notch signaling can help make predictions about how changes in protrusion length, dynamics and signal efficiency impacts spatiotemporal patterns at the level of the tissue. Future theoretical models can integrate diffusible transported factors that operate together with cellular protrusions, for example to examine the putative role of factors such as *Scabrous* in promoting Notch signaling at a distance.

These approaches offer valuable tools to aid the design of appropriate experimental perturbations to test the role of protrusions in Notch signaling and tissue patterning in vivo.

# 5. Case studies of long-range Notch signaling

## 5.1 Spatial and temporal control of branching angiogenesis

Angiogenesis is the process by which new blood vessels are formed. During this process, endothelial cells (ECs) in preexisting blood vessels undergo sprouting as a response to vascular endothelial growth factor (VEGF). The conversion of ECs into tip cells that propagate new branches occurs in a spatially heterogeneous pattern regulated via Notch mediated lateral inhibition (Herbert & Stainier, 2011; Potente, Gerhardt, & Carmeliet, 2011).

The process is initiated when levels of VEGF increase as a response to hypoxia and induce Delta expression in ECs. Activated VEGF receptors (VEGFR) also promote the formation of dynamic filopodia through rapid, local polymerization of actin (Rousseau, Houle, Landry, & Huot, 1997). As cells express increasing levels of Delta, they begin to activate Notch in neighboring cells and the process of lateral inhibition is underway. At the same time, filopodia continue to reach further into the VEGF gradient and VEGFR activation on the filopodia increases (Fig. 3A). Activated Notch inhibits the expression of VEFG receptors which in turn inhibits VEGF signaling and filopodia production. This generates sharp positive feedback without the need for multiple rounds of transcription (Bentley & Chakravartula, 2017). Importantly, filopodia are formed within seconds following VEGFR activation and so provide a fast mechanism for feedback amplification that ultimately speeds up cell fate determination. Eventually, cells with higher VEGF and Delta signaling will be selected as tip cells and become migratory (Fig. 3A).

The role of filopodia in this process is primarily to speed up the process of lateral inhibition and tip cell selection. Multiple rounds of transcriptional feedback are required by Notch-Delta lateral inhibition to amplify the initially small differences in neighboring ECs and select for the heterogenous tip cell pattern necessary for the branching network that forms blood vessels (Collier et al., 1996). Each round of transcriptional feedback is estimated to take 3–4h in vitro in mouse ECs and 2h in zebrafish (Leslie et al., 2007; Ubezio et al., 2016). This would predict a timescale for the selection of a

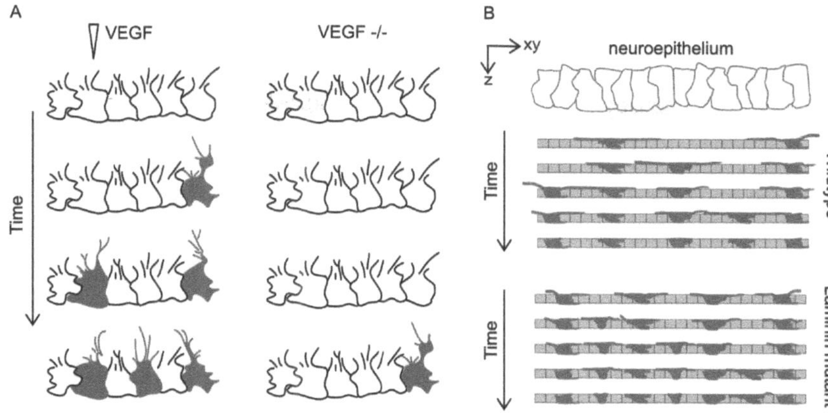


**Fig. 3** Examples of Notch signaling across length and timescales. (A) Endothelial cells during angiogenesis. At early time points, all cells have filopodia activity. In the presence of a VEGF gradient (left) that promotes Notch signaling, tip cells (green, Notch inactive) are patterned in a timely manner, with correct spatial distribution. In the absence of a VEGF gradient (right) tip cell selection and lateral inhibition is delayed. (B) The spacing of neurons (green) selected from initially unpatterned neuroepithelial cells depends on Notch signaling occurring along transient protrusions extending in the xy-plane. Neurons express Delta, present in the extended protrusions, which activates Notch in neighboring cells (gray). In laminin mutants, protrusions are shorter, leading to an decrease in the spacing between neurons and overall increase in density of neurons.

single tip cell that is of the order of a several hours to a day, much slower than the selection window observed experimentally within which all new tip cells are selected (~8 h in zebrafish) (Zakirov et al., 2021). As such, Notch mediated lateral inhibition alone cannot account for the spatiotemporal patterns of tip cell selection during angiogenesis. The sharp feedback generated through filopodia extension in this system offers a cell-based mechanism that overcomes this constraint. By physically expanding the length scale over which they can sense their environment and employing local feedback mechanisms cells effectively change the time-scale of the patterning process. The temporal dynamics of tip cell selection in turn impacts the spatial patterning of the branching network. Fast feedback implies a shorter timescale for the signaling dynamics, more tip cells being selected per unit time, and ultimately a denser branching network (Bentley & Chakravartula, 2017). Slower signaling dynamics delays lateral inhibition and tip cell selection and results in more sparse branching (Fig. 3A). Theoretical modeling of the process has shown than the positive feedback generated by filopodia sensing the VEGF signaling generate an ultrastable bistable switch that underlies fast and robust tip cell selection. Experimental perturbation of the VEGF levels in vivo and in vitro indicate that tip cell selection can be slowed down or sped up according to the levels of VEGF signaling. In agreement with theoretical predictions, higher levels of VEGF increased the number of ECs selected to sprout and, reversely, low level inhibition of VEGF signaling led to fewer tip cells being selected (Page et al., 2019).

This is an example whereby cell-based mechanisms, such as localized actin polymerization and filopodia, expand the range over which cells can access information, i.e., filopodia reaching further into the VEGF gradient. The fast, local feedback generated as a result impacts the dynamics of Notch signaling between neighboring cells allowing the timescale of fate determination to overcome the limitations set by the timescale of several rounds of gene transcription and activation required by the Notch pathway. Ultimately, the spatial and temporal scales achieved by these cellular processes impact the organization and length scale of patterning at the level of the tissue by specifying the duration of angiogenesis and density of the blood vessel branching network.

## 5.2 Spatiotemporal patterns of neurogenesis

During neurogenesis proliferative neuroepithelial cells gradually acquire proneural characteristics and generate specific neuron types. In the early stages of vertebrate neurogenesis, neurons of different subtype are born following a characteristic, non-random, spatiotemporal pattern in the spinal cord (Batista, Jacobstein, & Lewis, 2008; Dale, Roberts, Ottersen, & Storm-Mathisen, 1987; England, Batista, Mich, Chen, & Lewis, 2011; Higashijima, Mandel, & Fetcho, 2004; Roberts, Dale, Ottersen, & Storm-Mathisen, 1987). Quantitative analysis in vivo shows that neurons are rarely born close together in space and time. This pattern points to a local inhibitory mechanism that regulates the spatiotemporal appearance of new neurons. From a functional perspective, the sparse differentiation patterns observed experimentally may be important to allow for timely wiring of the central nervous system but this remains to be tested.

Live imaging in the zebrafish spinal cord has shown that neurons extend long protrusions along the apical-basal axis prior to differentiation (Fig. 3B) (Hadjivasiliou et al., 2019). The length of the protrusions resembles the spacing between neurons that are born at similar times. In addition, these cellular protrusions exhibit high levels of Delta, and Notch signaling is upregulated in their vicinity. Together these observations indicate that Notch mediated lateral inhibition may play a role in defining the spatiotemporal dynamics of neurogenesis. In agreement with this hypothesis, experimental perturbation of the length and dynamics of cellular protrusions through manipulation of the extracellular matrix protein laminin result in shorter protrusions and neurons that are born closer together in space and time (Fig. 3B).

The dynamics and length of the protrusions feed back into the timing of neurogenesis and spatial density of neurons. For example, theoretical modeling shows that the maximal length of the protrusions specifies an upper limit for the average spacing between neurogenesis events that occur at a similar time (Hadjivasiliou et al., 2019). This is an idealized state that emerges when protrusions are effectively extended instantaneously and quickly inhibit Delta production in cells within their reach. This limit is attained if the speed at which protrusions are extended is fast relative to the timescale of neural fate acquisition at the absence of Notch signaling. Conversely, if protrusions extend slowly or are inefficient at signaling, the spacing between neurons that are born close in time to one another is smaller than the maximal protrusion length (Fig. 3B). It follows that the interplay between the timescale of Notch signaling and protrusion dynamics together specify the spatiotemporal density of neuron differentiation.

A question that emerges from this work is the role of Notch signaling via the transient protrusions in neuron generation in higher dimensions. For example, protrusions extend additional filopodia along the Dorsal/Ventral axis suggesting that differentiating neurons may be able to influence their environment beyond the A/P axis they lie in. In addition, it is not known whether signaling interactions occur between the protrusions in the rare cases when protrusions from different neurons come in contact with one another, and how these could affect protrusion dynamics or retraction. Finally, although the presence of protrusions correlates with high levels of Delta, how extension is initiated as a response to Delta levels in this system is not clear. Further experimental work investigating possible feedback between Delta expression and the extension and dynamics of protrusions, and interactions between protrusions will help pin down the feedback mechanism in place.

## 6. Evolvability of patterns

Mechanisms that affect the ability of patterns to form, scale to body size, or to complete during the appropriate time window are likely under selective pressure (Curantz & Manceau, 2021). Using the previous example of airinemes during stripe formation in Zebrafish, a closely related species of fish (Pearl danio) does not feature stripes (Eom et al., 2015). Changes in the timing of xanthophore differentiation, due to the increased expression of a differentiation factor by neighboring interstripe cells, prohibits the formation of airinemes. Therefore, these cells cannot engage in long-range lateral inhibition, which is essential for stripe formation. These findings underline the importance of the interplay between gene regulation and the control of cell morphology for the evolution of patterning across species.

In another well-studied example, *Drosophila* species can exhibit a range of body sizes, and the organization of the thoracic sensory bristles varies with size (Simpson, Woehl, & Usui, 1999). There appears to be a mixture of scale invariant processes (e.g., the number and placement of large sensory bristles) in addition to the repeating processes that do not scale with tissue size (e.g., the number and organization of small sensory bristles). Scaling of the bristle patterns is important because the projection pattern of axons into the central nervous system is dependent on the location of the sensory bristle (Ghysen, 1980), and the behavioral response of the fly depends on the location of the stimulated bristles. Recent investigations of a mutant that alters both the spacing (Renaud & Simpson, 2001) and the timing of G2-exit in bristle precursors (Lacoste et al., 2022) indicates that mutations which disrupt the organization and timing of bristles and their neural projections lead to changes in cleaning reflexes. Despite the importance of the bristle organization for fly behavior, the regular array of sensory bristles on the dorsal thorax of Dipteran flies can be organized in a variety of ways, from randomly arranged small bristles to regularly spaced rows (Simpson et al., 1999). The genetic pathways and cell behaviors that control the placement of large and small sensory bristles was determined using Melanogaster, but in principle may extend to other Dipterans.

Both large and small sensory bristle patterns require Notch signaling, although in different contexts. The placement of the scale invariant large sensory bristles occurs during larval wing disc development and requires positional information generated by morphogen gradients (Yang, Hatton-Ellis, & Simpson, 2012). Notch signaling occurs after the placement of pro-neural clusters is determined, in order to generate the bristle cell lineages (Heitzler & Simpson, 1991). This is in contrast to the patterning of small sensory bristles, discussed previously, which is a more stochastic process that primarily relies on Notch signaling. Within the Drosophilidae family, larger flies tend to exhibit increased numbers of small bristles compared to the smaller Melanogaster, although the spacing between them does not always scale with size. There appears to be a conserved lateral inhibition process that drives a spaced pattern such that bristles do not occur adjacent to one another. However, these observations suggest that the cell based mechanisms for determining the length scale of Notch signaling may vary or in some cases may be absent. It is unknown if mechanisms that modify the timing of Notch signaling, as outlined above, contribute to the variety of bristle patterns observed.

The capacity of lateral inhibition for diversification, and the role of cell-based processes such as protrusions in evolution can be explored theoretically. For example, this can be addressed by exploring how mutations that impact Notch signaling and protrusion function affect patterning at the tissue level. It would be interesting to explore this scenario in tissues of varying size under selection to maintain or alter patterning proportions relative to size. Theoretical work can also address what is the capacity for evolvability and robustness in patterning systems that utilize local cell-cell interactions versus dispersed morphogens. This can help explain what the evolutionary benefits and constraints of alternative patterning mechanisms are.

## 7. Conclusion

Notch signaling is a highly conserved signaling pathway that regulates binary cell fate decisions in a variety of contexts—from controlling neural stem cell populations in the developing brain, differentiation of enterocytes in the intestinal epithelium, to the formation of spot and stripe patterns in Drosophila and Zebrafish. In many cases Notch signaling achieves this through lateral inhibition, but the range of patterns generated by this process should be limited to alternating cell types. Indeed, this is what we observe for initial mathematical models of lateral inhibition (Collier et al., 1996). However, the repertoire of Notch-mediated patterns includes those that are more sparse than would otherwise be predicted. As we develop better tools to observe the dynamics of Notch signaling over time, we also see that lateral inhibition occurs more rapidly than would be predicted by the feedback loops that dominate the timescale of the Notch pathway. Therefore, there must be mechanisms for modulating the length- and timescale of Notch signaling. In this review we have discussed how cellular mechanisms, from protrusions to ligand-receptor interactions, cell division and cellular rearrangements, as well as the coupling between secreted regulators and Notch signaling allow cells and tissues to overcome these spatial and temporal constraints.

In light of mounting experimental evidence that cellular protrusions contribute to Notch signaling across model systems, we have focused on the role of these processes in determining the spatiotemporal dynamics of patterning at the tissue level. Because protrusions allow cells to activate Notch signaling at a distance, the spacing between Delta expressing cells in the emerging patterns increases. At the same time, protrusions can be polarized allowing not only spots but also stripes and labyrinth like patterns to emerge through Notch signaling. The dynamic nature of protrusions that can grow and shrink, possibly as a response to Notch signaling, also introduces a temporal aspect to their function. During many differentiation processes, changes in cell morphology follow as a consequence of adopting a new cell state. Further work is needed to establish how cellular protrusions and Notch signaling dynamically modulate each other. A combination of theoretical and experimental approaches can offer the quantitative tools needed to address this challenge.

Modulating the length of protrusions may also allow patterns to become sparser in larger tissues, but as organ size undergoes changes of several orders of magnitude, cell based adjustments are unable to adapt to the tissue size. This suggests that the ways in which patterns transform between species of different size may vary depending on the mechanism that underlies pattern formation. Theoretical work together with comparative approaches can explore the scaling potential of patterns that depend on extracellular transport as well as of cell-based mechanisms across evolution.

## Acknowledgments

The authors would like to acknowledge the following support: G.H. received institutional support from Clarkson University. Z.H. was supported by the Francis Crick Institute, which receives its core funding from Cancer Research UK; the UK Medical Research Council and Wellcome Trust. We thank Buzz Baum (LMB, Cambridge) for critical and constructive comments on this manuscript.

## References

Aguilar-Hidalgo, D., Hadjivasiliou, Z., Romanova-Michaelides, M., González-Gaitán, M., & Jülicher, F. (2019). Dynamic modes of morphogen transport. *arXiv*. 1909. 13280 [physics.bio-ph].

Ali-Murthy, Z., & Kornberg, T. B. (2016). Bicoid gradient formation and function in the Drosophila pre-syncytial blastoderm. *eLife*, *5*, e13222. https://doi.org/10.7554/eLife. 13222.

Ayeni, J. O., Audibert, A., Fichelson, P., Srayko, M., Gho, M., & Campbell, S. D. (2016). G2 phase arrest prevents bristle progenitor self-renewal and synchronizes cell division with cell fate differentiation. *Development*, *143*, 1160–1169. https://doi.org/10.1242/ dev.134270.

Barad, O., Rosin, D., Hornstein, E., & Barkai, N. (2010). Error minimization in lateral inhibition circuits. *Science Signaling*, *3*(129), ra51. https://doi.org/10.1126/scisignal.2000857. PMID: 20606215.

Batista, M. F., Jacobstein, J., & Lewis, K. E. (2008). Zebrafish V2 cells develop into excitatory CiD and notch signalling dependent inhibitory VeLD interneurons. *Developmental Biology*, *322*, 263–275. https://doi.org/10.1016/j.ydbio.2008.07.015.

Bentley, K., & Chakravartula, S. (2017). The temporal basis of angiogenesis. *Philosophical Transactions of the Royal Society of LondonSeries B, Biological Sciences*, *372*, 20150522. https://doi.org/10.1098/rstb.2015.0522.

Ben-Zvi, D., Pyrowolakis, G., Barkai, N., & Shilo, B.-Z. (2011). Expansion-repression mechanism for scaling the Dpp activation gradient in Drosophila wing imaginal discs. *Current Biology*, *21*, 1391–1396. https://doi.org/10.1016/j.cub.2011.07.015.

Binshtok, U., & Sprinzak, D. (2018). *Modeling the Notch response, in: Molecular mechanisms of Notch signaling. Advances in experimental medicine and biology*. Cham: Springer.

Bischoff, M., Gradilla, A.-C., Seijo, I., Andrés, G., Rodríguez-Navas, C., González-Méndez, L., et al. (2013). Cytonemes are required for the establishment of a normal hedgehog morphogen gradient in Drosophila epithelia. *Nature Cell Biology*, *15*, 1269–1281. https://doi.org/10.1038/ncb2856.

Bocci, F., Onuchic, J. N., & Jolly, M. K. (2020). Understanding the principles of pattern formation driven by Notch signaling by integrating experiments and theoretical models. *Frontiers in Physiology*, *11*, 929. https://doi.org/10.3389/fphys.2020.00929.

Bosch, P. S., Ziukaite, R., Alexandre, C., Basler, K., & Vincent, J.-P. (2017). Dpp controls growth and patterning in Drosophila wing precursors through distinct modes of action. *eLife*, *6*, e22546. https://doi.org/10.7554/eLife.22546.

Bray, S. J. (2016). Notch signalling in context. *Nature Reviews. Molecular Cell Biology*, *17*, 722–735. https://doi.org/10.1038/nrm.2016.94.

Buser, D. P., Schleicher, K. D., Prescianotto-Baschong, C., & Spiess, M. (2018). A versatile nanobody-based toolkit to analyze retrograde transport from the cell surface. *Proceedings of the National Academy of Sciences of the United States of America*, *115*, E6227–E6236. https://doi.org/10.1073/pnas.1801865115.

Campbell, G., & Tomlinson, A. (1999). Transducing the Dpp morphogen gradient in the wing of Drosophila: Regulation of Dpp targets by brinker. *Cell*, *96*, 553–562. https://doi.org/10.1016/S0092-8674(00)80659-5.

Chen, N., & Greenwald, I. (2004). The lateral signal for LIN-12/notch in C. elegans vulval development comprises redundant secreted and transmembrane DSL proteins. *Developmental Cell*, *6*, 183–192. https://doi.org/10.1016/s1534-5807(04)00021-8.

Clark, E., & Akam, M. (2016). Odd-paired controls frequency doubling in Drosophila segmentation by altering the pair-rule gene regulatory network. *eLife*, *5*, e18215. https://doi.org/10.7554/eLife.18215.

Cohen, M., Baum, B., & Miodownik, M. (2011). The importance of structured noise in the generation of self-organizing tissue patterns through contact-mediated cell-cell signalling. *Journal of the Royal Society, Interface*, *8*, 787–798. https://doi.org/10.1098/rsif.2010.0488.

Cohen, M., Georgiou, M., Stevenson, N. L., Miodownik, M., & Baum, B. (2010). Dynamic Filopodia transmit intermittent delta-notch signaling to drive pattern refinement during lateral inhibition. *Developmental Cell*, *19*, 78–89. https://doi.org/10.1016/j.devcel.2010.06.006.

Collier, J. R., Monk, N. A., Maini, P. K., & Lewis, J. H. (1996). Pattern formation by lateral inhibition with feedback: A mathematical model of delta-notch intercellular signalling. *Journal of Theoretical Biology*, *183*, 429–446. https://doi.org/10.1006/jtbi.1996.0233.

Corson, F., Couturier, L., Rouault, H., Mazouni, K., & Schweisguth, F. (2017). Self-organized Notch dynamics generate stereotyped sensory organ patterns in Drosophila. *Science*, *356*, eaai7407. https://doi.org/10.1126/science.aai7407.

Curantz, C., & Manceau, M. (2021). Trends and variation in vertebrate patterns as outcomes of self-organization. *Current Opinion in Genetics & Development, 69*, 147–153. https://doi.org/10.1016/j.gde.2021.05.001.

Curran, S., Strandkvist, C., Bathmann, J., de Gennes, M., Kabla, A., Salbreux, G., et al. (2017). Myosin II controls junction fluctuations to guide epithelial tissue ordering. *Developmental Cell, 43*. https://doi.org/10.1016/j.devcel.2017.09.018. 480–492.e6.

Dale, N., Roberts, A., Ottersen, O. P., & Storm-Mathisen, J. (1987). The development of a population of spinal cord neurons and their axonal projections revealed by GABA immunocytochemistry in frog embryos. *Proceedings of the Royal Society of London - Series B: Biological Sciences, 232*, 205–215. https://doi.org/10.1098/rspb.1987.0069.

De Joussineau, C., Soulé, J., Martin, M., Anguille, C., Montcourrier, P., & Alexandre, D. (2003). Delta-promoted filopodia mediate long-range lateral inhibition in Drosophila. *Nature, 426*(6966), 555–559. https://doi.org/10.1038/nature02157. Erratum in: Nature 2004; 428(6981):445. PMID: 14654840.

del Álamo, D, Rouault, H, & Schweisguth, F. (2011). Mechanism and significance of cis-inhibition in Notch signalling. *Current Biology, 21*(1), R40–R47. https://doi.org/10.1016/j.cub.2010.10.034. PMID: 21215938.

Driever, W., & Nüsslein-Volhard, C. (1988). A gradient of bicoid protein in Drosophila embryos. *Cell, 54*, 83–93. https://doi.org/10.1016/0092-8674(88)90182-1.

Drocco, J. A., Grimm, O., Tank, D. W., & Wieschaus, E. (2011). Measurement and perturbation of morphogen lifetime: Effects on gradient shape. *Biophysical Journal, 101*, 1807–1815. https://doi.org/10.1016/j.bpj.2011.07.025.

Duque, J., & Gorfinkiel, N. (2016). Integration of actomyosin contractility with cell-cell adhesion during dorsal closure. *Development, 143*(24), 4676–4686. https://doi.org/10.1242/dev.136127. Epub 2016 Nov 11. PMID: 27836966.

Durrieu, L., Kirrmaier, D., Schneidt, T., Kats, I., Raghavan, S., Hufnagel, L., et al. (2018). Bicoid gradient formation mechanism and dynamics revealed by protein lifetime analysis. *Molecular Systems Biology, 14*, e8355. https://doi.org/10.15252/msb.20188355.

England, S., Batista, M. F., Mich, J. K., Chen, J. K., & Lewis, K. E. (2011). Roles of hedgehog pathway components and retinoic acid signalling in specifying zebrafish ventral spinal cord neurons. *Development, 138*, 5121–5134. https://doi.org/10.1242/dev.066159.

Eom, D. S., Bain, E. J., Patterson, L. B., Grout, M. E., & Parichy, D. M. (2015). Long-distance communication by specialized cellular projections during pigment pattern development and evolution. *eLife, 4*, e12401. https://doi.org/10.7554/eLife.12401.

Falo-Sanjuan, J., Lammers, N. C., Garcia, H. G., & Bray, S. J. (2019). Enhancer priming enables fast and sustained transcriptional responses to notch signaling. *Developmental Cell, 50*. https://doi.org/10.1016/j.devcel.2019.07.002. 411–425.e8.

Fancher, S., & Mugler, A. (2020). Diffusion vs. direct transport in the precision of morphogen readout. *eLife, 9*, e58981. https://doi.org/10.7554/eLife.58981.

Foe, V. E., & Alberts, B. M. (1983). Studies of nuclear and cytoplasmic behaviour during the five mitotic cycles that precede gastrulation in Drosophila embryogenesis. *Journal of Cell Science, 61*, 31–70.

Formosa-Jordan, P., & Ibañes, M. (2014). Competition in notch signaling with Cis enriches cell fate decisions. *PLoS One, 9*, e95744. https://doi.org/10.1371/journal.pone.0095744.

Fryer, C. J., White, J. B., & Jones, K. A. (2004). Mastermind recruits CycC:CDK8 to phosphorylate the notch ICD and coordinate activation with turnover. *Molecular Cell, 16*, 509–520. https://doi.org/10.1016/j.molcel.2004.10.014.

Gavish, A., Shwartz, A., Weizman, A., Schejter, E., Shilo, B. Z., & Barkai, N. (2016). Periodic patterning of the Drosophila eye is stabilized by the diffusible activator Scabrous. *Nature Communications, 7*, 10461. https://doi.org/10.1038/ncomms10461. PMID: 26876750; PMCID: PMC4756378.

Georgiou, M., & Baum, B. (2010). Polarity proteins and rho GTPases cooperate to spatially organise epithelial actin-based protrusions. *Journal of Cell Science*, *123*(Pt 7), 1089–1098. https://doi.org/10.1242/jcs.060772. Epub 2010 Mar 2. PMID: 20197404.

Ghysen, A. (1980). The projection of sensory neurons in the central nervous system of Drosophila: Choice of the appropriate pathway. *Developmental Biology*, *78*, 521–541. https://doi.org/10.1016/0012-1606(80)90351-6.

Glover, J. D., Wells, K. L., Matthäus, F., Painter, K. J., Ho, W., Riddell, J., et al. (2017). Hierarchical patterning modes orchestrate hair follicle morphogenesis. *PLoS Biology*, *15*, e2002117. https://doi.org/10.1371/journal.pbio.2002117.

González-Méndez, L., Gradilla, A.-C., & Guerrero, I. (2019). The cytoneme connection: Direct long-distance signal transfer during development. *Development*, *146*(9), dev174607. https://doi.org/10.1242/dev.174607. PMID: 31068374.

González-Méndez, L., Seijo-Barandiarán, I., & Guerrero, I. (2017). Cytoneme-mediated cell-cell contacts for Hedgehog reception. *eLife*, *6*, e24045. https://doi.org/10.7554/eLife.24045. PMID: 28825565; PMCID: PMC5565369.

Gordon, W. R., Zimmerman, B., He, L., Miles, L. J., Huang, J., Tiyanont, K., et al. (2015). Mechanical allostery: Evidence for a force requirement in the proteolytic activation of notch. *Developmental Cell*, *33*, 729–736. https://doi.org/10.1016/j.devcel.2015.05.004.

Green, J. B. A., & Sharpe, J. (2015). Positional information and reaction-diffusion: Two big ideas in developmental biology combine. *Development (Cambridge, England)*, *142*, 1203–1211. https://doi.org/10.1242/dev.114991.

Guo, M., Jan, L. Y., & Jan, Y. N. (1996). Control of daughter cell fates during asymmetric division: Interaction of numb and notch. *Neuron*, *17*, 27–41. https://doi.org/10.1016/s0896-6273(00)80278-0.

Hadjivasiliou, Z., Hunter, G. L., & Baum, B. (2016). A new mechanism for spatial pattern formation via lateral and protrusion-mediated lateral signalling. *Journal of the Royal Society, Interface*, *13*, 20160484. https://doi.org/10.1098/rsif.2016.0484.

Hadjivasiliou, Z., Moore, R. E., McIntosh, R., Galea, G. L., Clarke, J. D. W., & Alexandre, P. (2019). Basal protrusions mediate spatiotemporal patterns of spinal neuron differentiation. *Developmental Cell*, *49*. https://doi.org/10.1016/j.devcel.2019.05.035. 907–919.e10.

Hall, E. T., Dillard, M. E., Stewart, D. P., Zhang, Y., Wagner, B., Levine, R. M., et al. (2021). Cytoneme delivery of Sonic Hedgehog from ligand-producing cells requires Myosin 10 and a Dispatched-BOC/CDON co-receptor complex. *eLife*, *10*, e61432. https://doi.org/10.7554/eLife.61432.

Hamada, H., Watanabe, M., Lau, H. E., Nishida, T., Hasegawa, T., Parichy, D. M., et al. (2014). Involvement of Delta/Notch signaling in zebrafish adult pigment stripe patterning. *Development*, *141*, 318–324. https://doi.org/10.1242/dev.099804.

Hatori, R., Wood, B. M., Oliveira Barbosa, G., & Kornberg, T. B. (2021). Regulated delivery controls Drosophila Hedgehog, Wingless, and Decapentaplegic signaling. *eLife*, *10*, e71744. https://doi.org/10.7554/eLife.71744.

Heitzler, P., & Simpson, P. (1991). The choice of cell fate in the epidermis of Drosophila. *Cell*, *64*, 1083–1092. https://doi.org/10.1016/0092-8674(91)90263-x.

Herbert, S. P., & Stainier, D. Y. R. (2011). Molecular control of endothelial cell behaviour during blood vessel morphogenesis. *Nature Reviews. Molecular Cell Biology*, *12*, 551–564. https://doi.org/10.1038/nrm3176.

Hervé, J.-C., Derangeon, M., Bahbouhi, B., Mesnil, M., & Sarrouilhe, D. (2007). The Connexin turnover, an important modulating factor of the level of cell-to-cell junctional communication: Comparison with other integral membrane proteins. *The Journal of Membrane Biology*, *217*, 21–33. https://doi.org/10.1007/s00232-007-9054-8.

Higashijima, S.-I., Mandel, G., & Fetcho, J. R. (2004). Distribution of prospective glutamatergic, glycinergic, and GABAergic neurons in embryonic and larval zebrafish. *The Journal of Comparative Neurology*, *480*, 1–18. https://doi.org/10.1002/cne.20278.

Hsiung, F., Ramirez-Weber, F.-A., David Iwaki, D., & Kornberg, T. B. (2005). Dependence of Drosophila wing imaginal disc cytonemes on decapentaplegic. *Nature*, *437*, 560–563. https://doi.org/10.1038/nature03951.

Huang, H., & Kornberg, T. B. (2015). Myoblast cytonemes mediate Wg signaling from the wing imaginal disc and Delta-Notch signaling to the air sac primordium. *eLife*, *4*, e06114. https://doi.org/10.7554/eLife.06114. PMID: 25951303; PMCID: PMC4423120.

Huang, H., Liu, S., & Kornberg, T. B. (2019). Glutamate signaling at cytoneme synapses. *Science*, *363*, 948–955. https://doi.org/10.1126/science.aat5053.

Hunter, G. L., Hadjivasiliou, Z., Bonin, H., He, L., Perrimon, N., Charras, G., et al. (2016). Coordinated control of notch/Delta signalling and cell cycle progression drives lateral inhibition-mediated tissue patterning. *Development*, *143*, 2305–2310. https://doi.org/10.1242/dev.134213.

Hunter, G. L., He, L., Perrimon, N., Charras, G., Giniger, E., & Baum, B. (2019). A role for actomyosin contractility in notch signaling. *BMC Biology*, *17*(1), 12. https://doi.org/10.1186/s12915-019-0625-9. PMID: 30744634; PMCID: PMC6369551.

Inaba, M., Buszczak, M., & Yamashita, Y. M. (2015). Nanotubes mediate niche-stem-cell signalling in the *Drosophila testis*. *Nature*, *523*(7560), 329–332. https://doi.org/10.1038/nature14602. Epub 2015 Jul 1. PMID: 26131929; PMCID: PMC4586072.

Jaeger, J., Surkova, S., Blagov, M., Janssens, H., Kosman, D., Kozlov, K. N., et al. (2004). Dynamic control of positional information in the early Drosophila embryo. *Nature*, *430*, 368–371. https://doi.org/10.1038/nature02678.

Johnson, S. A., Zitserman, D., & Roegiers, F. (2016). Numb regulates the balance between notch recycling and late-endosome targeting in Drosophila neural progenitor cells. *Molecular Biology of the Cell*, *27*(18), 2857–2866. https://doi.org/10.1091/mbc.E15-11-0751. Epub 2016 Jul 27. PMID: 27466320; PMCID: PMC5025272.

Kicheva, A., Pantazis, P., Bollenbach, T., Kalaidzidis, Y., Bittig, T., Jülicher, F., et al. (2007). Kinetics of morphogen gradient formation. *Science*, *315*, 521–525. https://doi.org/10.1126/science.1135774.

Komatsu, H., Chao, M. Y., Larkins-Ford, J., Corkins, M. E., Somers, G. A., Tucey, T., et al. (2008). OSM-11 facilitates LIN-12 notch signaling during Caenorhabditis elegans vulval development. *PLoS Biology*, *6*, e196. https://doi.org/10.1371/journal.pbio.0060196.

Kondo, S., & Miura, T. (2010). Reaction-diffusion model as a framework for understanding biological pattern formation. *Science*, *329*, 1616–1620. https://doi.org/10.1126/science.1179047.

Kopan, R., & Ilagan, M. X. G. (2009). The canonical notch signaling pathway: Unfolding the activation mechanism. *Cell*, *137*, 216–233. https://doi.org/10.1016/j.cell.2009.03.045.

Kornberg, T. B. (2017). Distributing signaling proteins in space and time: The province of cytonemes. *Current Opinion in Genetics & Development*, *45*, 22–27. https://doi.org/10.1016/j.gde.2017.02.010.

Koto, A., Kuranaga, E., & Miura, M. (2011). Apoptosis ensures spacing pattern formation of Drosophila sensory organs. *Current Biology: CB*, *21*, 278–287. https://doi.org/10.1016/j.cub.2011.01.015.

Lacoste, J., Soula, H., Burg, A., Audibert, A., Darnat, P., Gho, M., et al. (2022). A neural progenitor mitotic wave is required for asynchronous axon outgrowth and morphology. *Elife*, *11*, e75746. https://doi.org/10.7554/eLife.75746. PMID: 35254258; PMCID: PMC8933001.

Lanford, P. J., Lan, Y., Jiang, R., Lindsell, C., Weinmaster, G., Gridley, T., et al. (1999). Notch signalling pathway mediates hair cell development in mammalian cochlea. *Nature Genetics*, *21*, 289–292. https://doi.org/10.1038/6804.

Langridge, P. D., & Struhl, G. (2017). Epsin-dependent ligand endocytosis activates notch by force. *Cell, 171*. https://doi.org/10.1016/j.cell.2017.10.048. 1383–1396.e12.

Lawton, A. K., Nandi, A., Stulberg, M. J., Dray, N., Sneddon, M. W., Pontius, W., et al. (2013). Regulated tissue fluidity steers zebrafish body elongation. *Development (Cambridge, England), 140*, 573–582. https://doi.org/10.1242/dev.090381.

Lee, E.-C., Yu, S.-Y., & Baker, N. E. (2000). The scabrous protein can act as an extracellular antagonist of notch signaling in the Drosophila wing. *Current Biology, 10*(15), 931–934. https://doi.org/10.1016/S0960-9822(00)00622-9. PMID: 10959842.

Leslie, J. D., Ariza-McNaughton, L., Bermange, A. L., McAdow, R., Johnson, S. L., & Lewis, J. (2007). Endothelial signalling by the notch ligand Delta-like 4 restricts angiogenesis. *Development (Cambridge, England), 134*, 839–844. https://doi.org/10.1242/dev.003244.

Liao, B.-K., & Oates, A. C. (2017). Delta-notch signalling in segmentation. *Arthropod Structure & Development Evolution of Segmentation, 46*, 429–447. https://doi.org/10.1016/j.asd.2016.11.007.

Maroto, M., Dale, J. K., Dequéant, M.-L., Petit, A.-C., & Pourquié, O. (2005). Synchronised cycling gene oscillations in presomitic mesoderm cells require cell-cell contact. *The International Journal of Developmental Biology, 49*, 309–315. https://doi.org/10.1387/ijdb.041958mm.

Mattes, B., Dang, Y., Greicius, G., Kaufmann, L. T., Prunsche, B., Rosenbauer, J., et al. (2018). Wnt/PCP controls spreading of Wnt/β-catenin signals by cytonemes in vertebrates. *eLife, 7*, e36953. https://doi.org/10.7554/eLife.36953.

McGough, I. J., & Vincent, J.-P. (2016). Exosomes in developmental signalling. *Development (Cambridge, England), 143*, 2482–2493. https://doi.org/10.1242/dev.126516.

Meinhardt, H., & Gierer, A. (2000). Pattern formation by local self-activation and lateral inhibition. *BioEssays, 22*, 753–760. https://doi.org/10.1002/1521-1878(200008)22:8<753::AID-BIES9>3.0.CO;2-Z.

Mlodzik, M., Baker, N. E., & Rubin, GM. (1990). Isolation and expression of scabrous, a gene regulating neurogenesis in Drosophila. *Genes & Development, 4*(11), 1848–1861. https://doi.org/10.1101/gad.4.11.1848. PMID: 2125959.

Moore, R., & Alexandre, P. (2020). Delta-notch signaling: The long and the short of a Neuron's influence on progenitor fates. *Journal of Developmental Biology, 8*, 8. https://doi.org/10.3390/jdb8020008.

Nègre, N., Ghysen, A., & Martinez, A. M. (2003). Mitotic G2-arrest is required for neural cell fate determination in Drosophila. *Mechanisms of Development, 120*, 253–265. https://doi.org/10.1016/s0925-4773(02)00419-7.

Ouyang, Q., Li, R., Li, G., & Swinney, H. L. (1995). Dependence of Turing pattern wavelength on diffusion rate. *The Journal of Chemical Physics, 102*, 2551–2555. https://doi.org/10.1063/1.468684.

Ozbudak, E. M., & Lewis, J. (2008). Notch signalling synchronizes the zebrafish segmentation clock but is not needed to create somite boundaries. *PLoS Genetics, 4*, e15. https://doi.org/10.1371/journal.pgen.0040015.

Page, D. J., Thuret, R., Venkatraman, L., Takahashi, T., Bentley, K., & Herbert, S. P. (2019). Positive feedback defines the timing, magnitude, and robustness of angiogenesis. *Cell Reports, 27*, 3139–3151.e5. https://doi.org/10.1016/j.celrep.2019.05.052.

Palmer, W. H., Jia, D., & Deng, W. M. (2014). Cis-interactions between Notch and its ligands block ligand-independent Notch activity. *Elife, 3*, e04415. https://doi.org/10.7554/eLife.04415. PMID: 25486593; PMCID: PMC4286723.

Petkova, M. D., Little, S. C., Liu, F., & Gregor, T. (2014). Maternal origins of developmental reproducibility. *Current Biology: CB, 24*, 1283–1288. https://doi.org/10.1016/j.cub.2014.04.028.

Potente, M., Gerhardt, H., & Carmeliet, P. (2011). Basic and therapeutic aspects of angiogenesis. *Cell, 146*, 873–887. https://doi.org/10.1016/j.cell.2011.08.039.

Powell, P. A., Wesley, C., Spencer, S., & Cagan, R. L. (2001). Scabrous complexes with Notch to mediate boundary formation. *Nature*, *409*(6820), 626–630. https://doi.org/10.1038/35054566. Erratum in: Nature 2001; 410(6829):718. PMID: 11214322.

Reilly, K. M., & Melton, D. A. (1996). Short-range signaling by candidate morphogens of the TGF beta family and evidence for a relay mechanism of induction. *Cell*, *86*, 743–754. https://doi.org/10.1016/s0092-8674(00)80149-x.

Renaud, O., & Simpson, P. (2001). Scabrous modifies epithelial cell adhesion and extends the range of lateral signalling during development of the spaced bristle pattern in Drosophila. *Developmental Biology*, *240*, 361–376. https://doi.org/10.1006/dbio.2001.0482.

Roberts, A., Dale, N., Ottersen, O. P., & Storm-Mathisen, J. (1987). The early development of neurons with GABA immunoreactivity in the CNS of Xenopus laevis embryos. *The Journal of Comparative Neurology*, *261*, 435–449. https://doi.org/10.1002/cne.902610308.

Romanova-Michaelides, M., Hadjivasiliou, Z., Aguilar-Hidalgo, D., Basagiannis, D., Seum, C., Dubois, M., et al. (2022). Morphogen gradient scaling by recycling of intracellular Dpp. *Nature*, *602*, 287–293. https://doi.org/10.1038/s41586-021-04346-w.

Rosa, A., Vlassaks, E., Pichaud, F., & Baum, B. (2015). Ect2/Pbl acts via rho and polarity proteins to direct the assembly of an isotropic actomyosin cortex upon mitotic entry. *Developmental Cell*, *32*, 604–616. https://doi.org/10.1016/j.devcel.2015.01.012.

Rousseau, S., Houle, F., Landry, J., & Huot, J. (1997). p38 MAP kinase activation by vascular endothelial growth factor mediates actin reorganization and cell migration in human endothelial cells. *Oncogene*, *15*, 2169–2177. https://doi.org/10.1038/sj.onc.1201380.

Sato, M., Kojima, T., Michiue, T., & Saigo, K. (1999). Bar homeobox genes are latitudinal prepattern genes in the developing Drosophila notum whose expression is regulated by the concerted functions of decapentaplegic and wingless. *Development (Cambridge, England)*, *126*, 1457–1466.

Serra-Picamal, X., Conte, V., Vincent, R., Anon, E., Tambe, D. T., Bazellieres, E., et al. (2012). Mechanical waves during tissue expansion. *Nature Physics*, *8*, 628–634. https://doi.org/10.1038/nphys2355.

Sharghi-Namini, S., Tan, E., Ong, L. L., Ge, R., & Asada, H. H. (2014). Dll4-containing exosomes induce capillary sprout retraction in a 3D microenvironment. *Scientific Reports*, *4*, 4031. https://doi.org/10.1038/srep04031. PMID: 24504253; PMCID: PMC3916896.

Shaya, O., Binshtok, U., Hersch, M., Rivkin, D., Weinreb, S., Amir-Zilberstein, L., et al. (2017). Cell-cell contact area affects notch signaling and notch-dependent patterning. *Developmental Cell*, *40*. https://doi.org/10.1016/j.devcel.2017.02.009. 505–511.e6.

Sheldon, H., Heikamp, E., Turley, H., Dragovic, R., Thomas, P., Oon, C. E., et al. (2010). New mechanism for Notch signaling to endothelium at a distance by Delta-like 4 incorporation into exosomes. *Blood*, *116*(13), 2385–2394. https://doi.org/10.1182/blood-2009-08-239228. Epub 2010 Jun 17. PMID: 20558614.

Shyer, A. E., Rodrigues, A. R., Schroeder, G. G., Kassianidou, E., Kumar, S., & Harland, R. M. (2017). Emergent cellular self-organization and mechanosensation initiate follicle pattern in the avian skin. *Science*, *357*, 811–815. https://doi.org/10.1126/science.aai7868.

Simpson, P. (1990). Lateral inhibition and the development of the sensory bristles of the adult peripheral nervous system of Drosophila. *Development*, *109*, 509–519.

Simpson, P., Woehl, R., & Usui, K. (1999). The development and evolution of bristle patterns in Diptera. *Development (Cambridge, England)*, *126*, 1349–1364.

Slowik, A. D., & Bermingham-McDonogh, O. (2016). A central to peripheral progression of cell cycle exit and hair cell differentiation in the developing mouse cristae. *Developmental Biology*, *411*(1), 1–14. https://doi.org/10.1016/j.ydbio.2016.01.033. Epub 2016 Jan 28. PMID: 26826497; PMCID: PMC4769996.

Spirov, A., Fahmy, K., Schneider, M., Frei, E., Noll, M., & Baumgartner, S. (2009). Formation of the bicoid morphogen gradient: An mRNA gradient dictates the protein gradient. *Development*, *136*, 605–614. https://doi.org/10.1242/dev.031195.

Sprinzak, D., Lakhanpal, A., LeBon, L., Garcia-Ojalvo, J., & Elowitz, M. B. (2011). Mutual inactivation of notch receptors and ligands facilitates developmental patterning. *PLoS Computational Biology*, 7, e1002069. https://doi.org/10.1371/journal.pcbi.1002069.

Sprinzak, D., Lakhanpal, A., LeBon, L., Santat, L. A., Fontes, M. E., Anderson, G. A., et al. (2010). Cis interactions between notch and Delta generate mutually exclusive signaling states. *Nature*, *465*, 86–90. https://doi.org/10.1038/nature08959.

Stapornwongkul, K. S., de Gennes, M., Cocconi, L., Salbreux, G., & Vincent, J.-P. (2020). Patterning and growth control in vivo by an engineered GFP gradient. *Science*, *370*(6514), 321–327. https://doi.org/10.1126/science.abb8205. PMID: 33060356; PMCID: PMC7611032.

Stapornwongkul, K. S., & Vincent, J.-P. (2021). Generation of extracellular morphogen gradients: The case for diffusion. *Nature Reviews. Genetics*, *22*, 393–411. https://doi.org/10.1038/s41576-021-00342-y.

Struhl, G., & Greenwald, I. (1999). Presenilin is required for activity and nuclear access of notch in Drosophila. *Nature*, *398*, 522. https://doi.org/10.1038/19091.

Tóth, B., Ben-Moshe, S., Gavish, A., Barkai, N., & Itzkovitz, S. (2017). Early commitment and robust differentiation in colonic crypts. *Molecular Systems Biology*, *13*, 902. https://doi.org/10.15252/msb.20167283.

Turing, A. M. (1952). The chemical basis of morphogenesis. *Philosophical Transactions of the Royal Society of London. Series B, Biological Sciences*, *237*, 37–72. https://doi.org/10.1007/BF02459572.

Ubezio, B., Blanco, R. A., Geudens, I., Stanchi, F., Mathivet, T., Jones, M. L., et al. (2016). Synchronization of endothelial Dll4-Notch dynamics switch blood vessels from branching to expansion. *eLife*, *5*, e12167. https://doi.org/10.7554/eLife.12167.

Uriu, K., Morishita, Y., & Iwasa, Y. (2010). Random cell movement promotes synchronization of the segmentation clock. *Proceedings of the National Academy of Sciences of the United States of America*, *107*, 4979–4984. https://doi.org/10.1073/pnas.0907122107.

Vasilopoulos, G., & Painter, K. J. (2016). Pattern formation in discrete cell tissues under long range filopodia-based direct cell to cell contact. *Mathematical Biosciences*, *273*, 1–15. https://doi.org/10.1016/j.mbs.2015.12.008.

Vasquez, C. G., Tworoger, M., & Martin, A. C. (2014). Dynamic myosin phosphorylation regulates contractile pulses and tissue integrity during epithelial morphogenesis. *The Journal of Cell Biology*, *206*, 435–450. https://doi.org/10.1083/jcb.201402004.

Wartlick, O., Mumcu, P., Kicheva, A., Bittig, T., Seum, C., Jülicher, F., et al. (2011). Dynamics of Dpp signaling and proliferation control. *Science*, *331*, 1154–1159. https://doi.org/10.1126/science.1200037.

Wolpert, L. (1969). Positional information and the spatial pattern of cellular differentiation. *Journal of Theoretical Biology*, *25*, 1–47. https://doi.org/10.1016/s0022-5193(69)80016-0.

Yamaguchi, M., Yoshimoto, E., & Kondo, S. (2007). Pattern regulation in the stripe of zebrafish suggests an underlying dynamic and autonomous mechanism. *Proceedings of the National Academy of Sciences of the United States of America*, *104*, 4790–4793. https://doi.org/10.1073/pnas.0607790104.

Yang, M., Hatton-Ellis, E., & Simpson, P. (2012). The kinase Sgg modulates temporal development of macrochaetes in Drosophila by phosphorylation of Scute and Pannier. *Development (Cambridge, England)*, *139*, 325–334. https://doi.org/10.1242/dev.074260.

Yatsenko, A. S., & Shcherbata, H. R. (2021). Distant activation of notch signaling induces stem cell niche assembly. *PLoS Genetics*, *17*, e1009489. https://doi.org/10.1371/journal.pgen.1009489.

Zakirov, B., Charalambous, G., Thuret, R., Aspalter, I. M., Van-Vuuren, K., Mead, T., et al. (2021). Active perception during angiogenesis: Filopodia speed up Notch selection of tip cells in silico and in vivo. *Philosophical Transactions of the Royal Society, B: Biological Sciences, 376*, 20190753. https://doi.org/10.1098/rstb.2019.0753.

Zhang, Z., Denans, N., Liu, Y., Zhulyn, O., Rosenblatt, H. D., Wernig, M., et al. (2021). Optogenetic manipulation of cellular communication using engineered myosin motors. *Nature Cell Biology, 23*, 198–208. https://doi.org/10.1038/s41556-020-00625-2.

Zhang, C., & Scholpp, S. (2019). Cytonemes in development. *Current Opinion in Genetics & Development, 57*, 25–30. https://doi.org/10.1016/j.gde.2019.06.005.

Zhou, S., Lo, W.-C., Suhalim, J. L., Digman, M. A., Gratton, E., Nie, Q., et al. (2012). Free extracellular diffusion creates the Dpp morphogen gradient of the Drosophila wing disc. *Current Biology: CB, 22*, 668–675. https://doi.org/10.1016/j.cub.2012.02.065.

Zhu, Y., Qiu, Y., Chen, W., Nie, Q., & Lander, A. D. (2020). Scaling a Dpp morphogen gradient through feedback control of receptors and co-receptors. *Developmental Cell, 53*, 724–739.e14. https://doi.org/10.1016/j.devcel.2020.05.029.

FSC
www.fsc.org
MIX
Papier | Fördert gute Waldnutzung
FSC® C083411

Kolibri 360 GmbH
Ferdinand-Jühlke-Straße 7
99095 Erfurt, Deutschland
produktsicherheit@kolibri360.de